SpringerWienNewYork

Studies in Space Policy
Volume 5

Edited by the European Space Policy Institute
Director: Kai-Uwe Schrogl

Ulrike Landfester, Nina-Louisa Remuss,
Kai-Uwe Schrogl, Jean-Claude Worms (eds.)

Humans in Outer Space – Interdisciplinary Perspectives

SpringerWienNewYork

Ulrike Landfester

Nina-Louisa Remuss

Kai-Uwe Schrogl

Jean-Claude Worms

Softcover re-print of the Hardcover 1st edition 2011

SpringerWienNewYork is a part of
Springer Science + Business Media
springer.at

Typesetting: Thomson Press (India) Ltd., Chennai

Cover: marsTotal – www.plani.ch/presse/MarsTotal.jpg;
spacewalk – www.ucl.ac.uk/news/news-articles/0911/09110501

Printed on acid-free and chlorine-free bleached paper

SPIN: 80016138

With 97 Figures

ISSN 1866-5307

ISBN 978-3-7091-1931-0 SpringerWienNewYork

DOI 10.1007/978-3-7091-0280-0 SpringerWienNewYork

Prefaces

As a child, as a little boy I always dreamt to fly. I was lucky enough to be born in the century when my dream could come true. I even had an opportunity to fly to space for six months. I believe it is inherent in the human nature to look out into the unknown, to be inspired by the remote horizons, to reach out into the sky. I am sure that many of the European citizens are fit and qualified, and dream to fly to space – by far more than there are flight opportunities today. I am convinced that one day it will change. I am looking forward to that day.

Europe has always been a society of explorers and visionaries. Columbus (whose name is given to the European space laboratory), Magellan, Marco Polo – just to name a few – with their life and work made a profound positive difference in the history of the whole world. I have always been proud of being a European and being connected to this heritage of exploration.

It was a particular honour for me to represent Europe as the first European commander of the International Space Station, today's only exploration outpost in space. It is also an honour and an achievement of the European Astronaut Corps and of the European Space Agency. It clearly shows what we can accomplish together as Europeans. It symbolises the success of the International Space Station in which all partners play an important role. Like Europe, the ISS is a true example of what humans can achieve when they decide to work together for a common goal leaving aside their differences.

Today we continue human exploration of outer space in order to push the frontiers of our knowledge and capability. In the 15th century Columbus travelled to the unknown destinations. While his voyages and discoveries barely had an impact on his contemporaries, they have shaped the world as we know it today. I am convinced that we owe it to the future generation, to the life on Earth to have a vision, which will take us to *Mundus Novus* that today lies outside of the boundaries of our planet.

With the words of Carl Sagan

> "In a cosmic perspective, most human concerns look insignificant, even petty. And yet our species is young and curious and brave and show much promise. In the last few millennia we have made the most astonishing and unexpected discoveries about the Cosmos and our place within it, explorations that are exhilarating to consider. They remind us that humans have evolved to wonder, that understanding is a joy, that knowledge is prerequisite to survival. I believe our future depends on how well we know this Cosmos in which we float like a mote of dust in the morning sky."

I strongly believe that a society that stops exploring is a society that stops progressing. Therefore, I hope that Europe also in the future will continue to explore and take up more and more responsibilities in space exploration. European ships and sailors sailed all the oceans of our planet. I hope to see European manned spaceships travel to the ISS and beyond.

As a child I had dreams and my dreams came true. I hope that tomorrow all children in Europe can have equal opportunities to realise their dreams, whatever they are, including an opportunity to travel to places far away from their continent, into the unknown of the universe. And I still dream of a world society continuing to explore, continuing to progress, by humanity and for humanity. This book gives the perspectives to realise that dream!

Frank De Winne
International Space Station Expedition XXI Commander
European Space Agency Astronaut

Interdisciplinarity, multidisciplinarity and transdisciplinarity are key notions in present-day research. It is practically impossible to envisage research that does not penetrate into new disciplines or does not cover more than one domain of science. The volume at hand goes beyond the traditional notions of interdisciplinarity by bringing together worlds which at first glance seem to be aeons apart. The contribution of the humanities and social sciences to space research opens up, without a doubt, new vistas of understanding of the Future, but also of the Present. The Future – in the sense of projecting possible ways of how humans will live in totally different environments than those that we are used to here on Earth. The Present – in the sense that through the process of conceptualising Future life in the universe, we in fact project a "mirror image" of humans and human behaviour on Earth. Both perspectives touch upon a multitude of issues, ranging from ethical to environmental, from aesthetic to scientific, from individual to global and in this case beyond global.

The papers assembled in this volume present a building block for future developments and approaches that the humanities and social sciences will provide for a better and a more thorough understanding of humans in their present context, but also for envisaging humans in their future contexts and environments.

The volume provides a "variation on the theme of dichotomy", and diversity. It builds upon a conference which was held at sea level and ended "on the Moon", or almost, i.e. in the isolated environment of the Roque de Los Muchachos astronomical Observatory, above the clouds, and above several layers of various vegetations characteristic of the island of La Palma in Spain.

As indicated, this book clearly illustrates that human space exploration is far more than simply technologies and (pure) science. The headings of the three parts encompass the very broad variety of topics that made the La Palma meeting lively and intellectually interesting: education, ethics, religion, history, aesthetics, governance, security even clothing and music, just to name a few, are subjects that scientists or engineers would not necessarily think of when considering manned spaceflight. Through our joint efforts, we paved the way for mutual assessment and analysis: interdisciplinarity, multidisciplinarity and transdisciplinarity at its best, and beyond traditional notions.

Milena Zic-Fuchs
Chair of the Standing Committee for the Humanities
European Science Foundation

Jean-Pierre Swings
Chairman European Space Science Committee of the European Science Foundation

Table of contents

CHAPTER 4 Annex

Introduction: from "odysseys" to "perspectives" – towards new interdisciplinary approaches to humans in outer space

Ulrike Landfester, Nina-Louisa Remuss, Kai-Uwe Schrogl and Jean-Claude Worms

This book is being published at a time when the subject of human space exploration appears to become very popular again. Most space-faring countries seem to agree on targets to reach (with robotic missions) in the solar system and in May 2007, 14 space agencies even agreed to a "Global Exploration Strategy" document,[1] delineating various options for the exploration of the solar system. In the USA, even though currently there is some uncertainty regarding the fate of the "Moon option", President Obama has committed his administration to continued exploration of the solar system with the goal of sending humans to Mars in the longer-term, and ambitious robotic missions to be implemented in the years to come through reinforced international cooperation. In Europe, most scientists believe that the – limited means of the continent should be directed towards strong participation in an international endeavour to explore the planet Mars, even though no one rules out the possibility of, or the interest in, returning to the Moon with strategic contributions from Europe's industrial base. European nations do not agree in full on the first steps (i.e. Moon or Mars) but there seems to be consensus that, beyond the application pro grammes Galileo and GMES, space exploration of targets that can ultimately be reached by humans could become one of Europe's next "Grand Challenges". What then of the human factor? Should it be a mere cog in the intricate gears of future missions or does it possess an intrinsic value? Should we leave the exploration of the solar system to robots or can *homo sapiens planetaris* bring added value to the adventure? This is what the European Science Foundation (ESF) and the European Space Policy Institute (ESPI) decided to explore through a series of events that culminated in a conference held on 2–3 April 2009 in La Palma, Canary Islands. This book is the result of the discussions that stemmed from this conference.

Humans in outer space – interdisciplinary odysseys

In 2006 the European Science Foundation (ESF) set out to organise the first trans-disciplinary comprehensive dialogue on humans in outer space. This dialogue goes further than regarding humans as better-than-robot tools for exploration. It investigates the human quest for odysseys beyond Earth's atmosphere and reflects on the implications of possibly finding extraterrestrial life. The inherent human curiosity for exploring the unknown is at the heart of this dialogue, which was addressed through collaboration between the ESF Standing Committee for the Humanities (SCH) and the ESF European Space Sciences Committee (ESSC), in cooperation with the European Space Agency (ESA) and the European Space Policy Institute in Vienna. From the ESF's perspective, the idea of involving humanities and social sciences in that debate was strengthened after completing the "Athens Declaration"[2] in May 2007, which was commissioned by ESA to ESF to establish a scientific framework for defining Europe's exploration programme.

A challenge was to bring together scholars who usually have few reasons to meet in scientific forums, and exchange views in a non-traditional fashion. Non-traditional because, beyond the technical aspects linked to human presence in space that have been studied by space scientists and engineers for the last five decades, humans in space pose challenges that go much further than the ability to survive. In March 2007, an ESF strategic workshop was therefore organised at the University of Genoa entitled "Humans in Space", addressing some of the issues identified above. The central theme was the role and situation of humans in orbit around the Earth, their place in exploration, and the search for life in the universe. Should humans explore space? Do the (cultural and economic) drivers for exploration require human participation? What are the human abilities and reasons to adapt to such extreme conditions as presented by the space environment beyond Earth? Are there scientific grounds that should lead humans to be prepared for the ethical and societal consequences of a possible encounter with extraterrestrial life? The interaction resulting from this workshop paved the way for a conference on Humans in Outer Space (HIOS), held on 11–12 October 2007 in Vienna, locally organised by ESPI with support from the Austrian Ministry for Transport, Innovation and Technology (BMVIT) and the financial support of the European Space Agency (ESA). Scholars from various disciplines and backgrounds, including history, cultural and religious studies, the arts, anthropology, policy, law, ethics and economics, but also space sciences and technology, presented their views. This conference resulted in a continued and further strengthening of the interdisciplinary European dialogue about human exploration of the Moon and ultimately Mars, with a particular emphasis on the human element.

Fig. 1. *Speakers at the conference on Humans in Outer Space (HIOS), held on 11–12 October 2007 in Vienna (source: ESPI).*

The presentations and discussions were structured around three "Odysseys" of humans leaving the Earth. The first odyssey dealt with humans in Earth's orbit and their effects on mankind. It addressed (i) the need to protect the Earth from natural and man-made threats and the role that humans in space can play in this context as "citizens of the planet Earth"; (ii) human space flight as a major source of scientific and technological innovation and improved international cooperation; (iii) the relationship between humans and machines, and the subsequent potential readjustment of our notion of "humanity"; (iv) the legal framework for peaceful space activities. The second odyssey tackled the effects that humans exploring the solar system will have on our societies and culture, starting with the rationale for sending humans out there, pursuing global cooperative endeavours in science, technology, search for resources, and cultural curiosity. The specific European cultural approach towards both scientific and moral issues was felt to be important in that endeavour. A key issue here is planetary protection that needs to be elaborated with international partners. Finally, the third odyssey discussed the last step: humans migrating off the Earth and the impact that this will have on human thought and culture. More exotic topics were thus discussed, such as: the search for settlements outside our planet and the first children to be born in space; the idea of possible encounters with other forms of life in outer space, either through the discovery of life in the solar system (extinct or extant) or through the reception of extraterrestrial radio signals, and whether this would cause the development of a new collective identity for humanity; how people's beliefs could change in the context of new living environments and in contact with other forms of life and societies, thus increasing the importance of the humanities and the social sciences; and finally, the adaptability of humans to new encounters was examined in the light of past encounters that took place on Earth, showing that human beings did eventually adapt to unforeseeable realities, although often at very great costs. While the first effects of an encounter between humans and extraterrestrial life are unpredictable, humans need to be aware that they will be held morally, economically and politically accountable for their choices.

The conference thus provided a unique European perspective by identifying various needs and interests of the humanities and social sciences that are linked with space exploration. The conference, which yielded the so-called "Vienna Vision on Humans in Outer Space", was a success not only on a scholarly level through discussions with colleagues in other disciplines with whom regular interaction is indeed not obvious, but also in demonstrating the necessity of productive contribution of humanities and social science disciplines in understanding the universe in which we live, or will live in the future. The proceedings from this conference were published by Springer in 2009[3] and led the ESF to publish a science position paper to guide future prospective reflexion and activities.[4]

Humans in outer space – from odysseys to perspectives

While the Genoa workshop and Vienna conference attendants were invited by ESF through its SCH and ESSC committees, the large interest in various communities triggered by these events led the ESF to launch in March 2008 a Call for Expressions of Interest open to researchers and scholars based in Europe, in order to identify key challenging topics from any discipline in this area and investigate the best ways to explore them. This consultation process was meant to provide ESF with the views of the European scientific community on these issues, to be synthesised in order to identify key topics of interest to be developed at the European level.

Twenty-one replies were selected out of 51 by the HIOS Steering Committee chaired by Luca Codignola, and their authors were invited to the April 2009 La Palma conference chaired by Ulrike Landfester. One topic recognised by ESF and ESA as interesting for further cross-disciplinary collaboration is the human impact of human spaceflight. Human spaceflight is a major endeavour that brings together many scientific and technical disciplines. Up to now, the emphasis in this context has mostly been on engineering, and physical and life sciences aspects, where major achievements have been made. However, with Europe preparing itself to make a decision on its ambition for future human spaceflight to further destinations other than a low orbit around the Earth, it becomes timely to also address the human and social aspects of having "some of us out there". The Vienna Vision on Humans in Outer Space already clearly indicated that this was a very interesting field to explore. Europe could take the lead in bringing this a step closer and provide a social sciences and humanities-based framework for decisions and events that are expected to happen in the next decades in such areas as:

- Psychology of isolation
- Ethical aspects of human spaceflight
- Socio-economic costs and benefits
- Space law
- Religious implications of leaving Earth
- Administrative and social structures in Lunar or Martian settlements
- Finding non-terrestrial life forms: social, psychological, religious implications
- Artistic expression as a means to share the human exploration experience

Fig. 2. *Speakers at the La Palma Conference (source: ESF).*

The programme of the La Palma conference was structured around the selected expressions of interest under five main themes: enlarging the horizon (cognitive sciences, culture and media); Philosophy, ethics and religious beliefs; Culture and society (arts, fashion, aesthetics); Space and education (school curricula, education, communication and outreach); and, Defining a legal, ethical, political and social framework (rights, governance, law, peaceful use of outer space, human impact on planetary bodies). The presentations at the conference attempted to go from the descriptive odysseys featured at the Vienna conference to a more perspective-based approach, defining aims, enabling science and stepping-stones. The broad discussion that ensued was summarised in a conclusion session, during which participants agreed to have proceedings published as a second "Humans in Outer Space" book.

Humans in outer space – interdisciplinary perspectives

The present book provides "Perspectives" related to governance, management of space exploration, space settlements, the role of astronauts in the future as well as possible encounters with extraterrestrial life. It is a source of insights and arguments for all who have a stake in human space activities. It is structured in three parts, covering the three main interdisciplinary perspectives dealt with in this volume.

The first section, Part I: Politics and Society, discusses the political context of human space exploration, trying to answer questions of ownership and governance as well as management and organisation as elaborated upon by Kai-Uwe Schrogl, Kurt Mills and Martin Parker, respectively. Nina-Louisa Remuss enumerates three scenarios for the potential role of astronauts in space security considerations and Adrian Belu highlights the search for habitable exo-planets as a global task. This section also deals with space-settlements and the encounter with extra-terrestrial life. Alan D. Britton sketches out a school curriculum for the children of space settlers, who have never known life on Earth, while Charles Cockell answers ethical questions related to microscopic extraterrestrial life. Michael T. Schetsche provides some exo-sociological considerations in encountering extraterrestrial life while Philippe Ailleris discusses encounters of extraterrestrial life on Earth in the form of UFOs and exogenous intelligence.

The second part, Part II: History and Religion, considers human space exploration from an historic or religious point of view. David Dunér outlines his concept of Astrocognition while the Jesuit Gustav Schörghofer describes his intentions behind the "Jesuitenkosmos" ("Cosmos of the Jesuits" in English), a reproduction of a International Space Station (ISS) photograph, which was spanned all over the nave of the Jesuit Church in Vienna. Thomas Brandstetter in turn discusses the history of terrestrial biology.

The third and final part of this volume, Part III: Culture and Psychology, deals with the influence of aesthetics on Earth and the shaping of an outer space aesthetics, discussing music as a means of communication in outer space, highlighting the question of clothing and fashion in the context of space travellers and space tourism, outlining the expectations of the next generation regarding contributions of humans in outer space to problem-solving on Earth and, finally, treating psychological, existential, social and ethical issues for crew members on long-term space missions beyond the Earth orbit. Ulrike Landfester, Anna Piotrowska, Christopher Mark Timmins, Agnieszka Lukaszcyk and Berna van Baarsen have contributed to this final section.

Humans in outer space – moving beyond

The case for a prospective activity supervised by the ESF was advocated and discussed at the end of the conference held in La Palma. ESF manages an instrument dealing with prospective actions called Forward Looks. It was therefore decided to prepare a corresponding proposal for the ESF Science Advisory Board and Governing Council. Recently the ESF Standing Committee for Social Sciences (SCSS) decided to take part in the reflexion concerning this proposal and to contribute to it. Dubbed "SpaceRoad", this Forward Look proposal concept includes such topics as:

- Socio-economic costs and benefits vs. philosophical and ethical aspects of human spaceflight: is it legitimate to support exploration or should we only take care of "down-to-Earth" problems?
- Man *or* Machine, or Man *and* Machine?
- Isolation: can we cope with the psychological and intellectual challenges facing the first crews to reach the planet Mars and the first isolated settlements on the Moon, Mars or the asteroids?
- Past examples: can we use historical and social sciences knowledge to infer future behaviour connected to human exploration of the solar system?
- Space law, already a fairly active field: do we need new paradigms to address the legal and administrative challenges of administrative and social structures in lunar or Martian settlements?

- Philosophy, psychology and religion: what are the implications for secular and religious structures and institutions of (a) leaving the Earth; (b) finding non-terrestrial life forms, whether primitive in the solar system (Mars, Titan, Europa...) or civilised in the Universe (hints of existence by radio contact, e.g. the SETI programme)?
- "Supra-globalisation": could the existence of colonies outside the Earth change the nature of international relations and the very perception that the human race has of itself? How can this process be guided?
- Arts: artistic expression as a means to share the human exploration experience.

Overall, humanities and social sciences questions relating to what will confront society if humans start to move and settle outside the Earth could be explored by such a foresight exercise. One goal then is to define a research "roadmap", identifying the various disciplines concerned, linking the scholars in these disciplines and allowing them to share their experience in support of this challenging activity and coordinate their research in the future. The initial discussions and exchanges that took place during the Genoa, Vienna and La Palma meetings clearly indicate that these are highly interesting fields to explore. Europe could thus take the lead in bringing this a step further and providing a humanities and social sciences-based framework (in addition to a scientific one) for decisions and events that are expected to happen in the next decades in solar system exploration by humans.

Many questions need to be addressed during such a prospective discussion, such as: why do we (want to) explore other worlds? Should/can we live elsewhere? How much do we need Earth? Can we settle on other bodies without local disruption? Will settlements diverge from mankind? Will that help mankind? Will state/planet-wide institutions be able to provide a legal framework beyond Low-Earth orbit or will it be the Wild West or Columbian Exchange once again? Can the search for life and extraterrestrial intelligence ever be more than "voices from the past"? Can/should we prepare for encounters? Can/should we meet? Will we be allowed to if we are less advanced? If we do, what happens to society? What will it teach us about being humans? Can decade- or century-long projects be sustained in our societies and, if so, through which mechanisms? What can this "cosmic perspective" tell humankind about itself? These are highly challenging questions that will require an appropriate structure for this foresight reflexion, and adequate participation from various sectors of academia, but also representatives from space agencies and related institutions, policy makers and, most probably, from the public itself.

The core of this activity will take place during strategic thematic workshops that will be structured around overarching questions based on the above. A general

Fig. 3. *Astronauts, cosmonauts and space experts from Canada, Europe, Japan, Russia, and the United States of America met at ESPI on 27 May 2010 to find common rationales and future perspectives for human spaceflight based on the respective cultural backgrounds of their countries and regions (Participants of the workshop (from left): Sergey Avdeev, Mamoru Mohri, Jean-Marc Comtois, Gerhard Thiele, Spyros Pagkratis, Jean-Francois Clervoy, Jeff Hoffman and Takao Doi) (source: ESPI).*

conference will then gather the participants of the individual workshops. The conclusions of these workshops and conference will be debated and integrated during a consensus conference. If accepted by the ESF governance, this SCH-SCSS-ESSC Forward Look could start in November 2010. This activity is foreseen for the 2010–2012 timeframe.

Within ESPI the topic of humans in outer space has also been developing further. In May a meeting of astronauts and cosmonauts from Europe, the U.S., Russia and Japan took place at ESPI, where the various cultural approaches to human spaceflight and the lessons to be drawn for the public and political debate were investigated.[5]

In this context, this book is a timely contribution towards providing foundations, inputs and introductions to the debate on humans in outer space. It is a voyage that has taken up speed. Trans-disciplinary promotion will certainly support and shape political decision-making. In 2011, the fiftieth anniversary of Yuri Gagarin's spaceflight is celebrated. This should certainly be regarded as another stimulus to think ahead.

[1] The Global Exploration Strategy: the framework for coordination, published jointly by ASI, BNSC, CNES, CNSA, CSA, CSIRO, DLR, ESA, ISRO, JAXA, KARI, NASA, NSAU, ROSKOSMOS, May 2007.

[2] Worms J. C. et al. "ESSC-ESF Position Paper – Science-Driven Scenario for Space Exploration: Report from the European Space Sciences Committee (ESSC)." Astrobiology 9 (2009): 23–41.

[3] Codignola Luca and Kai-Uwe Schrogl, eds. Humans in Outer Space – Interdisciplinary Odysseys. Vienna: SpringerWienNewYork, 2009. The "Vienna Vision on Humans in Outer Space" is contained in the Annex to this Volume.

[4] European Science Foundation. "Humans in Outer Space – Interdisciplinary Odysseys, SCH-ESSC Position Paper". Strasbourg: ESF, 2008.

[5] See "Astronauts join at ESPI to discuss cultural backgrounds of human spaceflight." 27 May 2010. European Space Policy Institute. 25 Jun. 2010. http://www.espi.or.at/index.php?option=com_content&view=article&id=499:27-may-2010-astronauts-join-at-espi-to-discuss-cultural-backgrounds-of-human-spaceflight&catid=39:news-archive&Itemid=37.

CHAPTER 1

POLITICS AND SOCIETY

1.1 The political context for human space exploration

Kai-Uwe Schrogl

1.1.1 Do humans in outer space solve any problem(s) on Earth? – An unfair, but politically relevant, question

Public and political discourses are often punctuated with unfair questions. Space activities seem to attract such unfair questions – in particular by the little minds. High-spirited activities such as humans in outer space and their visions risk being ridiculed or slammed. One such unfair question is: do astronauts solve any problems on Earth? It relates to the pressing questions our societies face in times of large financial and economic crises with a credit crunch, rising unemployment and lower living standards for many people. It also relates to global problems like securing peace, saving the environment or eradicating hunger and poverty in developing countries. Is there anything less likely to be a solution to these questions than sending humans into outer space? Certainly not, to be honest. So the strategy should be to refrain from entering into such debates, which are predestined to be lost.

Unfortunately, there were enough scientists, technocrats and politicians who thought they should exactly do that: argue for the impossible. The first lesson for them should be that it is not advisable to tell people that any problem on Earth can be *solved* by space activities. Not even the most useful and directly applicable space technology in the fields of telecommunications, navigation or Earth observation will solve such problems. They will simply – but certainly decisively – contribute to *managing* problems. Humans in outer space, on the other hand, will even less contribute to problem solving or even problem management. Therefore the need for an honest and open analysis of the rationale for humans in outer space is needed. This analysis will turn out to be more political than scientific or technological.

This is a highly relevant statement for the public discourse on humans in outer space in general, and for the engagement in space exploration in particular, which means the return of humans to the Moon and human exploration of Mars. The debate on this issue was initiated by former U.S. President George W. Bush in

2004[6]. Since then Europe has been deliberating its approach to joining this initiative. It will be quite important, How European politicians try to convince the European public to accept such participation and the resulting costs, or even become enthusiastic about them, will be quite important. If they start again to "sell" factories in outer space as they did in the 1980s, when they raised wrong expectations about space stations, they will certainly fail to guarantee sustained support. Therefore they have to revert to an honest and, by that, political approach.[7]

1.1.2 How to justify space activities – and what role do humans in outer space and exploration play?

Justification is necessary for any form of public engagement – and spending. Space activities are no exception, even if aficionados may be convinced that it should be out of the question that something as exciting as space should fall to such a down-to-earth regime as well. The ways that politicians and the space establishment have justified space activities throughout the past decades have varied. The same is true for the way space activities have been attacked by their opponents (which have varied as well and have not constituted a coherent and homogenous group of societies).

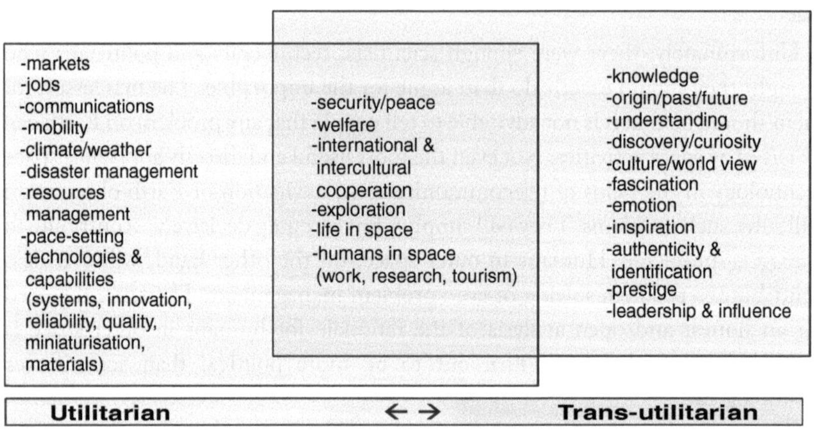

Fig. 1. *Approaches to justify space activities (source: Schrogl, Kai-Uwe and Nicola Rohner. "Für einen neuen Ansatz zur Begründung der Raumfahrt." Die Zukunft der Raumfahrt – Ihr Nutzen und Ihr Wert. Eds. Gethmann, Carl Friedrich, Nicola Rohner and Kai-Uwe Schrogl. Bad Neuenahr-Ahrweiler: Europäische Akademie GmbH, 2007, 139–42).*

The model (Figure 1) of the political justification of space activities[8] tries to provide a list of reasons for conducting space activities and it attempts to create a structured approach. Two kinds of "schools" have emerged to justify space activities: One focusses on tangible benefits like markets, jobs, technologies and applications. It can be labelled as the "utilitarian" approach. The other one builds on the less tangible benefits like the increase of knowledge, a better understanding of the universe and the Earth and the fascination and emotion space activities provide. This second school can be labelled as "trans-utilitarian".

In Figure 1 the two approaches are depicted at the opposite ends of a spectrum. This reflects the way in which the public debates have been conducted during the past decades in Europe. While in each country the public was confronted in different ways with space activities and humans in outer space (some of the countries have large, some have small, space programmes; some have astronauts, some have none), and while we cannot build on European-wide surveys nor can we compare single surveys in countries, where they have been conducted, it is clear from experience and communication with "space people" in European countries that the utilitarian vs. trans-utilitarian dichotomy has been present in all debates.

These debates have covered questions related to budgets, societal benefit, scientific achievements, technological development (spin-offs), peaceful uses, inspiration and soft power. Humans in outer space have been differently touched upon in countries like France (with numerous cosmonauts and astronauts and plans such as the Hermes project for human space transportation), in Austria (which had only one cosmonaut and no further human spaceflight ambitions since then) or in Finland, which has so far had no astronaut at all and which is interested in human spaceflight only from a technological standpoint. There has been, however, one common characteristic, which is that the debates have been conducted in the confrontation of utilitarian and trans-utilitarian views. The players have changed during the past decades (in Germany the Association of Physicists was for some time the most vocal enemy of human spaceflight until it was realised that this hurt the image of space activities as a whole and was detrimental to their interests in other space activities as well) but a dichotomised discourse was prevalent.

But is this confrontational discourse between these two positions really unavoidable and the true set of alternatives for decision-making? Figure 1 tries to show an answer to this question in that it opens a third box – or better – in that it depicts the justification of space activities as a continuum. Looking at it from this perspective, it is not an either-or but the benefits of space activities are spread from one end of the spectrum to the other, for which it is still useful to retain the terms of utilitarian and trans-utilitarian. In Figure 1 the most distinctive message is contained in the location of space for security and peace and human spaceflight. They are regarded as the most decisive exemplifications of space activities not being located at one or the other end of the spectrum, as it has been the case in the debates during the past decades.

This line of argumentation, i.e. to see human spaceflight in the centre of the justification continuum, has not yet been consciously understood. As presented here, it is primarily a proposal to decision-makers to avoid the trap of the "either-or". Scrutinising the recent public discourse on humans in outer space, it might be observed that a trend actually leads in that direction. The term "might" is used, since we have not seen a thorough debate in Europe on human spaceflight but can only observe how the publics in various countries react to their astronauts and to single plans for space exploration. We can also see from the way the recent 40th anniversary of the first landing on the Moon by humans on 20 (21) July 1969 was mainly reflected by the media and received by the public in Europe. The main story was not about heroism, or waste of money, or return on investment[9] (although how unimaginable amounts of money could be sunk into corrupted banks now makes investments in space look quite reasonable). The main story was about understanding political soft power, technological development, inspiration and the human exploratory nature.

Europe seems to have arrived at a point where human spaceflight is neither an enthusiastic hype nor a red flag for heated confrontation. Human spaceflight is located in the centre of the justification continuum between utilitarian and trans-utilitarian. Decision-makers tend to this view, media support this and the public is following. What has to develop from this is a European position and European decisions

1.1.3 What is, or could be, specifically European with regard to humans in outer space?

Europe is a late-comer with regard to human space activities. So far it has only provided "passengers" on U.S. or Soviet/Russian spaceflights. The U.S. and Russia are therefore uncontested senior partners in this field and Europe acknowledges their leadership.[10] This is why European decision-makers and the public do accept – without (political) envy – the display of leadership through iconographic images such as the first humans, U.S. citizens, on the Moon (Figure 2) and the first human, a Soviet citizen, in outer space (Figure 3).

We accept Russia and the U.S., but not China.

Fig. 2. *What Europe can accept: the first human (an American) on the Moon (source: "The eagle has landed." 20 July 2009. Tchironet. 31 May 2010. http://www.tichiro.net/wp-content/uploads/Mondlandung.jpg).*

Fig. 3. *What Europe can also accept: The first human (a Soviet) in outer space . . . (source: "The history of human spaceflight at a glance: from V-2 to Voyager, Gagarin to Melvill." 2004 Space Today online. 31 May 2010. http://www.spacetoday.org/History/MannedSpcFltHistory.html).*

Harder to accept for Europe is, however, the fact that not everybody waited to let Europe become the third power to autonomously conduct human spaceflight. Plans, which were prepared and adopted at the ESA Councils at

Ministerial Level of 1985 in Rome and 1987 in The Hague and comprised Ariane V, participation in the International Space Station with an own module (both of which materialised) plus a human space transportation system, Hermes, and a Man-Tended Free Flyer (both of which did not materialise) led in the direction of Europe becoming the number 3. But the funding problems and quarrels described above led to a generally negative political and public view of space activities.

So it was no surprise that eventually another contender became number 3 in human spaceflight: China. With its first Taikonaut flight in 2003 it created a new ranking in global space activities. Pointing to its supreme technology, Europe does not appreciate such "ranking" at all. But this logic is not shared by the world community, which sees that China has the ability to conduct autonomous human spaceflight while Europe does not. Figure 4 with the Taikonaut showing the V sign is emblematic of that. This simplicity, combined with the gain in soft power for China, is making Europe (in particular the decision-makers but also the public) think whether it is well positioned in the race to become the second (or even first?) superpower in 2020. If Europe is overtaken by China in high-technology spaceflight, where can we still be ahead in ten years?

Fig. 4. *What Europe cannot accept: Being second to China.*

So what is the answer? The current moderately positive political and public attitude combined with the competitive setting against China has to lead to decisions. Such decisions have to build on the current situation, which comprises, inter alia, a European astronauts corps and substantive participation in the International Space Station (ISS) with the Columbus module and the ATV (Automated Transfer Vehicle).[11] The current status, however, is also dominated by the lack of an autonomous human space transportation capability. Such a capability could build on Ariane V and ATV. Technology development should not be the problem. Investment and political will are required. Public interest is demonstrated again and again, most recently with regard to the selection of new ESA astronauts in May 2010 (Figure 5) and the European-Russian long-duration isolation test in June 2010.

When it comes to visions and plans, Europe is often enough number one. The U.S.' space exploration initiative of 2004 had been preceded by the European Aurora programme, which already in 2001 outlined the exploration of Moon, Mars and the planetary system. Europe is also ahead in systematically screening the scientific potential of space exploration.[12] Europe is also applying innovative approaches to space exploration, taking seriously contributions from other disciplines such as the humanities. One example is the trans-disciplinary investigation conducted by the European Science Foundation (ESF), ESA and ESPI, which is the basis for this book and which had its first output in 2009[13]

Fig. 5. *The new ESA astronauts selected from more than 8000 valid applications (source: ESA).*

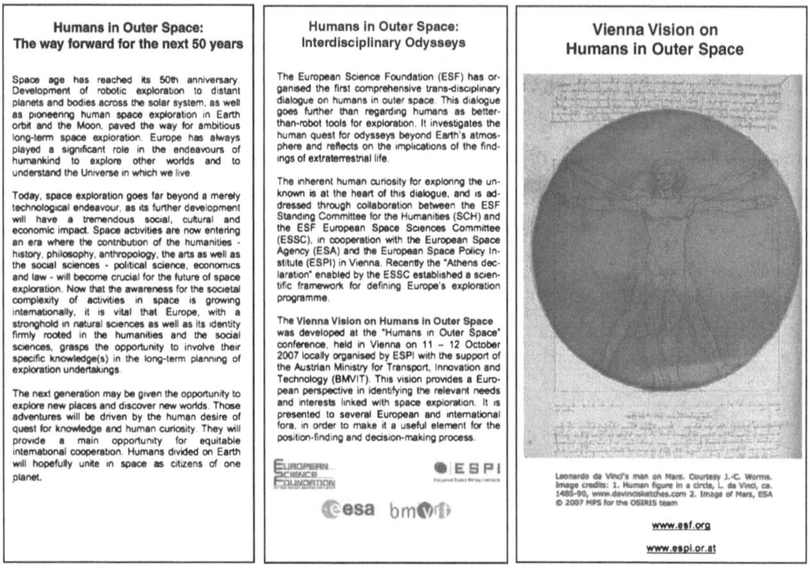

Humans in Outer Space:
The way forward for the next 50 years

Space age has reached its 50th anniversary. Development of robotic exploration to distant planets and bodies across the solar system, as well as pioneering human space exploration in Earth orbit and the Moon, paved the way for ambitious long-term space exploration. Europe has always played a significant role in the endeavours of humankind to explore other worlds and to understand the Universe in which we live.

Today, space exploration goes far beyond a merely technological endeavour, as its further development will have a tremendous social, cultural and economic impact. Space activities are now entering an era where the contribution of the humanities - history, philosophy, anthropology, the arts as well as the social sciences - political science, economics and law - will become crucial for the future of space exploration. Now that the awareness for the societal complexity of activities in space is growing internationally, it is vital that Europe, with a stronghold in natural sciences as well as its identity firmly rooted in the humanities and the social sciences, grasps the opportunity to involve their specific knowledge(s) in the long-term planning of exploration undertakings.

The next generation may be given the opportunity to explore new places and discover new worlds. Those adventures will be driven by the human desire of quest for knowledge and human curiosity. They will provide a main opportunity for equitable international cooperation. Humans divided on Earth will hopefully unite in space as citizens of one planet.

Humans in Outer Space:
Interdisciplinary Odysseys

The European Science Foundation (ESF) has organised the first comprehensive trans-disciplinary dialogue on humans in outer space. This dialogue goes further than regarding humans as better-than-robot tools for exploration. It investigates the human quest for odysseys beyond Earth's atmosphere and reflects on the implications of the findings of extraterrestrial life.

The inherent human curiosity for exploring the unknown is at the heart of this dialogue, and is addressed through collaboration between the ESF Standing Committee for the Humanities (SCH) and the ESF European Space Sciences Committee (ESSC), in cooperation with the European Space Agency (ESA) and the European Space Policy Institute (ESPI) in Vienna. Recently the "Athens declaration" enabled by the ESSC established a scientific framework for defining Europe's exploration programme.

The Vienna Vision on Humans in Outer Space was developed at the "Humans in Outer Space" conference, held in Vienna on 11 – 12 October 2007 locally organised by ESPI with the support of the Austrian Ministry for Transport, Innovation and Technology (BMVIT). This vision provides a European perspective in identifying the relevant needs and interests linked with space exploration. It is presented to several European and international fora, in order to make it a useful element for the position-finding and decision-making process.

European Science Foundation ESPI esa bmvit

Vienna Vision on
Humans in Outer Space

Leonardo da Vinci's man on Mars. Courtesy J.-C. Worms. Image credits: 1. Human figure in a circle, L. da Vinci, ca. 1485-90, www.davincisketches.com 2. Image of Mars, ESA © 2007 MPS for the OSIRIS team

www.esf.org

www.espi.or.at

Fig. 6. *Investigation by Europe in trans-disciplinary aspects of human space exploration (source: "Vienna Vision on Humans in Outer Space." Oct. 2007. European Space Policy Institute 15 Jan. 2010. http://www.espi.or.at/images/stories/dokumente/leaflet/humansinouterspace.pdf).*

following a conference that drafted the "Vienna Vision on Humans in Outer Space" as a manifest for such a trans-disciplinary perspective[14] (Figure 6). Through this exercise, history, human nature and culture are becoming additional keywords for human space exploration in addition to power, technology and science.

1.1.4 So what would be the (political) benefits from humans in outer space?

The setting seems to be good for decisions, propelling Europe into the club of powers conducting human space activities autonomously and shaping the global space exploration agenda.[15] Still, decisions have to based on an analysis of potential benefits justifying the investment.[16] Figure 1 already provided a systematic setting of categories. At this point, these categories shall be illustrated by a small bouquet of concrete benefits.

First of all, space exploration activities, including human spaceflight, are challenging endeavours that offer opportunities to further strengthen European ties and foster the European identity. Furthermore, visible and ambitious space exploration activities can help to motivate European students at all levels to embrace science and engineering disciplines, as well as attract foreign high-skilled students, scientists and investors. A strong European involvement or even leadership in space exploration would enable the enhancement of the European cultural sphere in the 21st century and promote European values.[17]

Through investing in space exploration, new markets and products could be generated, providing benefits for the European knowledge-based economy of the future and European society at large. Exploration could also facilitate the emergence of a dynamic private sector in Europe by creating the conditions for a sustainable space industry, encouraging the emergence of entrepreneurial activities, and the involvement of small and medium-sized enterprises as well as attract large, non-space industrial companies from many traditional fields. This potential should, however, not be overestimated (compare the promises for "factories in outer space" of three decades ago), but there are not too many thrilling areas for start-ups today.

Trans-utilitarian space programmes could be developed by bridging the gap between inspirational and utilitarian space programmes in order to provide increased benefits to European citizens. Synergies with wider European policies goals such as energy security should be sought, since this would broaden the acceptance and relevance of these initiatives. Finally, Europe now has the ability to shape its external policies into multi-dimensional comprehensive cooperation strategies that could prove beneficial to developing a substantive space exploration strategy.

As indicated, this is just a list of potential benefits that can be derived from a European engagement in human space exploration. It demonstrates the validity of positioning human space exploration in the centre of the continuum for the justification of space activities (Figure 1) with the ability to stretch from the utilitarian to the trans-utilitarian end of the spectrum.

1.1.5 Policy lessons for Europe

The current political context for humans in outer space and human space exploration provides a good basis for European decisions regarding the establishment of an autonomous human spaceflight capability and a leading role in shaping robotic as well as human exploration of our solar system. The justification for these has to be honest and should avoid an either-or debate with regard to the benefits of this area.

What is at stake for Europe as a future leading power in the world is the following: Europe has to gain the agenda-setting power in the international system.[18] It has to get the abilities to shape the priorities and timing of events. It also has to find the abilities to attract the best partners and remain the "partner of choice" particularly for the U.S. Europe also has to find ways to reap the broadly spread benefits from human space exploration objectives and put these into the context of other European policy objectives. Europe's human space flight policy should include moral obligations and be in line with the European values already established.[19]

1.1.6 Epilogue: and don't forget, even "serious" philosophers can't escape the big vision

The policy analysis of human space exploration may tend to be too little inspirational, since it primarily has to focus on interests and benefits. The trans-disciplinary approach, however, should bolster also the arguments of the trans-utilitarian elements of the justification for humans in outer space. Talking about discovery, curiosity, inspiration and emotion may sound displaced sometimes in political discourse. But it may help to show that even one of the most "serious" (some may say dry) philosophers, Immanuel Kant, two and a half centuries ago could not escape the fascination of imagining humans settling on other celestial bodies.

> "Wer weiss, ist es ihr [der edlen Seele] nicht zugedacht, dass sie dereinst jene entfernte Kugeln des Weltgebäudes und die Trefflichkeit ihrer Anstalten, die schon von weitem ihre Neugierde so reizen, von nahem soll kennen lernen? Vielleicht bilden sich darum noch einige Kugeln des Planetensystems aus, um nach vollendetem Ablaufe der Zeit, die unserem Aufenthalte allhier vorgeschrieben ist, uns in andern Himmeln neue Wohnplätze zu bereiten. Wer weiss, laufen nicht jene Trabanten um den Jupiter, um uns dereinst zu leuchten?
>
> ... wenn man mit solchen Betrachtungen und mit den vorhergehenden sein Gemüth erfüllt hat: so giebt der Anblick eines bestirnten Himmels bei einer heitern Nacht eine Art des Vergnügens, welches nur edle Seelen empfinden."[20]
>
> "Who knows whether it is not determined that in future the soul will get to know at close quarters those distant spheres of the cosmic structure and

the excellence of their dwelling places, which already attract its curiosity from far away? Perhaps that is why some spheres of the planetary system are already developing, in order to prepare for us in other heavens new places to live after the completion of the time prescribed for our stay here on Earth. Who knows whether those satellites do not circle around Jupiter so as to provide light for us in the future?

... when we have completely filled our dispositions with such observations and with what has been brought out previously, then the sight of a starry heaven on a clear night gives a kind of pleasure which only noble souls experience."[21]

This almost poetic (of course "pre-critical") thought should comfort decision-makers to admit to the fact that besides interests and concrete benefits, there is something more, which people feel when they reflect about humans in outer space. It is present when we see the magic of astronauts use for advertisements, in film and (pop) culture and whenever European astronauts speak to breathless crowds of all age groups in all parts of the continent. Humans in outer space and human space exploration have a solid political justification, and they also are an unmatched inspirational source for humans.

[6] Bush, George W. "A Renewed Spirit of Discovery. The President's Vision for U.S. Space Exploration." Jan. 2004 White House 5 Jan. 2010. http://www.ostp.gov/pdf/renewedspiritofdiscovery.pdf.

[7] An interesting ethical approach is provided by Arnould, Jacques. La Seconde Chance d'Icare: Pour une Éthique de l'Espace. Paris: Le Cerf, 2001.

[8] See Schrogl, Kai-Uwe and Nicola Rohner. "Für einen neuen Ansatz zur Begründung der Raumfahrt." Die Zukunft der Raumfahrt – Ihr Nutzen und Ihr Wert. Eds. Gethmann, Carl Friedrich, Nicola Rohner and Kai-Uwe Schrogl. Bad Neuenahr-Ahrweiler: Europäische Akademie GmbH, 2007: 139–42.

[9] For a special view on this question see: Baum, Seth D. "Cost-Benefit Analysis of Space Exploration: Some Ethical Considerations." Space Policy 25.2 (2009): 75–80.

[10] An exemplary expression of this leadership is contained in the book by former Moon astronaut and Senator Schmitt, Harrison H. Return to the Moon. Exploration, Enterprise, and Energy in the Human Settlement of Space. New York: Copernicus Books, 2006.

[11] See Hansel, Mischa. "The Political Dimension of Europe's New Spaceflight Capabilities." Yearbook on Space Policy 2007/2008: From Policies to Programmes. Eds. Kai-Uwe Schrogl, Nicolas Peter, Charlotte Mathieu. Vienna: SpringerWienNewYork, 2009: 188–195.

[12] See Worms, Jean-Claude. "Exploration – How Science finds its Way in Europe." Yearbook on Space Policy 2007/2008: From Policies to Programmes. Eds. Kai-Uwe Schrogl, Nicolas Peter, Charlotte Mathieu. Vienna: SpringerWienNewYork, 2009: 196–209.

[13] Codignola, Luca and Kai-Uwe Schrogl, eds. Humans in Outer Space – Interdisciplinary Odysseys. Vienna: SpringerWienNewYork, 2009.

[14] "Vienna Vision on Humans in Outer Space." Oct. 2007. European Space Policy Institute 5 Jan. 2010. http://www.espi.or.at/images/stories/dokumente/leaflet/humansinouterspace.pdf.
[15] This and the following section are drawing from two recent ESPI publications, both prepared by Peter, Nicolas. "Space Exploration 2025: Global Perspectives and Options for Europe." ESPI Report 14. Vienna: ESPI, 2008. http://www.espi.or.at/images/stories/dokumente/studies/espi_report_14.pdf; "Issues Underlying Space Exploration in Europe." Briefing Note for the European Parliament. Vienna: ESPI, 2009. http://www.espi.or.at/images/stories/dokumente/leaflet/itre%202009-021%20issues%20underlying%20space%20exploration%20europe.pdf.
[16] Martin, Parker. "Managing Space, Organising the Sublime." (in this volume) gives interesting insights in the managment structure preceeding the U.S. Apollo Moon landing. Europe could draw on the lessons learned from the U.S. experience.
[17] N. B. Schrogl is here referring to students on Earth and not the students of space settlers as described by Alan Britton. "A School Curriculum for the Children of Space Settlers." (in this volume).
[18] On the competition between the space powers see: Schaffer, Audrey M. "What do Nations Want from International Collaboration for Space Exploration?" Space Policy 24.2 (2008): 95–103.
[19] Cockell, Charles. "Ethics and extraterrestrial life." (in this volume).
[20] Kant, Immanuel. Allgemeine Naturgeschichte und Theorie Des Himmels oder Versuch von der Verfassung und dem mechanischen Ursprunge des ganzen Weltgebäudes, nach Newtonischen Grundsätzen abgehandelt. Stadt: Verlad, 1755.
[21] Kant, Immanuel. Universal Natural History and Theory of the Heavens. Translated by Ian Johnston. Canada: Vancouver Island University, Nanaimo, British Columbia, 1755.

1.2 Who will own outer space? Governance over space resources in the age of human space exploration

Kurt Mills

1.2.1 Introduction

Questions of ownership and control are key issues in international relations and international law. They have been addressed to a certain extent in the context of outer space. However, as humanity expands into space it will become obvious that there are still many unanswered questions regarding who should control the vast resources of space.

The main argument of this chapter is that the current concept of sovereignty, which is a core norm and organising principle of international relations, and the international law which underpins and is based upon it, are inadequate to fully understand the possible dynamics amongst various actors in space. In a somewhat speculative manner, it raises a number of questions about control of various aspects of outer space. While using current outer space law as one starting point for discussion, it raises the question of whether political dynamics and the actual actions of States and other actors may matter as much as contemporary law, which appears more unsettled than many observers contend.[22]

1.2.2 Sovereignty and law in outer space

The Outer Space Treaty and the Moon Treaty forbid States from making sovereignty claims on the Moon and other planets.[23] However, there are no enforcement mechanisms. States may decide not to recognise a claim, but *de facto* ownership may be established with little recourse against the offending State.[24] Additionally, international law is relatively silent on non-State actors in outer space.[25]

Could an extra-Earth corporation (EEC) decide it wants to set up a shop on the Moon to mine resources? What about asteroids?[26] Or what if a country wanted to capture a comet to provide water for its people? Or if a corporation wanted to do the same thing to sell the water?[27] On Earth, no State may make sovereignty claims in Antarctica and on the high seas and there is a right of access and passage. Further, any entity wishing to mine the seabed must comply with the Law of the Sea and share some of the resources it acquires. The entire world is supposed to benefit from the resources on the seabed. Should the same principles apply to Outer Space?

The Outer Space Treaty declared outer space, including all celestial bodies – indeed the entire universe (one wonders what extra-terrestrial beings might think about that . . .) – to be an interplanetary (indeed, intergalactic) commons, available to all of humanity. Should a portion of the resources mined from an asteroid be reserved for the benefit of all countries? If only a few countries have the technological and financial wherewithal to access these resources, would they agree to such a regime? What exactly would be the basis for a claim to a share of these resources on the part of Earth-bound countries? Developing countries make claims to share in resources on Earth on the basis of a right to development and other claims having to do with economic rights and a New International Economic Order. Could these claims also be extended to those resources off the Earth?

The Moon Treaty shifted the debate over who should benefit from the Moon. It states that "the Moon and its natural resources are the common heritage of mankind"[28] and binds the State Parties to create an international regime "to govern the exploitation of the natural resources".[29] It was an attempt to move the status of the Moon from the *res extra commercium*[30] of the Outer Space Treaty – everybody has access to Outer Space and nobody can make sovereignty claims to any part of Outer Space, including celestial bodies like the Moon – to *res communis humanitatis* – "common heritage of mankind". Everybody should benefit from resource extraction activities on the Moon, including those States who have no ability to get to the Moon and gain access to its resources.[31] Given that the U.S. and the then USSR did not sign, and, indeed, only a relatively few States have signed and ratified, it is currently in a State of flux.[32]

The Outer Space Treaty and the Moon Treaty draw from their terrestrial antecedents, the Antarctic Treaty System (ATS) and the UN Convention on the Law of the Sea (UNCLOS). The ATS, on which the Outer Space Treaty was modeled, designated the Antarctic as a preserve, suspending sovereignty claims. Further, since a convention to govern mineral extraction has not been ratified by the requisite number of States, resource extraction has been effectively prohibited, while scientific investigation has been encouraged. Like the Outer Space Treaty does for space, UNCLOS provides for free use of the high seas, both recognising the impossibility of controlling the vast expanses involved. A significant development in the law of the sea came with development of the idea of the "common heritage of mankind" with its requirement that the economic benefits derived from the deep seabed should be shared with all States. This requirement provided the basis for the claims in the Moon Treaty for such sharing. However, it also prevented the United States from becoming a party to UNCLOS, in the same way that none of the States who might be most impacted by this requirement in the Moon Treaty have an incentive to become parties to the treaty.[33] Both of these regimes have come under strain. Some States have begun to make claims to the Antarctic seabed.[34] Russia has attempted to claim almost half of the Arctic seabed.[35] These regimes are likely to come under more pressure as the search for resources intensifies. In the same way, the legal prohibitions in the outer space regime may come under pressure as States acquire the capabilities to have a more sustained presence beyond low Earth orbit.

1.2.3 Space as a commons?

What is the moral status of outer space? Does it occupy the same moral "space" as Earth? Orbital space has obvious benefits for developing countries – remote sensing for weather and crop development, for example – and is relatively accessible. These countries are able to benefit from cooperative activities between States, the UN, etc. It is relatively easy to make a direct connection to this area of space and Earth – call it the Greater Earth Metropolitan Area (GEMA). It is part of the Earth's environs. Is the Moon, which is intimately tied to Earth, also part of the GEMA? The international community has seemed to say "yes". Going beyond the Moon, it is harder to make that claim, although perhaps this has more to do with current transportation capabilities than anything inherent about outer space. Yet, this is what the international community has decreed. Does this make sense or, as some have suggested, should hard celestial bodies – planets, asteroids, etc. – be considered *res nullius* – able to be claimed – and the vast stretches of vacuum in between *res communis* (owned by all – or *res extra commercium* not able to be owned by anybody) – equivalent to the high seas where all have a right to go?[36]

The question is whether all of outer space, including planets and other hard celestial bodies, are to be considered commons, and if so, what kind of commons. The outer space regime designates all of outer space as a commons, but as a type of commons, which is susceptible to the tragedy of the commons.[37] In its current status as *res extra commercium* outer space, including planets, is open to all. No State can claim sovereignty, but there is a possibility for private access and removal of resources. In such an anarchical situation, as with the classic scenario of farmers overgrazing a town common, private actors might be able to exploit the Moon's resources, for example, without any regard for the effects they are having on the Moon itself.[38] The Moon Treaty's *res communis humanitatus* would collectivise the Moon, in the same way that the deep sea bed has been collectivised, allowing humanity as a whole to control access and benefit. This was the intent of the Moon Treaty. However, given that the State powers most likely to be able to access the Moon are not parties and have not agreed to this principle, this principle likely will not constrain these States[39] although, since some European States are parties to the treaty, the European Space Agency (ESA) may be constrained in its activities.[40]

1.2.4 Territorial claims?

States were originally able to claim territorial waters out to three miles. This was because this was how far cannons could shoot and thus provide protection. Today, however, States' reach is much further, but it is also much more variable – there is greater disparity in the extent to which States have the ability to reach and effectively control the space radiating out from their territory. During the Cold War States decided that it was not worth trying to control space in the way they controlled terrestrial territory. Could this change? In the same way that territorial claims have moved from three miles to twelve miles (and, in terms of exclusive economic zones, 200 miles) offshore, might the possible extent of sovereignty claims extend upward and outward? This becomes a problematic question when one considers that traditional sovereignty claims are based upon territory – *terra firma*.

According to international law, a State is defined as having a territory, a permanent population, a government, and independence. If these rules are applied to outer space, then obviously no State can make such claims. Aside from the few celestial bodies, there is nothing – certainly nothing to put a flag on. States cannot make sovereignty claims in orbit, but what if a country puts a permanent human settlement in geostationary orbit? This part of GEMA has been the sole preserve of satellites. But if a population is added to the mix, could that bit of space 35,785 km above the equator be claimed as sovereign territory? Say this was above Gabon. Gabon obviously has no way of physically challenging such a claim. Taking the territorial waters analogy, 35,785 km is far beyond the range of its cannons (although it may not be beyond the control of anti-satellite technology – of course the issue of whether Gabon might have access to such technology is another question). It is also beyond the height at which airplanes can fly, and thus would be outside of air space, over which States do have sovereignty.[41] Beyond this question of effective control,[42] however, is a much more fundamental one which points to the problematic assumptions of the Westphalian sovereignty discourse.[43] That bit of vacuum above Gabon may be a *place* – one could chart its GPS coordinates – but it is not territory in the way we conceive of it. Can we export Westphalian sovereignty beyond Earth where fundamental assumptions underpinning sovereignty – land, control, dominion – do not apply? Further, geostationary orbit is considered *res extra commercium* – no State may own it. However, the International Telecommunications Union (ITU) allocates the frequencies satellites (many of which are privately owned) may use (although the frequencies are not able to be owned by States or private corporations).[44] An international body regulates the frequency resource common, something even more intangible than geostationary orbit itself. Such an incorporeal common, access to which has very important security and commercial implications, further de-stabilises traditional understandings of sovereignty.

1.2.5 Problematic sovereignty

1.2.5.1 Sovereignty in flux: the International Space Station

The International Space Station Intergovernmental Agreement (IGA) of 1998, along with associated bilateral memoranda of understanding and implementing arrangements, governs legal claims and responsibilities on the ISS. According to Article 5 of the IGA, the main rule is that "each partner shall retain jurisdiction and control over the elements it registers and over personnel in or on the Space Station who are its nationals."[45] Thus, different States own – enjoy sovereignty over – different parts of the ISS, and different laws apply to different parts of the ISS. In one sense, this is analogous to diplomatic missions where the mission is conceived, through diplomatic fiction, of being the sovereign territory of the country being represented and not subject to the sovereign control of the country in which the mission exists. Thus, a series of diplomatic fictions is created to try to impose a problematic Westphalian sovereignty onto an even more problematic space, which defies categorisation under Westphalian sovereignty. Each State or entity (i.e. the European Union[46]) owns a part of the station. The different parts are bound together into a whole unit where, unlike most confluences of sovereignty, but

Fig. 7. *International Space Station: Fictional, extraterritorial sovereignty in flux (source: National Geographic 8 Jan. 2010. http://s.ngeo.com/wpf/media-live/photologue/photos/2009/09/02/custom/5448_1280x1024-wallpaper-cb1262885893.jpg).*

perhaps like the Schengen area in Europe, there is freedom of movement – one assumes that there is no need to show a passport when moving from the Russian to the U.S. section of the ISS. Perhaps it is like post-War Berlin, which was divided into four zones, each controlled by a different country. Yet, unlike Berlin, the ISS has no fixed territorial referent.

Unlike the station in geostationary orbit, it has no fixed coordinates. We can predict where it will be at any moment, but that is a far cry from the Westphalian insistence on fixed territory. The ISS drags bits of sovereignty through the *res extra commercium* GEMA. It is an unfixed, magical sovereignty – now you see it, now you don't – or quantum sovereignty – now it's here, now it's there. The space in which the sovereignty is claimed is not fixed and thus the sovereignty claims are in constant – if predictable – flux. Although the IGA is an attempt to export Westphalian sovereignty into a realm and situation that could not be conceived of by the parties to the Peace of Westphalia which helped to solidify the contemporary concept of sovereignty, it is a highly problematic export. To refer back to the practice of diplomatic missions, Ruggie refers to this extraterritoriality as an "'unbundling' of territoriality."[47] The ISS would seem to be an example of extraterritoriality, but can territory be "unbundled" if there is no territory to begin with?

1.2.5.2 How tall is sovereignty: space elevators

Or, what if a country builds a space elevator? The end will be in geosynchronous orbit but it will be tethered to the Earth in some way – either on land or, perhaps, on a floating platform. It is in orbit but one might also think of it as a very tall building. It is not launched in the way that spacecraft are launched. Its end may resemble the previously imagined permanent space station in that it inhabits a space devoid of Westphalian-imagined territory, but it is attached to territory (although not if based on a platform floating on the high seas). Is it just an extension of that State's territory? How could this be if States cannot make permanent claims off the Earth? And what about the possible dangers of such a structure? Given the recent collision of two satellites,[48] the possibilities for running into something "up there" is real. They would be magnified by such a large structure. Could a State legally build such a thing on its own without consulting its neighbors (in this case all of humanity)? Or, would governments be able to deny it planning permission on the grounds that it impacts their ability to freely navigate near-Earth space? What body could adjudicate such a dispute?

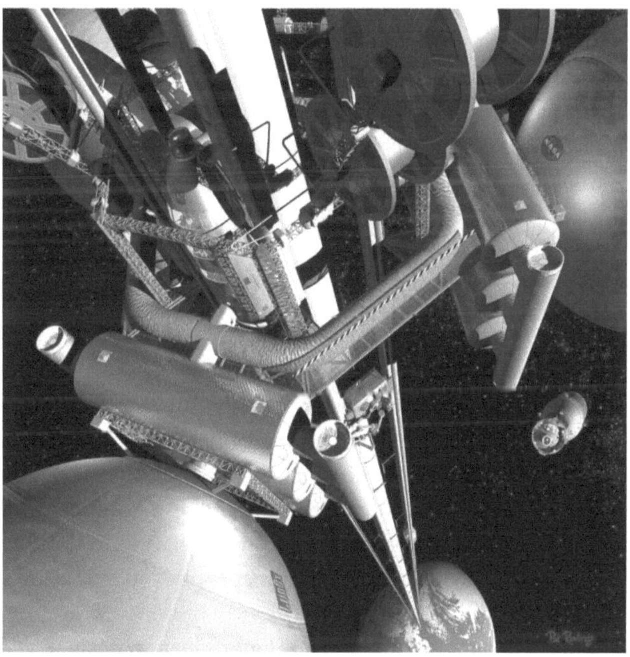

Fig. 8. *Space elevator: How tall is sovereignty? (source: NASA (Pat Rawling) 8 Jan. 2010. http://upload.wikimedia.org/wikipedia/commons/f/f0/Nasa_space_elev.jpg).*

The International Court of Justice (ICJ)? This would be rather different from the boundary disputes it usually handles. Would arguments about cross boundary environmental effects provide any type of precedent? If so, what boundary is being crossed if the space elevator goes straight up from a State's territory? The Westphalian sovereignty discourse fails us again.[49]

1.2.6 Changing sovereignty claims

1.2.6.1 De jure meets *de facto*

What happens when States are able to travel to the Moon or other bodies and establish permanent populations? Will they have *de facto* sovereignty? Will State practice change international law in such a way that these claims can be made and recognised? It is easy to foreswear such claims when you have no way of effectively making such claims – and, indeed, fear that others might make them before you. Once you have the capabilities and the interests – strategic advantage, resources, etc. – it becomes harder to sustain your earlier position. It becomes harder, still, when there are no material penalties for violation. What will happen when the U.S. or Russia or China make it back to the Moon or to Mars? The incentives to withdraw from the outer space regime – or at least to try to radically amend its provisions – would be great.

1.2.6.2 Non-State actors and appropriation

Or, consider a situation where a non-State actor – an EEC – sets up an operation. It establishes an ongoing operation on what is practically *res nullius* (even if legally *res extra commercium*). It establishes itself as the authority and claims that it fulfills all the criteria for being a State – it occupies the asteroid or some portion of the Moon, there is a continued human presence, it acts as a government (although perhaps not according to some of the recent criteria having to do with popular representation), and it is independent of any other entity.[50] Can it claim that it is a State?

The legal status of outer space and the non-appropriation norm would seem to undermine the claim. However, if such a claim might be considered could an EEC make that claim on behalf of its workers? On what moral basis could this claim be made? While previous State-practice has been to recognise whoever has effective control of a territory, this is no longer sufficient as States have begun demanding that such claimants are democratic and protect human rights. While the logic of neo-liberal economics might accord a more independent governance role to EECs, it would be hard to find a moral basis for a company to claim sovereignty over its workers. A corporate function does not – and cannot – equal a governance function. In fact, as we have seen many times the logic of the corporation can dramatically undermine governance – and certainly any type of democratic governance.

1.2.6.3 Self-Determination in space

Could the workers themselves make such a claim? Setting aside for the moment the legal status of the territory involved, they would certainly have a greater moral claim. It would be a real claim to self-determination. The people themselves would be deciding to determine their own future and how to govern themselves. Certainly the EEC would have a few things to say about that. And this would raise the issue of the connection between the EEC, the workers, the parent company back on Earth, and the State where the EEC is based. What laws would apply? What laws could apply? Would it be able to use force to keep the workers under control and prevent such a claim from being put into effect? The answer would seem to be no, given that the State is supposed to be the only entity with a claim to the legitimate use of violence. But what if there is no State – or at least no effective State control? The EEC would almost certainly use force to protect its interests. In the Wild West of outer space, legal niceties are not going to get in the way of profit. Could it come under legal sanction back on Earth? What if such an operation had headquarters in a country, that did not care too much about corporate conduct and human rights?[51] There would be little that could be done.

1.2.7 Can sovereignty be exported to space?

Westphalian sovereignty assumes every bit of territory is – or is liable to be – under the control of States. The Outer Space Treaty fundamentally alters this assumption, decreeing that States may not make sovereignty claims on celestial bodies.[52] As States – and other entities – gain greater abilities to access such places and spaces, there may be pressure to change this legal

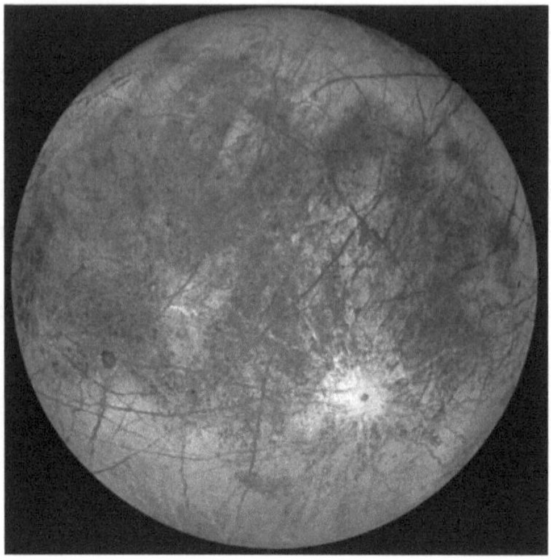

Fig. 9. *Could the European Union claim sovereignty over the Jovian moon Europa? (source: Western Washington University. http://www.wwu.edu/depts/skywise/planets/europa.jpg).*

understanding of outer space, although the question will remain of the extent to which non-spacefaring States are entitled to the fruits of outer space.

What will happen when individual States, groupings of States, and private commercial enterprises all move into, and compete in, space? Will States be able to extend their authority beyond the earthly realm? Will the UN be able to claim any relevance? If multinational expeditions make it to another planet and want to claim control, on whose behalf would this be? Could a European expedition claim Europa for Europe? What would this mean, given that Europe itself is made up of 27 sovereign entities? While these States pool their sovereignty to a certain extent, there is still no sovereign "European" territory. Could such pooled sovereignty be extended to a celestial body? If so, how could this be encompassed within existing international political arrangements and assumptions? Current outer space law says that outer space – all of it (a rather hubristic assertion) – belongs to everybody and no one simultaneously. It is still extremely unclear what this may mean, and whether our current legal and political concepts are adequate to help us envision how to apportion and govern the vastness of outer space.

[22] Quinn, Adam G. "The New Age of Space Law: The Outer Space Treaty and the Weaponization of Space." Minnesota Journal of International Law 17.2 (2008): 487–96.
[23] Indeed, Freeland and Jakhu argue that the principle of nonappropriation found in Article II of the Outer Space Treaty has "become a significant norm of general international law (in fact a *jus cogens*

norm)". Steven Freedland and Ram Jakhu. "Article II of the Outer Space Treaty" Cologne Commentary on Space Law, Vol. I. Eds. Stephan Hobe, Bernhard Schmidt-Tedd, and Kai-Uwe Schrogl. Cologne: Carl Heymanns Verlag, 2009: 55.

[24] Just look at the actions of Israel in the Occupied Territories. It has made sovereignty claims over certain parts and has *de facto* annexed other parts through settlements, all in clear violation of international law. The international community has been powerless to stop these violations. What recourse could the international community possibly have in outer space when it cannot enforce its own laws on Earth?

[25] Except to say that they are accountable to their home States, and such States are responsible for their actions, about which more will be said later on in this chapter. See Bohlmann, Ulrike M. "The Need of a Legal Framework for Space Exploration." Humans in Outer Space – Interdisciplinary Odysseys. Eds. Luca Codignola and Kai-Uwe Schrogl. Vienna: SpringerWienNewYork, 2008: 182–95.

[26] Surely a potentially lucrative activity, given that the metals contained in one small asteroid could have a market value of U.S.$20 trillion. Fountain, Lynn M. "Creating Momentum in Space: Ending the Paralysis Produced by the 'Common Heritage of Mankind' Doctrine." Connecticut Law Review 35 (2003): 1785. In fact, as Quinn points out, '[t]he resources in the [asteroid] belt alone have an estimated value of $100 billion for every person on Earth.' Quinn, Adam G. op. cit. 487.

[27] Further scenarios that will trigger further questions for international law are described by Remuss, Nina-Louisa. "From Envoys of Mankind to Combatants?" (this volume).

[28] Article XI(1).

[29] Article XI(5).

[30] "A thing outside commerce". Many authors use the term *res communis* to define the status of outer space. This also denotes a commons, and there seems to be confusion in terminology.

[31] However, the International Institute of Space Law appears to claim that the current regime, as outlined in the Outer Space Treaty, mandates a common heritage of mankind approach: "The clear goal of such a regime is to preserve outer space, including the Moon and other celestial bodies, for the exploration and use of all mankind, *not only for those States and private enterprises that are capable of doing so at any particular time*" [emphasis added]. "Statement of the Board of Directors of the International Institute of Space Law (IISL)" 22 Mar. 2009. International Institute of Space Law. 22 July 2009. http://www.iislweb.org/docs/Statement%20BoD.pdf. In fact, there seems to be split between spacefaring nations – who see a clear distinction between the prohibition on States making sovereignty claims, which is clearly prohibited in the Outer Space Treaty, and removal of resources by non-State actors, which is not mentioned – and developing States, who support the common heritage principle, thus preventing any actor from taking resources for purely private gain. Quinn, Adam G. op. cit. 480–1. Although, according to Freeland and Jakhu, "[t]here is no room for classification in terms of 'national' (State or public) or non-governmental (private) beyond the requirements of Article VI of the Outer Space Treaty" (which says that States parties are responsible for non governmental actors) (Steven Freedland and Ram Jakhu. op. cit. 51).

[32] Laver argues that the treaty has little effect, and von der Dunk argues that "its relevance in legal terms is therefore to be severely doubted," while Peterson argues that the principle of *res communis humanitatus* "has been expressed and has enough supporters to make a political difference, though the remoteness of lunar or other space mining makes judging the principle's practical effects difficult. The principle stands as a point of appeal . . . ." Laver, Michael. "Public, Private and Common in Outer Space: *Res Extra Commercium* or *Res Communis Humanitatis* Beyond the High Frontier?" Political Studies XXXIV (1986): 359–73; von der Dunk, Frans G. "Space Law in the Age of the International Space Station." Humans in Outer Space – Interdisciplinary Odysseys. Eds. Luca Codignola and Kai-Uwe Schrogl. Vienna: SpringerWienNewYork, 2008: 148–61; Peterson, M. J. "The use of analogies in developing outer space law." International Organization 51.2 (1997): 265.

[33] Quinn, Adam G. op. cit. 483–6; Fountain, Lynn M. op. cit. 1757–9; 1769–74. On UNCLOS see also Steven Freedland and Ram Jakhu. op. cit. 61–3. They argue that "the creation of the international regime similar to that established under UNCLOS is, at least at this stage, premature for the governance of regular and widespread exploitation of these natural resources of outer space" (Ibid. 63).

[34] Doyle, Alister. "Antarctica: The next North Pole?" The New York Times. 1 Feb. 2008.
[35] Chivers, C. J. "Russians Plant Flag on the Arctic Seabed." The New York Times 3 Aug. 2007.
[36] Hickman, John. "Still Crazy after Four Decades: The Case for Withdrawing from the 1967 Outer Space Treaty." 24 Sept. 2007. The Space Review: Essays and Commentary about the Final Frontier. 12 Feb. 2009. http://www.thespacereview.com/article/960/1. Fountain suggests a regulated free-market approach. Fountain, Lynn M. op. cit. 1174–782. Quinn advocates a similarly regulative framework to encourage the development of resources in outer space. Quinn, Adam G. op. cit. 498–500.
[37] Hardin, Garrett. "The Tragedy of the Commons." 3 Dec. 1968. Science 162 (3859): 1243–8. 3 Mar. 2009. http://www.sciencemag.org/cgi/content/full/162/3859/1243. For a critique of the tragedy of the commons see Dolman, Everett C. Astropolitik: Classical Geopolitics in the Space Age. London: Frank Cass, 2002: 101–5.
[38] Although, as Freeland and Jakhu point out, a celestial body, such as a small asteroid, could not be "exploited 'out of existence'," since the interests of other States and future generations must be taken into account. The question might then become where to draw the line between appropriate and inappropriate use of resources. Steven Freedland and Ram Jakhu. op. cit. 53–4.
[39] Laver, Michael. op. cit.
[40] Carswell, Bill. "The Outer Space and Moon Treaties and the Coming Moon Rush." 18 Apr. 2002. Space Daily. 3 Mar. 2009. http://spacedaily.com/news/oped-02c.html.
[41] In 1976 eight equatorial States tried to claim that "synchronous geostationary orbit, being a natural resource, is under the sovereignty of the equatorial States." However, this claim was not taken seriously. Laver, Michael. op. cit. 369. Alexander Harris and Ray Harris argue that there is need to demarcate where air space ends and outer space begins. Harris, Alexander and Ray Harris. "The Need for Air Space and Outer Space Demarcation." Space Policy 22 (2006): 3–7. Should it be a vertical demarcation – so many miles above sea level – or a functional demarcation – perhaps the height at which airplanes can no longer fly because they cannot get the respective lift under their wings from the air? The space shuttle renders this definition problematic since it is a spacecraft when it launches, even though it is going through air space through which airplanes may fly, and functions as an airplane, using the lift from the air under its wings to facilitate its landing. Further, what if there is a space between the upper limit of aircraft and the lower limit of satellites?
[42] And Freeland and Jakhu are clear that effective control has been clearly forbidden by Article II of the Outer Space Treaty as a legal argument for sovereignty (Steven Freedland and Ram Jakhu. op. cit. 54).
[43] Camilleri and Falk define the 'sovereignty discourse' as 'a way of describing and thinking about the world in which nation-States are the principle actors, the principle centres of power, and the principle objects of interest.' Camilleri, Joseph A. and Jim Falk. The End of Sovereignty? The Politics of a Shrinking and Fragmenting World. Brookfield, VT: Ashgate Publishing Company, 1992: 2.
[44] Fountain, Lynn M. op. cit. 1756–67. See also Steven Freedland and Ram Jakhu. op. cit. 61.
[45] "International Space Station Legal Framework." 24 Oct. 2008. European Space Agency. 3 Mar. 2009. http://www.esa.int/esaHS/ESAH7O0VMOC_iss_0.html. See also Dunk, Frans von der. op. cit. on issues of criminal and other legal jurisdiction on the ISS. He points to the novel questions of "the nationality of space-born babies, the applicability of human rights to outer space, and the validity of contracts drawn up in outer space on outer space matters," as well as the question of whether jurisdiction, which is currently not possible on a territorial basis, should be structured so as to ensure that law will actually follow man into outer space. Or will jurisdiction based on the nationality of the humans involved suffice – but then, what about these future space-born humans? Von der Dunk also points out that previous space stations owned by a single country 'qualified as their quasi-territory for legal purposes,' while the multi-national character of the ISS makes things much more complicated legally.
[46] Von der Dunk notes the unique situation of the European Space Agency – an intergovernmental organisation – being a party to the agreement and acting as a registration "State".
[47] Ruggie, John Gerard. "Territoriality and Beyond: Problematizing Modernity in International Relations." International Organization 47.1 (1993): 165. Jill Stuart further discusses this idea of unbundling sovereignty in the context of outer space. Stuart, Jill. "Unbundling Sovereignty, Territory

and the State in Outer Space." Securing Outer Space. Eds. Natalie Boormann and Michael Sheehan. London: Routledge, 2009.
[48] Broad, William J. "Debris Spills into Space After Satellites Collide." The New York Times. 11 Feb. 2009.
[49] Stuart argues that "Westphalian sovereignty as a concept is inadequate for analysing outer space politics. The concept does not provide a language through which to understand spaces outside of the traditional territorial state." Stuart, Jill. op. cit. 9.
[50] Although, given that States are supposedly responsible for any non-State entity based in its territory which goes into outer space, this might be a somewhat problematic claim to make.
[51] This could be a radical extension of the concept of "offshore."
[52] Indeed, Freeland and Jakhu argue that "the primary intent of Article II [of the Outer Space Treaty] was to... [confirm] that principles of territorial sovereignty do not apply to outer space" (Steven Freedland and Ram Jakhu. op. cit. 48).

1.3 Managing space, organising the sublime

Martin Parker

1.3.1 Beginnings

In 1968, James Edwin Webb, the ex-administrator of NASA, delivered a series of McKinsey Foundation lectures at the Graduate School of Business, Columbia University, which were published under the title *Space Age Management* the next year, but before the first Moon landing of July 1969. At that time, "space age" was a prefix that was being applied to everything. Effectively, it meant something like "modern", with connotations of being streamlined, efficient and fashionable. The coupling of "space age" and "management" combines this modernity with a particularly technocratic sense of control. The space age was an age of mass organisation, of new products that saved time, of automated factories and paperless offices and launch pads connected by freeways and telephones. It was a world of harmonious organization, managed by wise and well-qualified elders who would eradicate the problems that beset humanity in its Earth-bound dark ages. The paradox, for Jim Webb,[53] was how to retain the advantages of this technocracy without eclipsing democracy. After all, there would be no point in winning the space race against the Soviets if the form of social organisation that enabled it was effectively reproducing what was understood to be the centralised repression of communism.

This chapter will engage with Webb's paradox in two ways: one, by exploring the sense in which the space age was necessarily the age of management. In order to achieve Kennedy's goal, huge numbers of people, things and places needed to be co-ordinated and made controllable. In 1966 NASA directly employed 36,000 people, with another 360,000 people working for 20,000 contractors and 200 universities in 80 countries.[54] The total cost, at that time, for the Apollo programme was 24 billion dollars, ninety percent of which ended up in the private sector. This meant accounting for people and things, chains of command, scheduled meetings with determinate agendas, as well as plans, graphs, reports and deadlines. The Moon landing was a triumph of organisation, of project management and control of a complex socio-technical system.[55] It was also a huge exercise in new deal economics, with a wide variety of State subsidies being channelled through NASA to aerospace corporations, universities, regional development organisations and so on. The contradiction between the ideology of the free market and this command and control centralisation did not go unnoticed, and meant that Webb and others needed to make constant reference to a dead President and a new frontier in order to keep NASA's funding rolling in.

But the rhetoric wore thin because, in management terms, going to the Moon made little rational sense. It was criticised then and ever since as a monumental distraction from the problems of the Earth, a subsidy for the military-industrial complex, and (in cultural terms) a white, middle class, male fantasy, in which heroes ride rockets whilst the rest of the world just watches.[56] Yet, Apollo was also one of the iconic moments of the twentieth century, and one that inspired feelings of awe in many people, which still resonate forty years later. It has also encouraged many people to make connections between science, fiction, and possibility that are substantially at odds with the capitalist and militarist political and financial interests that were key to driving the programme.[57] On the one hand we have the possibility of a new world for all humankind, on the other the predictable calculus of a twentieth century empire. Given its cold war roots, and its managerialist methods, how did Apollo produce something so incomprehensibly strange, and excessive?

In this paper, I will deal with these questions in turn. The first section looks at the state-based forms of organisation that made the space age possible. It also deals with some of the political paradoxes that needed to be negotiated.[58] I then move on to suggest that complex organisation is necessary in order to produce an event that exceeds it, that suggests the sublime, and opens up possibilities that have not yet been imagined. *Space Age Management*[59] is an account of how something astonishing was made, but it never approaches its object because it does not have the language to do so. It seems to me that, symmetrically, the only way that we can understand how that incomprehensible event happened is as the outcome of space age management and of the rather mundane political and administrative contexts that made it possible.

1.3.2 Means

1.3.2.1 Administration and management

James Webb was, from 1962 to 1968, the Chief Administrator who understood that making things happen required mundane politics. Webb had

Fig. 10. *James Edwin Webb, NASA Administrator, February 14, 1961–October 7, 1968 (source: "James E. Webb." NASA. 10 Jan. 2010. http://www.hq.nasa.gov/office/pao/History/Biographies/webb2.gif).*

experience in Big Oil and Big Aerospace, as well as various posts in a series of administrations, including being Director of the Bureau of the Budget under Truman. According to most sources,[60] Webb's real skill was in Washington, where he bamboozled politicians with jobs, universities with research money, and journalists with wholesome stories and an unstoppable torrent of words. He also kept a direct line to two presidents, ensuring that their interests did not diverge too much from his and spent a considerable amount of time organising and re-organising different lines of communication, job descriptions and spans of control, restlessly trying to find an organisational form that he thought would be both stable and dynamic.

The lessons for "management" in *Space Age Management* are less about "managing employees", and more often about managing other executives, as well as politicians. Webb's line is essentially a Machiavellian one. If you wish to keep a big organisation going, you need to ensure that your friends are close and your enemies even closer. But beyond the tactical pragmatism, Webb was also selling a "new deal" and "big government" line. He wanted NASA to be an example of large-scale interventionism that required "rapid advances in so many disciplines – engineering, physics, astronomy, mathematics, economics, political science, psychology, public administration – the whole list of the physical, behavioural and social sciences".[61] He wanted the space programme to change the U.S.

This is a long way from the idea that space investment merely resulted in Teflon and pens that can write upside-down. Webb wanted to claim that his Keynesian approach to spreading gigantic NASA contracts around the private and public sector was not merely political pragmatism, but also an example of large scale social engineering which by 1966 had resulted in 420,000 people owing their jobs to the organisation. It is this, rather than any sniff of the sublime, or starry-eyed belief in exploring frontiers, that provides his rationale for spending so many billions of dollars. Webb is clear that much of what NASA did had nothing to do directly with the Moon landing, but was basic science and engineering aimed at developing long-term national capacity-unmanned probes, new materials, better missiles but also a more sophisticated knowledge of the sort of large scale administration and management upon which all these sort of projects depended. Not very far in the background here is a response to one of the most perplexing questions left by Apollo – "*if we can put a man on the Moon, why can't we . . . ?*" And whether the question ends with education, poverty, housing, healthcare or whatever, Webb suggests that big projects from big government are the answer.

1.3.2.2 Bureaucratisation and freedom

Big aggregations of resources and power, like big rockets, need sophisticated control systems that can deal with turbulent conditions and provide the sort of "*dynamic equilibrium*" required to keep it moving forward.[62] So whilst Webb has to appreciate the importance of a certain sort of bureaucratisation in order to achieve co-ordination of a large number of people and things, he is continually also aware of the dangers of command and control, in terms of political authoritarianism: "Otherwise we might well destroy the values we are trying hardest to preserve and promote".[63] Getting to the Moon first will be pointless if it merely demonstrated that the Russians were right, so "freedom" needed to be both a means and an end. "Goals that depend on undesirable systems are undesirable goals".[64] "Bigness", whilst often needed to create great nations and great events, is also something to be feared, something that needs to be treated with a certain suspicion. But, suspicion or not, Webb has enough faith in science and research to suggest that much can be learned for the future by comparing NASA with other large forms of organisation and innovation. The railroads, U.S. Steel, General Motors, Du Pont chemical and others are compared with huge State-sponsored projects, particularly that paragon of new deal Keynesianism, the Tennessee Valley Authority. These comparisons are not systematic, though certainly insightful but it is the wartime organisation of the State that is most useful in understanding Webb's perspective on what the State can do.

Comparing NASA to the War Production Board, the Manhattan Engineering District Project and the Office of Price Administration allows Webb to show the modern State as necessarily entangled with the private and public sector in all sorts of complex ways. During the war, the partial control of large corporations and the direction of universities were effectively ceded to State bureaucracies whilst the State printed money and attempted to control supply and demand. In 1958, in the middle of the cold war, this was the institutional history which administrators like Webb understood, at the same time that the more perceptive amongst them also understood that demonstrating the virtues of the free market through state control was a rather contradictory project. Webb, and NASA, were the culmination of an understanding of government interventionism that begins with Roosevelt, runs through Truman and Kennedy, and ends with Johnson's "great society" initiatives. Even Eisenhower, the cold warrior in the middle who reluctantly began the agency, was explicitly and famously suspicious of the ways in which the "military-industrial complex" could influence state policy. This was a generation of politicians who were shaped by the deep suspicion of business that led to new deal progressivism in the 1930s, but also the suspicion of state control and the suppression of liberties that were at the heart of the cold war understanding of what the U.S. was fighting for.

1.3.2.3 Elitism and public responsibility

This has some interesting consequences for Webb's account of organisation. He knows Taylor, Weber, and Fayol, as well as "the newer doctrine of the behavioural or participative school",[65] but is most impressed by the new sciences of systems, with their emphasis on operations, forecasting, and quantitative methods. Here the paradox once again surfaces, because Webb wants clear limitations on the freedom of the many, because participation in goal setting simply is not feasible or desirable for all of them, at the same time as the freedom of the few is to be enhanced and recognised as a special skill that must be discovered and nurtured – "leadership".[66] So we have computers and systemic control for middle management and below, but "*the art of administration*" for those at the top, because the computer might become "the master of the systems disciple rather than a useful tool in his hands".[67] Later on, in a chapter on leadership, he suggests that "executives" (not managers now) should not be constrained by defined areas of responsibility and control, or centralisation, or plans and objectives.[68] Neither do such people need to be "psychologically coddled" in the way that the participative school suggests.[69] These executives, and by implication Webb himself, are not the sort of vulnerable children who need to be looked after by personnel managers, or given instructions by their superiors.

The clear sense of elitism here is underpinned and legitimised by an assumption of "public responsibility". Whether the leaders are politicians, administrators, university presidents or CEOs, they all share the burden for their society, their nation. This means that it is not enough for any leaders, particularly business managers, to limit themselves to looking after their own organisation. Their views

> "must be expanded to include school dropouts, crime rates, the prevalence of poverty, the number of university graduate students, effectiveness of government policies, incidence of group violence, and even indexes of the nation's willingness to act like a great power on the international scene."[70]

There is a systemic view of the social at work here, which echoes the systems view of the complex organisation and adds a certain moralism. Organisational self-interest is not enough and, by implication, market mechanisms do not necessarily produce the greatest good for the greatest number. For example, the Office of Price Administration prevented profiteering and hence protected both consumer and taxpayer. The view of society as in some sort of dynamic balance, or equilibrium, also suggests that there will need to be continual vigilance about the unintended consequences of large organisations and new technologies.

Reading *Space Age Management* gives you the impression of a man who is more interested in public policy and administration than space. Like Eisenhower, Kennedy and Nixon (though perhaps not Johnson), there is no particular sense of Apollo being interesting in itself, but only as a form of politics by other means. And the means were organisational, led by "men of rare abilities".[71] Or, as Bizony expresses it, Webb believed in a "perfectible society, led by selfless philosopher-kings".[72] Webb left NASA in 1968, worn down by continual budget cuts due to the increasing costs of Vietnam, and bruised by the damaging enquiry into NASA's relationship with North American Aviation, following the Apollo 1 launch pad fire.[73] Ultimately, Webb's vision of great leaders shouldering great burdens required an agreement concerning what the aims of the social system were. As De Groot acidly concludes "only in outer space could consensus be reached"[74], simply because the rest of the great society projects – education, housing, healthcare – were too contentious. In the background to Webb's lectures was the Tet Offensive, the My Lai massacre, the Soviet invasion of Czechoslovakia, the assassinations of Martin Luther King and Robert Kennedy, student protests and Olympic athletes making black power salutes on the podium. Even space age management could not solve all those problems.

1.3.3 Ends

The end of Apollo seems trivial compared to the political and organisational complexity of the state organised means. A few rocks, some photos, a litter of urine bags, trampled dust and a collapsed flag. No-one has been back, and three of the astronauts who stood on the Moon are already dead. This is now history, a science fiction future fading into the past. In a sense, the space age was the end of the age of industry, its steaming roaring metal now replaced by the marketing of information, miniaturisation, and communication.[75] But it certainly wasn't the end of management. Indeed we could explain the ends in terms of the management means, as James Webb does. He, like Kennedy, was not really interested in space, but in what the state could do if enough organisations and resources were pointing in the right direction.[76] Apollo was intended to be a demonstration of this point, but I think it demonstrated something else too. The paradox was ultimately not about the state versus the market, but that NASA's complex organisation intimated something else, something entirely other, something that could result from organisation but never be explained by it.

Fig. 11. *The Apollo Moon Landing (source: "Moonwalk One: Apollo 11 landing film to be released on DVD." Telegraph UK 26 May 2009).*

In his book *American Technological Sublime*,[77] David Nye writes about railroads, bridges and skyscrapers as examples of the sublime response that can be created by awesome forms of industrial engineering. Or, to be more precise, he is writing about a particularly American sense of these as modern collective projects that are worthy of being tourist attractions in their own right. If the classical sublime was concerned with the emotions produced by peaks and storms, then this modern sublime is concerned with great energies, heights and distances. Nye devotes half a chapter to Apollo, and nicely demonstrates its aesthetic affinity with the atomic bomb, the World's Fair and the electrified cityscape. They were illustrations of the control of power, and of the power that can result from control, from being the most technologically advanced nation on the planet. Still, I think that Apollo exceeds all of these descriptions too. It is the end point of a hugely ramified and complex network of people, things and controls that inscribe the Earth in order to make space possible. But all this does not contain Apollo, does not organise and

Fig. 12. *David Nye's "American Technological Sublime" (source: Google Books).*

manage its meaning for everyone, as if the ends of management were consumed by its means.

In one of the stories in Italo Calvino's *Invisible Cities (1974)*, there is one city that rests on the Earth only by long flamingo legs, whilst the city itself is high above the clouds.

> "There are three hypotheses about the inhabitants of Baucis: that they hate the earth; that they respect it so much they avoid all contact; that they love it as it was before they existed and with spyglasses and telescopes aimed downward they never tire of examining it, leaf by leaf, stone by stone, ant by ant, contemplating with fascination their own absence."[78]

Space is this absence in a much more profound sense – the absence of human organisation, and the absence of any value that might mark an enduring difference between flesh and Moon dust. It is not sublime in Nye's sense, though the machines that took people there might be, but space itself is an absolute zero of meaning. It seems evident that astronauts would have been fascinated by the Earth and took endless photographs of it, as it receded and then grew through their triple glazed portholes. Outside was vacuum and very little, almost nothing. Certainly nothing that could be inscribed with the varieties of humanism that made the flight possible, and allowed an astronaut to look misty eyed with wonder at his "home". Outer space is nothing, a nothing that diminishes all our best-laid plots to insignificance. A desolation that (like Buzz Aldrin) we might call magnificent if we wish, but it is still only desolation, however we inscribe it with value.

Frederic Jameson says something similar, with reference to "science fiction" films, noting the difficulty of representing space when it is so easy to "paradoxically abandon space for the Earth-bound world of competition, government funding, male bonding, patriotism, science, bureaucracy, and technological innovation".[79] In order to fit space into a plot, of manifest destiny, or market managerialism, or humanism, space needs to be ignored, and become the stage set for a story. That is why Webb's book is really about the age of management and space is never given any space at all.

In Andrew Smith's touching book about the remaining Moon walkers, he suggests two rather interesting, and perhaps contradictory, ideas. One is that the book was motivated by one question, the question that everyone has always wanted to ask Armstrong – "*What was it like?*" Embedded in that question is the idea that there must be something else, something we do not yet know, despite the millions of words that have been spent on Apollo since. Perhaps a glint in his eye or some reassurance about infinity, from the man who was there, and could have blotted out the Earth with his thumb. But what could he say to us, this man who "never admits surprise"[80], the product of a space age management that was designed to fill space with organisation, and reduce chaos to prediction within normal parameters. The very person, who might open the Otherness that the Apollo programme had as its end, is incapable, because of its means, to tell us about anything but engineering.[81] His is the language of management, and the calculus of the engineer, all suited and booted in the body of a military man.

Forty years ago, William Burroughs captured the problem nicely – Dr. Paine of the Space Center in Houston says:

> "This flight was a triumph for the squares of this world who aren't ashamed to say a prayer now and then. Is this the great adventure of space? Are these men going to take the step into regions literally unthinkable in verbal terms? To travel in space you must leave the old verbal garbage behind: God talk, country talk, mother talk, love talk, party talk. You must learn to exist with no religion, no country, no allies. You must learn to live alone in silence. Anyone who prays in space is not there."[82]

Burroughs, the drug addled homosexual who would never be chosen by NASA, expresses a very important truth here. Getting humans to outer space requires space age management, it requires large scale organisation. But dreaming of humans being in outer space expresses a rather different desire, a sense of the sublime, or even the inhuman, that could make little sense to someone like Webb. The means may be organisational, but the ends could only make sense to people who do not see the present, or the Earth, as their home – utopians, and lunatics. For future adventures, it might be better to let the space age managers do their work of preparation, construction and scheduling, but then put a poet on top of the rocket to Mars.

---

[53] Webb, James Edwin. Space Age Management: The Large Scale Approach. New York: McGraw Hill, 1969.

[54] Bizony, Piers. The Man Who Ran the Moon. Cambridge: Icon Books, 2006. 79; Johnson, Stephen B. The Secret of Apollo: Systems Management in American and European Space Programs. Baltimore: Johns Hopkins University Press, 2002: 5.

[55] Johnson, Stephen B. op. cit.

[56] DeGroot, Gerard. Dark Side of the Moon: The Magnificent Madness of the American Lunar Quest. London: Jonathan Cape, 2007; Dickens, Peter and James Ormrod. Cosmic Society: Towards a Sociology of the Universe. London: Routledge, 2007; Parker, Martin. "Capitalists in Space." Space Travel and Culture: From Apollo to Space Tourism. Eds. David Bell and Martin Parker. Oxford: Blackwell, 2009: 83–97.

[57] Mailer, Norman. A Fire on the Moon. London: Pan Books, 1971; Nye, David E. American Technological Sublime. Cambridge, MA: MIT Press, 1994; Jameson, Frederic. Archaeologies of the Future. London: Verso, 2005; Parker, Martin. "After the Space Age: Science, Fiction and Possibility."

SciFi in the Mind's Eye: Reading Science Through Science Fiction. Ed. Margret Grebowicz. Chicago: Open Court, 2007: 275–88; Shukaitis, S. "Space is the (non)Place." Space Travel and Culture: From Apollo to Space Tourism. Eds. David Bell and Martin Parker. Oxford: Blackwell, 2009: 98–113.

[58] In this essay, I am assuming that any large-scale humans in outer space programme will not come from the private sector. The scale of the undertaking, and the unlikely short-term economic pay-off seem to make that a safe assumption for the time being.

[59] This chapter draws on part of Parker, Martin. "Space Age Management." Management and Organizational History 4.3 (2009): 317–32.

[60] Lambright, W. Henri. Powering Apollo: James E. Webb of NASA. Baltimore, MD: Johns Hopkins University Press, 1995; Bizony, Piers. op. cit.

[61] Webb, James Edwin. Space Age Management: The Large Scale Approach. New York: McGraw Hill, 1969: x.

[62] Ibid. 11.

[63] Ibid. 15.

[64] Ibid. 27.

[65] Ibid. 65.

[66] Not as common a term at that time, when "management" was still trying to distinguish itself from "administration".

[67] Ibid. 67–8.

[68] Though a few pages later he appears to contradict himself again, suggesting that "the large complex endeavour cannot allow the executive such freedom of personal choice" (Ibid. 145.). In context, he seems referring to middle managers, not "leaders".

[69] Ibid. 136.

[70] Ibid. 76.

[71] Ibid. 169.

[72] Bizony, Piers. op. cit. 68.

[73] Ibid. 154; DeGroot, Gerard. op. cit. 212.

[74] DeGroot, Gerard. op. cit. 259.

[75] Parker, Martin. "Memories of the Space Age: From Apollo to Cyberspace." Information, Communication and Society 11.6 (2008): 846–60.

[76] Klerkx, Greg. Lost in Space. London: Secker and Warburg, 2004: 134.

[77] Nye, David. American Technological Sublime. Boston, MA: MIT Press, 1994.

[78] Calvino, Italo. Invisible Cities. London: Secker and Warburg, 1974: 77.

[79] Jameson, Frederic. "The Square Peg in the Round Hole or the History of Spaceflight." Critical Enquiry 34.S2 (2008): 172–83. 183.

[80] Smith, Andrew. Moondust. In Search of the Men Who Fell to Earth. London: Bloomsbury, 2005: 349.

[81] Mailer, Norman. op. cit.

[82] Burroughs, William S. Word Virus. The Williams Burroughs Reader. Eds. James Grauerholz, and Ira Silverberg. London: Flamingo, 1999: 320.

1.4 Astronauts: from envoys of mankind to combatants

Nina-Louisa Remuss

With Europe aiming at becoming a leader in human spaceflight, the question of what impact astronauts might have on the development of space security and particularly, whether they might become involved in the weaponisation of outer space, becomes relevant. The international debate on "weaponisation" versus "peaceful uses" is the context for this analysis.[83]

It is possible to distinguish three potential future scenarios in this regard: first, astronauts indeed become involved and become an element of the weaponisation of outer space; secondly, humans engage for the purposes of verification and other peaceful security uses; and third, humans refrain from getting involved in this field (which does not rule out that military personnel – as has been quite common – are "peaceful" astronauts). While drawing on these three possibilities, this essay aims at providing a path analysis as well as a normative recommendation, thereby also considering the societal implications involved such as the prevalent image of the astronaut in Europe. Moreover, the current international legal framework – with the notion of astronauts as "envoys of mankind" (Art. V of the 1967 Outer Space Treaty) as centrepiece – will be looked at in the light of the path analysis.

1.4.1 The setting

Generally speaking, exploration is perceived as inherent to humans, with space-walks providing unforgettable images to the whole world.[84] Exploration as such is not to be understood as purely the realm of scientists but rather a societal enterprise mandating, defining and enforcing rules and ethics. Exploration can also be understood as "to adapt to situations you did not plan for"[85] or "to boldly go where no man has gone before".[86]

Today space is increasingly used militarily. While initially only the U.S. and Russia used space for military purposes, Europe, China, India and the developing countries are now increasingly using space for dual (i.e. civil and military) purposes. In addition, China has recently shown its anti-satellite capability through its Anti-Satellite Test (ASAT) of January 2008. Using weapons against satellites in such an ASAT-test proves the risk of a weaponisation of space.

In the context of increasing military use of outer space; Europe is looking for its own formative role and principled identity[87] and should therein carefully consider the role of military personnel as astronauts and the possible resulting scenarios.

1.4.2 Legal considerations

From its very beginning the issue of astronauts in distress was central in the discussion on international space law due to the fact that it became clear that the space race between the U.S. and the Soviet Union would concentrate on human spaceflight.[88] These considerations resulted among other things in the "Declaration of Legal Principles Governing the Activities of States"[89] in the Exploration and Use of Outer Space, whose first three principles are equivalent to Art. V of the Outer Space Treaty of 1967 (OST).

Art. V of the OST defined astronauts as the envoys of mankind:

> "States Parties to the Treaty shall regard astronauts as *envoys of mankind* in outer space and shall render to them all possible assistance in the event of accident, distress or emergency landing on the territory of another State Party or on the high seas. When astronauts make such a landing, they shall be safely and promptly returned to the State of registry of their space vehicle.
>
> In carrying on activities in outer space and on celestial bodies, the astronauts of one State Party shall render all possible assistance to the astronauts of other State Parties.
>
> States Parties to the Treaty shall immediately inform the other States parties to the Treaty or the Secretary-General of the United Nations of any phenomena they discover in outer space, including the Moon and other celestial bodies, which could constitute a danger to the life or health of astronauts." [emphasis added]

The *Agreement on the Rescue of Astronauts* of 1968, which complemented the regulations on the rescue of astronauts as contained in the OST some forty years ago, underlined the importance of astronauts. It aimed at moving away from nationalist thinking and instead underlined the importance of understanding astronauts as the envoys of mankind embedded in the peaceful uses of outer space as it advocates that astronauts are to be understood as being on mission for all mankind.

Additionally, each year around 140 nations vote for the resolution entitled "*Prevention of an Arms Race in Outer Space*", which is commonly known as PAROS. It recognises "the common interest of all mankind in the exploration and use of outer space for peaceful purposes", reaffirms the will of all States that the exploration and use of outer space "shall be for peaceful purposes and shall be carried out for the benefit and in the interest of all countries", and declares "that prevention of an arms race in outer space would avert a grave danger for international peace and security".[90]

In addition to the legal framework laying down the rights and duties of astronauts, discussions on the term "astronaut" have recently emerged. With the rise of commercial space flight as well as the increasing use of scientific personnel involved in the conduct of experiments on space stations such as the ISS, it seems that different categories of participants in spaceflight are being established. This article uses van der Dunk and Groh's definition, seeing astronauts as "human beings travelling into outer space for professional reasons of a non-commercial, non-private nature".[91]

1.4.3 Astronauts: envoys of mankind or combatants

Thus, while the existing legal international framework states that astronauts are "envoys of mankind", in practice the questions arises as to whether astronauts are really the envoys of mankind and apostles of peace as international law tries to convey? Art. V (OST) was originally meant to address the Cold War fears of both the U.S. and the Soviet Union that their respective astronauts might not be

Fig. 13. *Astronauts with a Military Background: from left to right Malcolm Scott Carpenter, Yuri Gargarin, Rakesh Sharma, Thomas Reiter (from l. to r.) (source: Absolute Astronomy. Nov. 2009. absoluteastronomy.com; "India seeks Russia's Help in Space Pilot Training." 26 Mar. 2008. Space Travel Nov. 2009. http://www.space-travel.com/reports/India_Seeks_Russia_Help_In_Space_Pilot_Training_999.htmlSpaceDaily; "Luftwaffenpilot Thomas Reiter im All." 5 Jul. 2007. Bundesregierung.de. Nov. 2009. http://www.bundesregierung.de/nn_914534/Content/DE/Archiv16/Artikel/2006/07/2006-07-05-luftwaffenpilot-thomas-reiter-im-all.html).*

treated correctly in case of unforeseen emergency landings in the lands of the other.[92]

With exploration being inherent to humans, nothing is probably more strongly associated with national pride than human spaceflight. The military and astronauts have historically always been connected to a great degree. While initially only the superpowers recruited astronauts or cosmonauts with a military background, even the Indian astronaut Rakesh Sharma and German astronaut Thomas Reiter were previously pilots for their respective national armies.

However, the fact that having a military background does not necessarily conflict with the principle of peaceful uses was enshrined in the Art. IV of the OST:

> "States Parties to the Treaty undertake not to place in orbit around the Earth any objects carrying nuclear weapons or any other kinds of weapons of mass destruction, *install such weapons on celestial bodies*, or station such weapons in outer space in any other manner.
>
> The Moon and other celestial bodies shall be used by all States Parties to the Treaty exclusively for peaceful purposes. *The establishment of military bases, installations and fortifications, the testing of any type of weapons and the conduct of military manoeuvres on celestial bodies shall be forbidden.* The *use of military personnel* for scientific research or for any other peaceful purposes *shall not be prohibited.* (. . .)" [emphasis added][93]

While Art. IV paragraph 2 prohibits the establishment of military bases, installations and fortifications, prohibits the testing of any type of weapons and emphasises that the conduct of military manoeuvres on celestial bodies shall be forbidden, it explicitly states that the use of military personnel for peaceful purposes shall not be prohibited. The wording of Art. IV was strongly influenced by the Antarctic Treaty of 1959, the Treaty Banning Nuclear Weapon Tests in the Atmosphere, in Outer Space and Under Water (Partial Test Ban Treaty) of 1963 and UNGA Resolution 1884 (XVIII) of 1963.[94] Specifically, the Antarctic Treaty (Art. 1 paragraph 1) contains a very similar provision on the use of military personnel, which was copied to the OST with only stylistic changes.[95]

In practice until now, astronauts or cosmonauts have not acted in a non-peaceful manner. Yet, recalling that according to Mike Horn, exploration is to adapt to situations you did not plan for, one can identify three possible future scenarios:

(1) Astronauts become involved in the weaponisation of space,
(2) Astronauts engage for the purposes of verification,
(3) Astronauts remain or actually become the envoys of mankind.

These will now be further elaborated upon.

1.4.3.1 Scenario 1: "Space Cowboys"

The first scenario is what can be referred to as the Space Cowboy scenario in which astronauts actively operate space weapons aimed against satellites, thereby engaging in warfare, and actively operate sensors detecting and intercepting ASATs.

While Art. IV paragraph 2 of the OST as well as the respective Art. 3 paragraph 4 of the Moon Treaty prohibit the establishment of military bases, installations and fortifications, the testing of any type of weapons and forbid the conduct of military manoeuvres on celestial bodies, both explicitly state that the use of military personnel for peaceful purposes shall not be prohibited. Given that not all States have ratified the Moon Treaty, its provisions might not be respected.

Art. IV paragraph 1 only prohibits nuclear or any other weapons of mass destruction. The formulation "any objects carrying nuclear weapons or any other kinds of weapons of mass destruction" can also be interpreted as prohibiting only weapons with mass destructive effects. Such narrow interpretations however do not find support in the language of Art. IV and its ratio legis.[96] The word "install" means "to place (an apparatus, a system of ventilation, lighting, heating or the like) in position for service or use". Thus, such a wording only prohibits the "installation" of weapons but not their actual use.[97] Art. IV does however prohibit the testing of all conventional weapons. In turn it explicitly allows the use of military personnel. Yet, it limits their use to "scientific research" and other "peaceful purposes".[98] Art. IX is of relevance in the context of Art. IV. It underlines that States Parties to the treaty "shall be guided by the principle of co-operation and mutual assistance and shall conduct all their activities in outer space (. . .) with due regard to the corresponding interests of all other States Parties to the Treaty". It also obliges States to conduct international consultations:

> "[i]f a State Party to the Treaty has reason to believe that an activity or experiment planned by it or its nationals in outer space including the Moon and other celestial bodies, would cause potentially harmful interference with activities of other State Parties in the peaceful exploration".

This provision may clearly affect the military uses of outer space.[99] In this context, Art. XII is also of relevance, providing that "[a]ll stations, installations, equipment and space vehicles on the Moon and other celestial bodies shall be open to representatives of other States Parties to the Treaty on a basis of reciprocity".

Fig. 14. *Fictitious Picture of a "Space Cowboy" (source: "Space Cowboy." Media Photobucket. Mar. 2009. http://media.photobucket.com/image/space%20cowboy/HurricaneGene/Texas.jpg?o=137).*

A change in the role of astronauts also has an impact on the perception of astronauts and, in more general terms, the perception of human exploration. As a consequence, they will no longer be perceived as the apostles of peace but rather will become combatants. Exploration as such will also be perceived differently resulting from a changed role of astronauts. It becomes a tool of the weaponisation of space. In the next section some examples of such a scenario will be outlined.

Between 1952 and 1954, as part of a campaign to sell space flight in a series of articles in Collier's magazine, the German-American rocket engineer, Wernher von Braun, presented a manned space station which would not only serve as a base for further exploration but also as an orbiting reconnaissance platform and battle station for achieving "space superiority" over the USSR. As one of its roles he mentioned the launching of nuclear missiles. In order for the space station to be protected, von Braun even proposed pre-emptive atomic strikes from space as a response to the development of a hostile anti-satellite capability. Von Braun described the space station as the "ultimate weapon" to impose a pax Americana on the USSR. Based on some previous work he referred to it as "Lunetta".[100]

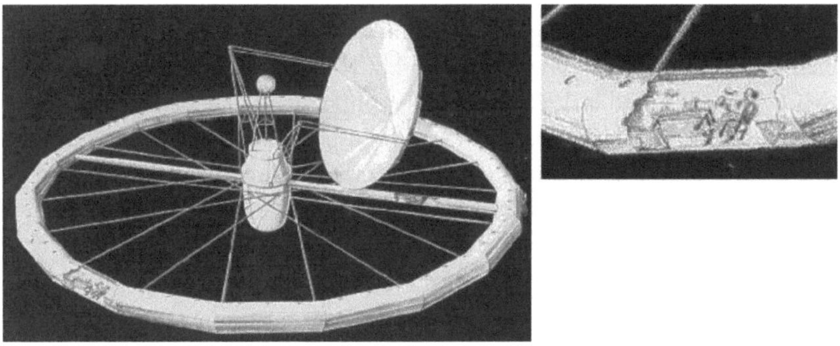

Fig. 15. *von Braun's manned Space Station (source: "Von Braun Station." Astronautix. Nov. 2009. http://www.astronautix.com/craft/vonation.htm).*

While von Braun initially presented its uses "as an observation station for both military and civilian purposed" and as a base for space astronomy, he went on to present the geostrategic argument to the wider public: "Our space station could be utilized as a very effective bomb carrier and for all present-day means of defense, a non-interceptible one". Von Braun foresaw a battle station for dominating Earth, dropping atomic missiles on "Soviet industrial and military facilities until the USSR collapsed."[101] Von Braun based his ideas most probably on the work of the German-Romanian space visionary Hermann Oberth, who mentioned reconnaissance from orbit and first broached the idea of a giant mirror, up to 100 km in diameter, that could be used to modify the climate, light the polar regions, or set enemy ammunition dumps, troop concentrations and cities on fire. He might also have used the work of Herman Noordung, actually Herman Potocnik, a former Austro-Hungarian officer, who published a book on the challenges of exploration, in which he provides an elaborated description of an inflatable, wheel-shaped space station that uses a solar concentrator mirror to generate electricity which proves largely similar to von Braun's space station.[102]

Since the very start of the space endeavour, the U.S. had been seeking to demonstrate that having military crews operating in orbit could contribute to U.S. military power.[103] In fact, from the visionary proposals of General Bernard A. Schriever during the 1950s for a major U.S. Air Force role in space, through the cancellation of the Manned Orbital Laboratory (MOL) programme in 1969, there were many suggestions for human space fight activities under dedicated military auspices. Both the Dynasor X-20 programme and the Manned Orbiting Laboratory (MOL) programme got close to fight testing.

NASA's Dynasor X 20 programme could be perceived as the first step in the military use of outer space. Numerous different versions of the ship were planned, including satellite inspection and electronic and photographic intelligence gathering. Later versions included plans for a mini space station and "orbital bombers" which could carry nuclear weapons into orbit.[104]

MOL, which started in 1963, was highly confidential. One of its objectives as mentioned in a confidential document was "to inspect satellites of the enemy", i.e. of the USSR, from close distance, to put them off the right orbit or even to destroy them. Given the scepticism that arose in the context of the Vietnam War and given the introduction of a new NRO programme for reconnaissance satellites that proved less expensive, none of the fourteen trained MOL astronauts ever went on this mission.[105] Ultimately the lack of a clearly defined mission that could not be performed either more cheaply or more effectively by other means doomed these programmes.[106]

From 1971 to 1986, military planning for human space fight focused on the Space Shuttle, to be operated by the National Aeronautics and Space Administration (NASA) and, once declared operational, to be used by the national security community as its sole means of access to space.

Military officers were trained as shuttle pilots making use of security payloads for dedicated military and national security missions. In the context of this endeavour, Pete Altridge issued a guidance, saying:

> "The Air Force has been examining the potential role of military man in space for over two decades. Thus far, our military space missions have not required man's presence in space. Thus, there has not been an identified role for the military man in space. However, with the advent of the space shuttle and man's routine presence in space, there is a greater opportunity to exploit man's unique capabilities." Accordingly, the following policy should be used in the planning of future space systems by the Air Force: "The Air Force policy is to ensure that the unique capabilities that can be derived from the presence of military man in space shall be utilized to the extent feasible and practical to enhance existing and future missions in the interest of national security objectives".[107]

While this guidance quickly became obsolete, it nonetheless proves that the Space Cowboys scenario is not that unrealistic after all. Considering the U.S. Vision 2020 one gets a similar impression. Already the title page is striking: a multicoloured cover showing a weapon shooting a laser beam from space. The Vision 2020 also calls for space superiority by saying: "Now it is time to rule space".[108]

Looking at different U.S.' documents, space is depicted as "the fourth medium of warfare-along with land, sea, and air"[109] and "the ultimate high ground". In 1996 Joseph W. Ashy, former commander-in-chief of the U.S. Space Command told Aviation Week and Space Technology:

> "It's politically sensitive, but it's going to happen. Some people don't want to hear this, and it sure isn't in vogue, but-absolutely – we're going to fight in space (. . .) We will engage terrestrial targets someday – ships, airplanes, land targets – from space".[110]

Furthermore, the Space Cowboy might become of relevance in the context of encountering extraterrestrial life.[111]

A Little Digression
Along similar lines Robert Zubrin in his book entitled "Entering Space – Creating a Spacefaring Civilization" of 1999, proposes a high degree of freedom for Martian settlers. Accordingly, higher levels of freedom would encourage people to move to Mars. In this context he says that Mars will only succeed if it can retain and innovate further the bet forms of law, culture and society Earth has to offer and leave the worst behind. Referring to the U.S. Bill of Rights as the "best one" he suggests an Art. 2 for a Martian constitution on the right to bear arms. Welcome to the Wild Wild Space.

1.4.3.2 Scenario 2: "Space Sheriffs"

The Space Sheriff Scenario; in contrast to the previous scenario is the scenario of military use but peaceful. In this scenario, astronauts engage as a means for verification and other peaceful but military/security uses. Such "Space Sheriffs" could check the purposes and claimed orbits of satellites, operate sensors to detect ASATs, counter space terrorism, verify the purpose of missions to Mars and Moon and conduct verification and treaty monitoring. Such a scenario can be understood as an answer to the current suspicion of China and other space newcomers.

With a increasing need for transparency and confidence building measures, such a scenario seems ever more possible in the near future. Currently EO satellites, so-called image intelligence (IMINT) or photo surveillance satellite and signal intelligence (SIGINT) satellites, help in terms of treaty monitoring and verification through, for example, the detection of underground nuclear tests revealed in satellite radar interferometry.[112] Pure imagery data is however of limited use. Derived imagery has to be placed in the relevant context, analysed and assessed. This is what is referred to as GEOspatial INTelligence (GEOINT), combining disciplines such as mapping, charting, imagery analysis and imagery intelligence. GEOINT organises and combines all available data around a particular geographical location and then exploits it in order to prepare products useable by

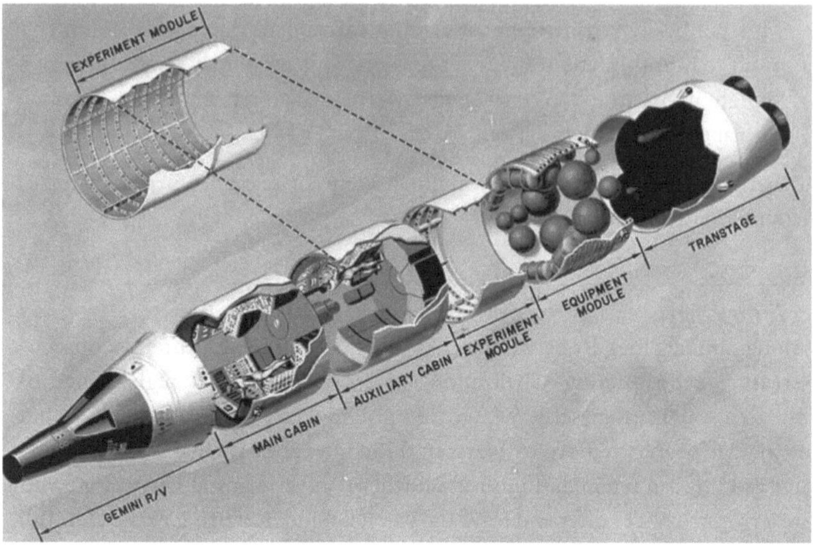

Fig. 16. *The MOL Project (source: Raumfahrer.Net. Nov. 2009. http://www.raumfahrer.net/raumfahrt/raumstationen/images/molcutusaf.jpg).*

decision makers, intelligence agencies and emergency responders.[113] The question arises as to whether the "Space Sheriffs" could contribute to treaty monitoring and verification missions as well. Such verification missions do not necessarily need to be limited to military missions but could also include civil ones such as missions in the context of climate change.

As already explained above, both the U.S. and the USSR pursued projects to develop spy space stations. The MOL programme described above also contained plans for astronauts to inspect USSR' satellites from close distances.[114] Similar to the U.S.' MOL programme, plans for a secret military spy space station were secretly developed in the USSR from 1964 with the project actually starting in 1967. In contrast to the U.S. Airforce project, the USSR Almaz Programme and its space stations called OPS "obritalnaja pilotirujemaja stanzija" were put into practice. They were 16 m long and weighted 20 tonnes. One of their instruments was a huge telescope, being directed at the Earth. It gave astronauts the possibility to get high resolution images showing types of planes positioned on U.S. aircraft carriers and nuclear missile silos as well as submarines. After some technical difficulties, the USSR managed to successfully launch the Almaz-2 (also known as Saljut 3) in June 1974. Pawel Popowitch and Juri Artjuchin were thus the first space spies, being able to even identify brands of cars. The USSR Almaz programme "was peaceful" in the sense that the astronauts did not shoot anything but were just taking pictures.[115]

Fig. 17. *In Verhoeven's "Starship Troopers" military astronauts are fighting bugs (extra-terrestrials) (source: MyVideo. 27 Jan. 2010. http://is1.myvideo.de/de/user_pics/59/pic_3980059_1206274440.jpg).*

Satellite imagery is today increasingly also used for monitoring of climate change. The analysis of the imagery is done on Earth by scientists. The same holds for the interpretation of satellite imagery for other purposes, e.g. maritime security and the fight against piracy or critical infrastructure protection. When thinking about larger space missions, settlements on Mars and increasing time constraints, one could however also think of involving astronauts in space in this process of interpretation.

With respect to possible space settlements, "Space Cowboys" could be tasked to enforce law and order in the settlement. Thinking of the "encounter of the unknown", i.e. the contact of humans with extra-terrestrials, "Space Sheriffs" might be tasked to mediate between humans and extra-terrestrials.

Given these possible activities, astronauts will be perceived as acting in some kind of policing function. Consequently, exploration as such is no longer of a purely civilian nature.

Another Little Digression

The idea of making use of special military forces against hostile extra-terrestrials has been the frequent subject of Science Fiction movies. In Paul Verhoeven's "Starship Troopers" (1997) volunteer high school students after their final year to become soldiers in the fight against bugs. The movie defines soldiers as someone who "takes personal responsibility for the security of the

(*Continued*)

Earth and defends it with his life". In the movie soldiers only fight in space. A new form of astronauts is depicted. Apart from scientific personnel, commercial staff, space tourists and today's ISS astronauts, a whole army consisting of mobile infantry, secret service and pilots has been created. Similarly, in James Cameron's movie Alien-2 (1986) a special marine troop, trained for fights against extra-terrestrials, using the newest technology is called to help a group of land settlers in the fight against hostile extra-terrestrials. Both movies give reasons for using military personnel as astronauts not only because of their pilot training but because of their tactical and training with weapons.[116]

1.4.3.3 Scenario 3: "envoys of mankind"

Scenario 3 is the most optimist of the three possible scenarios. Accordingly, all remains as it is. Astronauts continue to be recruited from the military staff only due to their experience as pilots. An arms race in space can be avoided. Astronauts continue to be perceived as the peaceful apostles and envoys of mankind and the concept of space exploration remains peaceful and civilian.

The "envoys of mankind", "working (...) fundamentally for a greater public common good"[117] even if this common good might be limited to one single nation, explore outer space bringing scientific knowledge to the Earth.

Discussions on the key phrase "envoys of mankind" in Art. V OST centred on notions such as "diplomatic envoy" and the meaning of "envoy" in more general terms. The original interpretation of astronauts "working (...) fundamentally for a greater public common good"[118] are today becoming questioned given the evolution of commercial space flight and the increase in space tourism. The international community rejected reading any implications into this term beyond the agreement and thus did not convey jurisdictional immunity to astronauts on the basis of this phrase.[119]

Such a scenario is in line with Kofi Annan's warning to not make space the next high ground:

> "Above all, we must guard against the misuse of outer space (...) We must not allow this century, so plagued with war and suffering, to pass on its legacy, when the technology at our disposal will be even more awesome. We cannot view the expanse of space as another battleground for our Earthly conflicts".[120]

When thinking about a possible encounter with extra-terrestrials, these "envoys of mankind" could act as some kind of diplomatic force representing earthly-humans to extra-terrestrials. They could further make sure that ethical reasoning in such an encounter is applied, making sure that microbes in outer space are treated in an ethical way.[121]

With more and more States participating in human exploration and manned spaceflight, the idea of the "envoy of mankind" will become further manifested as more and more States benefit from the "common good" that astronauts work for. A particularly interesting aspect will be the participation of developing countries in human exploration. At the same time, however, the increasing numbers of space tourists will undermine the idea of the astronaut as an "envoy of mankind".

The elaboration on the previous two scenarios showed that none of them is exaggerated or unrealistic. On the contrary, the analysis of U.S. and Soviet Union human space flight programmes has shown that similar scenarios have been actually planned. Thus, in order for astronauts to remain "the envoys of

Fig. 18. *James Cameron's "Avatar" (2009) uses military personnel for a mission to resettle the extra-terrestrials "Na'vis" on Pandora, a planet where humans exploit resources to solve the energy crisis on Earth (source: "Images." Official Avatar Movie Website. 26 Jan. 2010. http://www.avatarmovie.com).*

mankind" and for human exploration to remain as it is today, political precautionary measures have to be taken. The main question that arises is whether the international space law established 40 years ago still adequately regulates humans in outer space? Which concepts could be used to further regulate humans in outer space?

The Last Little Digression

In James Cameron's recent movie "Avatar" (2009), the former marine Jake Sully (Figure 18 sitting), is recruited to travel light years to the human outpost of Pandora, where humans are mining minerals to solve the Earth's energy crisis. Due to the toxic atmosphere of Pandora, the Avatar programme has been created. Avatars are genetically engineered hybrids of human DNA mixed with DNA from the native of Pandora the Na'vi (Figure 18 right).

Most humans participating in the Avatar Programme are scientists. They have the mission to learn more about the Na'vis who did not follow the human plans to move to a new settlement. Jake was initially chosen because his twin brother died (he has the same DNA and can thus use the expensive Avatar of his brother). Arriving on Pandora however Jake is also getting an additional mission by Col. Miles Quaritch (Figure 18 left). He is given the mission to infiltrate the Na'vi and inform the military based on Pandora about their vulnerabilities. This happens solely on the basis of his previous career and experience as a marine. Similarly to other Science Fiction movies the use of military personnel as some kind of astronauts is not questioned. Also, it is no longer justified on the basis of their pilots training but rather depicted as a logical consequence based on their previous experiences in conflicts.

When Jake however becomes part of the Na'vi community he learns to respect their culture and tradition and is impressed by their form of lifestyle and their knowledge. However, the marines abuse his knowledge and attack the Na'vis at their very core. Jake is fighting at the site of the Na'vis. Eventually, he manages to overrun the humans with the help of all creatures living on Pandora. As a result he and his friends (mainly the scientist community can stay on Pandora) most other humans have to leave the planet.

By entering the Na'vi community he can bring their knowledge, tradition and culture to the rest of the human world, thereby becoming a true "envoy of mankind".[122]

1.4.4 Conclusions and recommendations

"I am not an advocate for frequent changes in laws and constitutions, but laws and institutions must go hand in hand with the progress of the human mind. As that becomes more developed, more enlightened, as new discoveries are made, new truths discovered and manners and opinions change, with the change of circumstances, institutions must advance also to keep pace with the times".

(Thomas Jefferson)

This quote by Thomas Jefferson highlights that laws and institutions must go hand in hand with the progress of the human mind. Thus a change of circumstances also requires a change in the related laws and regulations.

Given the deadlock in relevant CD discussions, the U.S. National Space Policy of 2006 under the Bush Administration that opposed the development of new legal regimes infringing its right to use and access space, and the renewed focus on Transparency and Confidence Building Measures (TCBMs), the idea of a code of conduct for outer space activities has gained ground. As a non-legally binding instrument, a code of conduct, through which adhering States voluntarily commit themselves to rules of the road can be seen as an ultimate goal in itself, or as a building block towards a legally binding treaty, or as part of the more all encompassing concept of space traffic management.

Building upon discussions around an Italian food-for-thought paper, the Portuguese EU Council Presidency drafted a first version of a European Union Code of Conduct (CoC) in the second half of 2007. An updated version, entitled "Best Practice Guidelines for/Code of Conduct on Outer Space Activities" was circulated in the beginning of 2008, under the Slovenian Presidency. After taking several comments into account, the document was agreed upon in June 2008. France took over the EU Council Presidency in July 2008 making the proposal a priority. In December 2008 the Council of the European Union officially released its Draft Code of Conduct for Outer Space Activities.[123]

On the one hand the EU Draft Code of Conduct aims to strengthen existing United Nations treaties, principles and other arrangements, as subscribing States commit to make progress toward adhering to them, implementing them and promoting their universality. On the other hand, it aims to complement the United Nations treaties, principles and other arrangements by codifying new best practices in space operations, including notification and consultation. This should strengthen confidence and transparency among space actors and contribute to developing good faith solutions that allow access to space and the carrying out of space activities for all.[124]

The Draft Code of Conduct comprises a preamble and 12 articles, subdivided into four sections: Core Principles and Objectives, General Measures, Cooperation Mechanisms and Organisational Aspects. In the preamble, the EU recognises the "need for the widest possible adherence to relevant existing international instruments". The preamble further states that a comprehensive approach to safety and security in space should be based on freedom of access to space for all for peaceful purposes, preservation of the security and integrity of space objects in orbit, and due consideration for the legitimate defence interests of States. Additional general principles to be followed by the subscribing States are laid down in Article 2, such as "the freedom of access to, exploration and use of outer space and exploitation of space objects for peaceful purposes without interference, fully respecting the security, safety and integrity of space objects in orbit". Sections II and III introduce the rules of the road, augmented by relevant provisions covering space debris and notification of manoeuvring. Compliance and verification are ensured through a consultation mechanism and an investigation mechanism.

While repetitively strengthening the importance of the peaceful uses of outer space, the EU Draft CoC does not include any provision on human exploration. Given the above described possible scenarios it is thus of great importance that measures related to astronauts are incorporated in the EU CoC.

The EU Draft CoC stresses that it does not intend to replace other initiatives and that it complements and contributes to those initiatives by underlining the importance of taking all measures to prevent space from becoming an area of conflict. Thus it can be understood as a first step a future comprehensive regulation of space activities: space traffic management (STM). The term has been around for a number of years, but only a study by the International Academy of Astronautics (IAA), prepared between 2001 and 2006, looked into STM in an interdisciplinary and fundamental way.[125] The International Space University (ISU) and the International Association for the Advancement of Space Safety (IAASS), among others, have also undertaken work on space traffic management since this initial study. STM is defined as "a set of technical and regulatory provisions for guaranteeing safe access to outer space, operation in outer space and return from outer space to Earth free from physical or radio-frequency interference".[126] STM is not about tackling single issues, but regards the regulation of space activities as a comprehensive concept, based on the idea of regarding space activities "as a traffic system and not as disconnected activities of States".[127] It is a permanent solution to the issues of safety and security in space.

The comprehensive STM regime aims at drawing up rules in the areas of safety provisions for launches; safety provisions for human spaceflight; zoning (selection of orbits); right of way rules for in-orbit phases; prioritisation with regard to manoeuvre; specific provisions for geostationary and low Earth orbit, respectively; debris mitigation mechanisms; safety provisions for re-entry; and environmental provisions. While it is already planned to include safety provisions for human exploration, the STM needs to go even further in taking into account questions related to the military use of astronauts. Special protection and/or prioritisation of human spaceflight becomes necessary. While it should stress the importance of maintaining the existing restrictions, it should include further limitations on military activities by humans. In order to do so, there is a need for implementation of notification, control and enforcement mechanisms.

Given that the international space law established forty years ago does not seem to adequately regulate humans in outer space, the EU Draft Code of Conduct as well as proposals to establish a comprehensive STM should take up this issue and provide the necessary precautionary regulations to ensure that astronauts do not get involved in the weaponisation of outer space but remain the "envoys of mankind".

---

[83] For an analysis on the history of the discussion on the peaceful uses of outer space, the legal context and the current state see Rathgeber, Wolfgang, and Nina-Louisa Remuss. "Space Security – A Formative Role and Principled Identity for Europe." ESPI Report 16. Vienna: European Space Policy Institute, 2009.

[84] "Walking in the Void." NASA. 9 Mar. 2004. 19 Mar. 2009. www.nasa.gov/vision/space/workinginspace/eva_stats.html.

[85] Mike Horn in Worms, Jean-Claude. "Preface – Humans and Space." HIOS 1. vii.

[86] This is a quote from title sequence of most episodes of the original *Star Trek* science fiction television series that was used in Worms, Jean-Claude. op. cit. vii.

[87] Rathgeber, Wolfgang and Nina-Louisa Remuß. "Space Security: A Formative Role and Principled Identity for Europe." ESPI Report 16. Vienna: ESPI, 2009.

[88] Dunk, Frans von der, and Géradine Meishan Goh. "Article V." Cologne Commentary on Space Law: Volume 1 Outer Space Treaty. Eds. Hobe, Stephan, Benhard Schmidt-Tedd, and Kai-Uwe Schrogl. Cologne: Carl Heymanns Verlag, 2009: 94–102. 95.

[89] United Nations General Assembly. Declaration of Legal Principles Governing the Activities of States in the Exploration and Use of Outer Space. UNGA Resolution 1962 of 13 Dec. 1963. New York: United Nations.

[90] Grossman, Karl. "Master of Space." Third World Traveler. Jan. 2000. 19 Mar. 2009. http://www.thirdworldtraveler.com/Pentagon_military/MasterofSpace.html.

[91] Dunk, Frans von der, and Géradine Meishan Goh. op. cit. 97.

[92] Ibid.

[93] Art. IV para. 2.

[94] Schrogl, Kai-Uwe and Julia Neumann. "Article IV." Cologne Commentary on Space Law: Volume 1 Outer Space Treaty. Eds. Hobe, Stephan, Benhard Schmidt-Tedd, and Kai-Uwe Schrogl. Cologne: Carl Heymanns Verlag, 2009: 70–88. 73.

[95] Ibid.

[96] Ibid. 76.

[97] Ibid. 79–80.

[98] Ibid. 85.

[99] Ibid.

[100] Lunetta is a Science Fiction story von Braun had written at the age of 17.

[101] Neufeld, Michael J. "'Space Superiority': Wenrher von Braun's Campaign for a Nuclear-Armed Space Station, 1946–1956." Space Policy 22 (2006): 52–62. 52–5.
[102] Ibid.
[103] Logsdon, John M. "Human Space Flight and National Power." High Frontier 3.2 (2007): 10–13. 12 Oct. 2009. http://www.scribd.com/doc/1448693/US-Air-Force-AFD070322103.
[104] http://www.aerospaceguide.net/dynasoar.html.
[105] Hertenberger, Gerhard. Aufbruch in den Weltraum – Geheime Raumfahrtprogramme, dramatische Pannen und faszinierende Erlebnisse russischer Kosmonauten. Vienna: Seifert Verlag, 2009: 107.
[106] Logsdon, John M. "Human Space Flight and National Power." High Frontier 3.2 (2007): 10–13. 12 Oct. 2009. http://www.scribd.com/doc/1448693/US-Air-Force-AFD070322103.
[107] Ibid.
[108] Grossman, Karl. "Master of Space." Third World Traveler. Jan. 2000. 19 Mar. 2009. http://www.thirdworldtraveler.com/Pentagon_military/MasterofSpace.html.
[109] Ibid.
[110] Ibid.
[111] Schetsche, Michael T. "Encounters among the stars – exosociological considerations." (in this volume).
[112] Liebig, Volker. "Remote Sensing for Civil and Security Applications." Presentation. An AFCEA International Symposium on AEROSPACE TECHNOLOGIES and APPLICATIONS for DUAL USE. Hotel Parco del Principe: Rome, Italy. 13 Sep. 2007. http://www.afcearoma.it/presentazioni/internazionali/AFCEA_VLiebig_Final.pdf.
[113] cf. Hettling, Jana. "The Use of Remote Sensing Satellites for Verification in International Law." Space Policy 19.1 (2003): 33–9; For information on technical capabilities and existing R&D projects at both European and national level cf. Remuss, Nina-Louisa. "Space and Internal Security – Developing a Concept for the Use of Space Assets to Assure a Secure Europe." Report 20. Vienna: European Space Policy Institute, 2009. http://www.espi.or.at/images/stories/dokumente/studies/espi%20report%2020_final.pdf.
[114] Hertenberger, Gerhard. Aufbruch in den Weltraum – Geheime Raumfahrtprogramme, dramatische Pannen und faszinierende Erlebnisse russischer Kosmonauten. Vienna: Seifert Verlag, 2009: 107.
[115] Ibid.
[116] In Alien-2 moral considerations related to killing extra-terrestrials are touched upon as well. This has been the subject of analysis of Cockell, Charles. "Ethics and Extraterrestrial Life." (in this volume).
[117] Ibid. 98.
[118] Ibid. 98.
[119] cf. Ibid. 98.
[120] Grossman, Karl. "Master of Space." Third World Traveler. Jan. 2000. 19 Mar. 2009. http://www.thirdworldtraveler.com/Pentagon_military/MasterofSpace.html.
[121] cf. Cockell, Charles. op. cit.
[122] Official Avatar Movie Website http://www.avatarmovie.com/index.html.
[123] Draft Code of Conduct for Outer Space Activities, EU Council, document 17175/08, PESC 1697, CODUN 61, Brussels, 17 December 2008, Annex II, at http://register.consilium.europa.eu/pdf/en/08/st17/st17175.en08.pdf.
[124] A summary of the EU report to COPUOS can be found in Report of the Committee on the Peaceful Uses of Outer Space, UN document A/64/20, Vienna, 2009, paragraph 45.
[125] Corinne Contant-Jorgensen, Petr Lála and Kai-Uwe Schrogl (eds), 2006, Cosmic Study on Space Traffic Management, Paris, International Academy of Astronautics, at http://iaaweb.org/iaa/Studies/spacetraffic.pdf.
[126] Ibid.
[127] Schrogl, Kai-Uwe. "Space Traffic Management: The New Comprehensive Approach for Regulating the Use of Outer Space." ESPI Flash Report No. 3. Vienna: European Space Policy Institute, 2007.

1.5 Space inclusiveness and empowerment, or how The Frontier becomes a mirror

Adrian Belu

1.5.1 The current values of inclusiveness and empowerment in political and organisational endeavours

At the beginnings of the space age, belief in the world-harmonising virtue of the space frontier is likely to have been a widely shared belief, at least in the leading circles of the developed world. Over the decades, however, the magnitude of the task required such de-humanisation[128] that optimism about such an outcome for space endeavours is likely to have dwindled from its former levels.[129]

In political and organisational endeavours, one of the currently most active principles, or values, appears to be "inclusiveness". It could be loosely defined as: ensuring that no particular groups are left out of a given action. This may require wide-perimeter participation or involvement at the earliest stages of a given project. Organised by the European Space Agency-ESA, through the European Space Policy Institute (ESPI), and the European Science Foundation (ESF), the present Humans in Outer Space (HIOS) initiative has been active for three years now.[130] One of the main focuses of HIOS (as well as of its participating bodies in their respective missions) is access to space. Unfortunately, evolution has neither shaped our naïve physics, nor our naïve psychology, or for that matter none of our other untrained, natural skills, to the scale of outer space. It is interesting that at the opposite end of the "scene of operations" scale, nanosciences and nanotechnologies face a similar inclusiveness challenge.[131]

Of course, dematerialisation of interactions and exchanges has a role to play: outreach sector, event organising, school teachers flying, and more recently gaming[132] (also for training and talent spotting). However, individual access to space represents access to tangible resources.[133] The private sector will contribute in widening the access, and may even be able to provide for an inclusive component. The question remains: What structural initiative can be made at a political level towards inclusiveness in space access?

Related to "inclusiveness", another key term is "empowerment". It can be loosely defined as enabling individuals or groups to act, on issues they define as important. We start here from the argument that perspective is a universal empowerment asset ("teach a man to fish" Chinese proverb). More importantly, perspective sharing can be cost-effective (in the long term), and definitely is environmentally responsible. What perspective and representations (concerning space) can/should we ensure sharing of today, to contribute to a harmonious social tomorrow? Can we bridge the "scale gap" between the daily experience of homo sapiens and outer space?

1.5.2 The mirror of space: The Common mirror

It appears that we are living in happy times.[134] For a decade and a half now, space is no longer exponentially alien with distance, but is starting to form a mirror for us. Each month or so, exoplanets increasingly Earth-like are being detected and make the headlines.[135] With the successor of the Hubble Space Telescope scheduled to

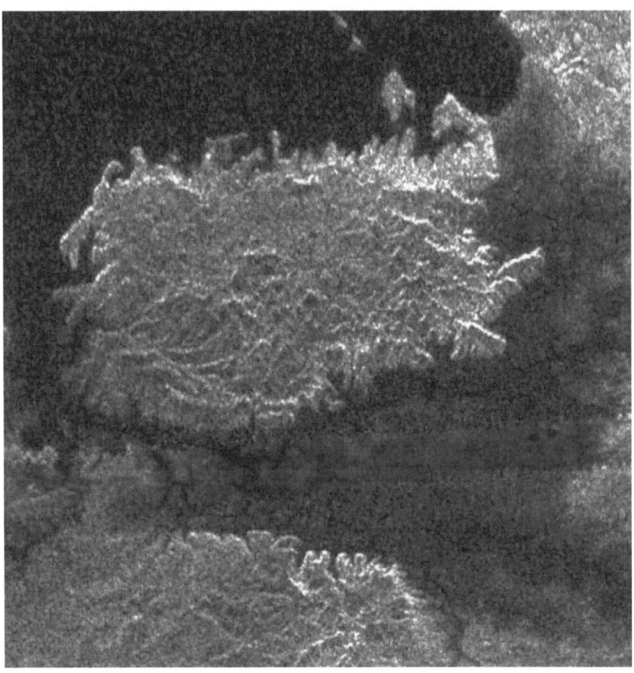

Fig. 19. *Radar image of Saturn's moon Titan, clearly showing "hydrological" networks above and beneath (through) a sitting liquid surface (lake). The liquid is natural gas (methane) (Source: NASA).*

fly just four years from now, research papers[136] on spectroscopic characterisation of habitable exomoons are being published in the same time as the premieres of the probably most viewed cinematographic entertainment ever.[137] Goodbye the scale gap problem. No more decades-worth of schooling, and above all social milieu encouraging, to train your perception in quantum physics or relativity. Atmospheres, oceans, rivers (Figure 19) are at Our common fingertips.

As first pointed out more than half a century ago, extrasolar planet detection is an endeavour that has been accessible to mankind for some time now.[138] So don't start getting all set up for field trip arrangements when your kid announces his class is going to observe an "exoplanet transit" at the end of the week: it will most likely happen in the school's backyard. This incredibly effective technique consists simply of taking advantage of the fact that, out of all the planetary systems out there, some just happen to be edge-on from our point of view. So, from our point of view, at least one planet happens to pass (transit) in front of its star, blocking part of the light periodically. It is like looking from a great distance at a matte energy-saving light bulb, on which a big fly patiently but steadily walks round and round.

Little surprise therefore that big science was done and probably will still be done with amateur support. The first transiting neptune for instance. The transiting nature of the planet (missed by its professional discoverers) was secured by a transcontinental consortium involving amateurs.[139] The conclusion was just this year: the first analysis of spectroscopic data from this planet, acquired with the Spitzer Space Telescope, was released,[140] demonstrating the pathway leading from ground to space endeavours.[141]

Eleven years ago, David Charbonneau was one of the discoverers of the transiting nature of the first known such exoplanet, with a 10 cm-wide telescope. Presently a Harvard professor and Principal Investigator of exoplanet studies for both the Hubble and the Spitzer space telescopes, he has brought the technique to maximum efficiency while remaining in the same budget and infrastructure range. The announcement last December of the closest transiting "Earth relative"[142] (6.4 Earth masses) was secured by a single observatory, entirely automatic, but consisting only of 40 cm-wide telescopes (MEARTH project, Figure 20). Given the uncertainties, this planet may well be close to habitable conditions. No wonder it is the highlight of this summer's observations of the Hubble Space Telescope, in order to obtain spectroscopic information.[143] Also, naturally, the discovering team has priority in proposing the space-based follow-up observations, therefore once again demonstrating how off-the-shelf, low cost hardware effectively paves the way to accessing a multi-billion euro, 20-year spanning space project,[144] and its associated international community at large.

Fig. 20. *The eight, 40 cm-wide telescopes of the MEARTH project (operating together in coordination in a totally automatic manner over the two year project span) have been visible on the Internet through three webcams for more than a year now (source: "MEarth has discovered a super-Earth transiting a nearby low-mass star!" Harvard University May 2010. http://www.cfa.harvard.edu/~zberta/mearth).*

How far beyond the academic and pop culture circle does the impact of this current exploration path reach? One anecdote probably worth mentioning is New York Times bestseller on President G.W. Bush's war cabinet having its title reused for the opening review of the 2008 International Astronomical Union symposium on this new particular technique.[145] In the meanwhile the involvement of the private[146] and for-profit sector are real. A very recent example of the latter is *Over Sky*, a French-based company, aiming a "private exoplanet search project".[147] The company is backed by a regular corporation: such "business models" seem to be fairly common for these types of projects, and in general could be dubbed "show window-patron symbiosis".[148] Whatever the level, the main ingredients remain:

(a) low or moderate cost equipment,
(b) a "network over which the sun never rises".[xix]

While the northern hemisphere will be surveyed nearly completely by MEARTH by next year, the southern hemisphere remains a fresh ground for a similar endeavour. This is an opportunity to improve on the original design. By not concentrating all the telescopes into one spot, but rather distributing them in longitude, one enables an even better completeness of the search, in just a couple of

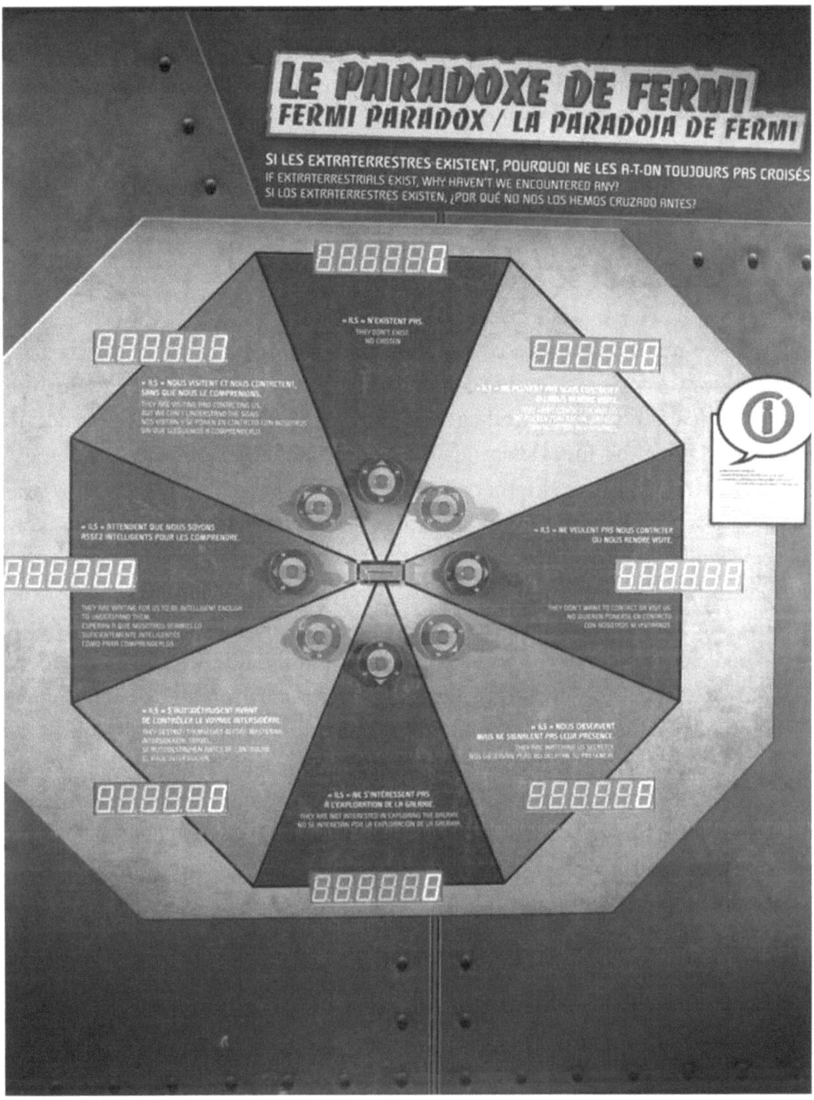

Fig. 21. *"Punch-your-opinion" wall at "Are you ready to meet Them?" exhibit running all 2010 at Cité de l'Espace, Toulouse, France. June 2010. http://www.pretalarencontre.com.*

months, while also fostering international cooperation within and with the Southern Hemisphere. Europe has the culture of such initiatives. It should not miss the chance. Robust and simple yet highly innovative hardware particularly adapted to this task is available today.[149]

It is not only about sharing of perspective and representations, but also effective citizen science[150] fostering. Social inequalities relative to space access rapidly vanish when one considers the scale accessible to simple observation. Observation, after all, is the first step of empowerment. How about a consortium of developing countries in the Southern Hemisphere discovering what will prove to be the first planet out there likely to bear Earth-like life?

1.5.3 Second payoffs of taking the risk to share exoplanet science

Finally, there may be a second payoff from such a unifying global project. This volume, as well as current scientific conference projects,[151] are gradually stating the need for "a scientific and societal agenda" for scenarios[152] of extraterrestrial life detection, whether this life is less, equally or more potent than ourselves. In the last case, sixty years ago, Enrico Fermi pointed out that these civilisations have had the means to 'discover' us several times over; however there seems to be no observational evidence that they have. One of the academically most agreed-upon[153] solutions to this paradox (Figure 21) is the so-called "zoo hypothesis", or "Prime Directive" (after the popular Star Trek series). It states that contact with an unprepared civilisation may prove harmful for it, and should therefore be avoided, in favour of discrete observation. Arguably, active participation of developing countries in exoplanet search is a major step forward in the global awareness that may be required for our recognition in a putative Galactic Club (even if at the beginning membership only consists of "you are not alone, but you are on your own"). However, this hypothetical consequence may also expose an inclusiveness and empowerment project such as the one outlined here to profound structural political anthropology limitations.[154]

[128] See Parker, Martin. "Managing Space, Organising the Sublime." (in this volume).

[129] Acknowledgements: The author wishes to thank A.-K. Martins, N. Baumard, and F. Selsis for fruitful discussions.

[130] "Inter-Disciplinary Activities at ESF – Humans in Outer Space (HiOS). "European Science Foundation. May 2010. http://www.esf.org/research-areas/humanities/strategic-activities/humans-in-outer-space.html.

[131] The creation of research consortia like the French *P2I-Physique des Deux Infinis* is a terrific opportunity to address this issue globally.
[132] Tarter, Jill, et al. [with contribution from the TED organisation and the Sapling Foundation]. "Getting the world actively involved in SETI searches." To appear in Astrobiology Science Conference, April 26–20, 2010, League City, Texas, Lunar & Planetary Institute. Also see study by the think tank MindArk PE AB (Sweden) for ESA's General Studies Program, 2010.
[133] On how to take you own photos from space for 150 U.S. dollars worth of hardware, see Lee & Yeh, Sept. 2009. Jul. 2010. http://edition.cnn.com/2009/TECH/09/21/space.camera.icarus.ireport/index.html.
[134] Also see the evolution of optimism relative to extraterrestrial life through the last three centuries: Cockell, Charles. "Habitability." Complete Course in Astrobiology. Eds. Gerda Horneck and Petra Rettberg. Weinheim: Wiley-VCH Verlag GmbH & Co. KGaA, 2007: 151–76.
[135] The exoplanet 'buzz', ignited fifteen years ago and showing no sign of curbing ever since, should not lead to forgetting the quiet and methodical progress of the whole astronomy field throughout the century, which is required for the proper and rapid understanding of current observations.
[136] Kaltenegger, Lisa. "Characterizing Habitable Exomoons." Astrophysical Journal 712.2 (2010): L125–L130.
[137] Cameron, James. "Avatar." 2009.
[138] Struve, Otto. "Proposal for a project of high-precision stellar radial velocity work." The Observatory 72 (1952): 199–200.
[139] Gillon, M. et al. "Detection of transits of the nearby hot Neptune GJ 436 b." Astronomy & Astrophysics 472.2.III (2007): L13–L16.
[140] Stevenson, Kevin, et al. "Possible thermochemical disequilibrium in the atmosphere of the exoplanet GJ 436b." Nature 464.7292 (2010): 1161–4.
[141] Besides their detectability with low cost equipment, the enormous interest of transiting planets is that they can be characterised spectroscopically with the highest sensitivity yet achieved. The principle is similar to the one behind the red colour of the Moon during a Moon eclipse: it comes from sun's white light being filtered through the blue-diffusing atmosphere of Earth. Already carbon dioxide, water and even an organic compound (methane) have been identified, alone or in subsets, in the atmospheres of several transiting jupiters and one neptune.
[142] Charbonneau, David; Berta, Zachory K., et al. "A super-Earth transiting a nearby low-mass star." Nature 462.7275 (2009): 891–4.
[143] "Hubble Space Telescope Approved Programs for Cycle 18." Space Telescope Science Institute. June 2010. http://www.stsci.edu/hst/proposing/exp_abstract-catalogs/cycle18-approved-programs.pdf
[144] See also Hubble Space Telescope's 'successor': the international James Webb Space Telescope. Scheduled for launch in 2014, the size of a tennis field when deployed, it will be situated five times further than the orbit of the Moon, and has a cost plan of approximately 4.5 billon U.S. dollars.
[145] Mann, James. "Rise of the Vulcans: The History of Bush's War Cabinet", New York: Viking Adult, 2004, cited by Carbonneau, David. "The Rise of the Vulcans", Opening review, International Astronomical Union Symposium 253; "Transiting Planets", May 2008, Boston, 2009: 1–8.
[146] For example in the Las Cumbres Observatory Global Telescope Network, founded by Wayne Rosing.
[147] cf. "Over Sky Corporation." Over Sky. May 2010. http://www.over-sky.fr.
[148] It must be noted that corporations are also a common founding choice for non-profit but financially well-off projects that wish to avoid the incumbent administrative and accounting hassles.
[149] Cf. the "20 inch Astro System Austria (ASA) Astrograph on ASA DDM 85 mount." Youtube. May 2010. http://www.youtube.com/watch?v=0Yhsiv2A_Po and Astro System Austria. May 2010. http://www.astrosysteme.at/eng/mount_ddm85.html.
[150] For the alternate perspective on how public constituency for space projects may arise from exoplanet citizen science, see Dasch, Pat. "Public involvement in extra-solar planet detection." Astrophysics and Space Sciences 241 (1996): 147–53.

[151] see for example the Dominik, Martin & Zarnecki, John eds. "The detection of extraterrestrial life and the consequences for science and society." 25–26 Jan. 2010, London; Royal Society. With amongst others Catherine Cesarsky as chair of one of the sessions. May 2010. http://royalsociety.org/Event.aspx?id=1887. Also Dominik, Martin and Zarnecki, John eds. "Towards a scientific and societal agenda on extra-terrestrial life." 4–5 Oct. 2010, Kalvi Royal Society International Centre, Buckinghamshire, London: Royal Society. July 2010. http://royalsociety.org/extra-terrestrial-life.
[152] For more on the methodology of scenario analysis see Schetsche, Michael. "Encounters among the Stars – Exosociological Considerations." (in this volume).
[153] Baumard, Nicolas. "Who can build a radio-telescope? For a reasoned anthropomorphism in evolution, in psychology and in social sciences." CNRS Workshop "Search for Extra-terrestrial life", Paris (France), June (2002). This approach relies on extending the biological concept of *convergence* (see Brandstetter, this volume).
[154] Wendt, Alexander and Duval, Raymond. "Sovereignty and the UFO." Political Theory 36 (2008): 607–33.

1.6 A school curriculum for the children of space settlers

Alan Britton

1.6.1 Introduction

If present day projections in relation to the future of space exploration are realised, at a point approximately midway through the current century humankind may witness the first generation of children to be born on another world. These young people will need to be nurtured and educated in a very different spatial, technological and cultural context. What knowledge, skills and values will have to be inculcated in these young people in order to equip them for life on another world, or in an orbiting habitation? In short, what school curriculum needs to be developed for the children of space settlers?[155]

At first glance the question of what to teach in space might appear to fall into the realm of idle speculation, if not outright science fiction. However on further reflection it is apparent that specific disciplinary interests might yet be relevant to such a question, given that they lie at the confluence of educational policy, pedagogy and citizenship; in other words the much – contested landscape where political, social and economic goals are translated into national curricular programmes for schools and other places of learning.

Like many people who were born during the high water mark of human spaceflight in the late 1960s, I have retained a passionate, albeit amateur interest in this field. I also believe strongly in the capacity of effective education to promote human progress. The British Astronomer Sir Patrick Moore has made the striking observation that the lives of two men; Orville Wright, the first to achieve powered flight, and Neil Armstrong, the first to walk on the Moon, substantially overlapped, indeed Moore had met them both.[156] The popular imagination has often struggled to keep pace with such rapid progress: the opening words of the 1956 science fiction film classic "*The Forbidden Planet*", were "[i]n the final decade of the 21st century, men and women had landed on the Moon".[157] The fact that the Moon landings followed only thirteen years after the release of this film, and only 66 years after Kitty Hawk, is a pertinent illustration of the extent of technological development during the 20th century, much of which can be attributed to the growing accessibility and effectiveness of public education in that era.

Fig. 22. *The first powered flight (1903). Only 66 years later, humans walked on the Moon (source: NASA).*

Education lies at the heart of human endeavour and ambition, not least in relation to space. Significant public education programmes around the world are devoted to fostering young people's interest in space, and in turn promoting careers in science and engineering, as well as bolstering support for space programmes themselves. Education about space is therefore generally well understood and reasonably well-resourced. The purpose of this chapter is to make the case for education in space, and to present a rationale for what it might look like. In embarking on such a journey, one can be both reassured and challenged by Brumbaugh's[158] observation that "(T)here are no future facts, but there are no past possibilities". In other words, while we may remain uncertain about particular aspects of the future of humans in space, we can build our speculation on solid foundations to anticipate situations and trends, and to arrive at the requisite curricular response. As with the best science fiction, when we undertake informed guesswork about the future, very often the greatest meaning can actually be derived about our present condition.

1.6.2 What is "curriculum"?

Curriculum can be understood from a variety of perspectives. The word is derived from the Latin for a *race course*, in other words symbolising the journey on which the student must embark through learning. Bernstein[159] suggested that "curriculum defines what counts as valid knowledge", in other words the knowledge that a particular state or society considers relevant, true and apposite. In pedagogical terms it might be regarded as "all the learning, which is planned and guided by the school, whether it is carried on in groups or individually, inside or outside the school".[160] This latter definition encompasses notions of both the overt and the "*hidden*" curriculum, that is, explicit categorised knowledge as well as the learning that is implicit (indeed sometimes inadvertent) in matters of teacher–pupil relationships, school and examinations structures, processes of socialisation, values and ideals, and so on.

The postmodern/poststructuralist turn in educational scholarship contributed to the identification of a whole series of underlying purposes being attributed to curriculum, which was now interpreted as a form of "*text*", including the view that it is never values-neutral. Rather it is in essence political, or racial, or gendered, or phenomenological, or aesthetic or even theological.[161] There is another strand of thought (stretching back as far as Dewey, if not to Plato) that sees curriculum and education more generally as a humanistic and idealistic undertaking that seeks to improve society. Bruner suggests that "education should serve as a means of training well-balanced citizens for a democracy".[162] Pring believes that curriculum should be "the forum or the vehicle through which young people are enabled to explore seriously (in the light of evidence and argument) what it is to be human".[163]

The notion of curriculum as public hope[164] or "utopian text" has recently been revisited.[165] It is this latter formulation that is perhaps most relevant to curriculum development in the context of space colonisation, given that it shares the often optimistic outlook of space and technological discourses and the astrofuturist movement. If we are to build an "ideal" society in future space communities from a *tabula rasa*, we must construct a curriculum that could contribute to that process. Before sketching out what I believe that curriculum might look like, I will briefly outline some of the ways in which educationalists have already begun to conceptualise and envisage the future of education.

1.6.3 Education, curriculum and the future

Many facets of schooling often tend to be unchanged vestiges of 19th or early 20th century practice (Cuban[166]), and most schools are still organised around particular ages and stages due to De La Salle's influence in the 18th century.[167] In response to this, over recent decades some educationalists have begun to examine *educational futures*, a field which seeks to predict the likely evolution of schooling in the medium- to long-term. In 2001 the Organisation for Economic Cooperation and Development (OECD)[168] developed six scenarios for future schools in 2020 that described a possible spectrum from status quo (minimum change and schools to look and feel largely familiar to the present time) through re-schooling (significant change that nonetheless sees schools as the on-going focal point for educational activity) to de-schooling (where the school becomes essentially an obsolete institution to be replaced by radical technologically-driven alternatives). This last vision is shared by influential commentators such as Ivan Illich[169] and Manuel Castells,[170] who have prescribed a radical repositioning of education, suggesting that schools be replaced by learning networks and learning webs, within a wider network society thus in part anticipating the (as yet not fully realised) potential of the internet as an educational tool within and beyond the classroom.

There is also an emerging literature in the specific area of curriculum futures, which has appeared as a response to factors such as globalisation and the rapid pace of change in relation to information and communication technologies. Many educationalists argue that we do not equip young people adequately for these new realities. For example Kress suggests that "the presently existing curriculum still assumes that it is educating young people into older dispositions, whereas the coming era demands an education for instability".[171] Others argue that curriculum will always tend to be behind the pace of contemporary social and economic movements and realities. This has been captured well in the fable of the Sabre-tooth curriculum,[172] which describes how a tribe of early humans persists in teaching a "classical curriculum" long after climatic and environmental changes have rendered it obsolete.

One challenge to the apparent ossification of the modern curriculum is a strand of thought that advocates education *about* the future. Hicks observes that "we live in a world of rapid change where yesterday's certainties are no longer an effective guide to tomorrow".[173] He suggests that the contemporary schools curriculum gives little or no consideration to the preparation of young people to anticipate, plan for, or cope with the future. The skills of clear and thoughtful foresight ought to be viewed increasingly as a "survival skill". Elsewhere Inayatullah suggests that if such forecasting within the curriculum is to be meaningful, it must be founded upon two basic epistemological considerations: "the technical concerned with predicting the future and the humanist concerned with developing a good society".[174] Such considerations help to form the framework by which we might construct a curriculum for the children of space settlers.

1.6.4 Education and space

It is a revealing exercise to explore the distinction between education *about* space and education *in* space. As suggested previously, there is substantial evidence of the former, whether in schools or through high profile competitions, public exhibitions and a variety of documentary media; while the latter appears barely to have been considered. Furthermore, there is a significant literature on the future of space exploration and the logistics of space colonisation, while there are only fleeting references, if at all, to the idea of the classroom or teacher in space. It is perhaps unsurprising that authors (whether of fiction or non-fiction) tend to focus on a cast list of heroic terra-forming engineers, humanoid robots, space marines[175] and spaceship pilots. In science fiction the narrative potential of education is largely unrealised and teachers are rarely encountered as *dramatis personae*. A famous exception is in the (very) short story "The Fun They Had" by Isaac Asimov (1951), set in a future where the delivery of education is highly personalised and delivered on a one to one basis, albeit by "mechanical teachers".[176]

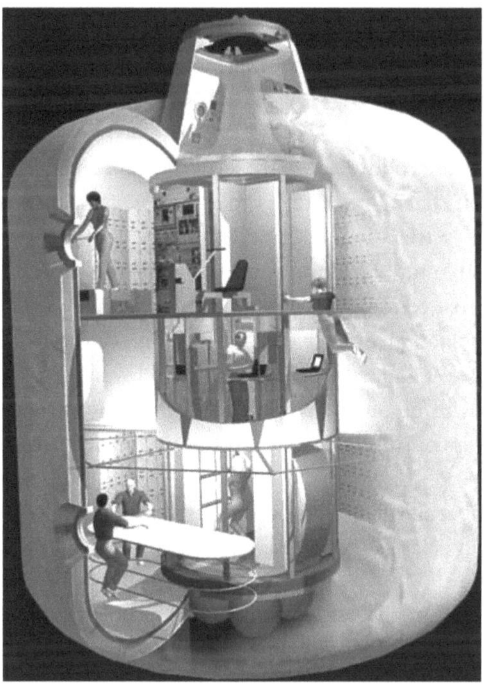

Fig. 23. *A contemporary school's vision of future habitation modules: "The Asten Project." (source: NASA).*

In non-fiction, there are sporadic references to the provision of educational facilities in space.[177] In other similar recent works, however, such as Schrunk et al.[178] as well as Crossman and Zubrin,[179] the provision of educational facilities is barely evident at all, despite otherwise highly detailed descriptions of the amenities required. The NASA Space Settlement Contest,[180] given that it is aimed at schoolchildren themselves, encourages entries that recognise the need for some form of educational provision in space colonies, although there tends to be limited consideration of the content of education in space. Due credit should be given to one of the 2009 Grand Prize entries, however, which suggests "three areas of focus: social development, academic development and linguistic development" together with a short rationale for these themes.[181] Elsewhere, educators sometimes prompt young people to imagine a future for humans in outer space, expressed through art or literature.[182] Such initiatives show that young people are able to engage with such concerns in creative and thoughtful ways when given the opportunity. However it is to be hoped that if and when the time comes to embark on detailed planning for real-life colonisation there will be a role for educationalists. Taking into account present knowledge, what might such a planned curriculum entail then?

1.6.5 A curriculum for the children of space settlers

The need to educate the children of space settlers would emerge during the second wave of colonisation that O'Neill refers to as the *humanisation* of space[183] – the point at which the "settlers will become less heroic and less exceptional".[184] In establishing a possible rationale and framework for this education, it should be noted firstly that most contemporary educational theorists argue that curriculum *content* cannot be planned or delivered in isolation and without simultaneous attention to the *method* and *materiel* of teaching. However in the context of this chapter, the focus will be on the subject areas that might form the core of the curriculum, as well as considering some of the objectives that the curriculum would have to fulfil.

I propose a number of illustrative considerations that ought to help justify the particular choices in establishing a space-based curriculum, including:

- The social, cultural, medical, emotional and psychological challenges that the young people will be confronted with.
- The knowledge, skills and dispositions that will be required, valued and rewarded by the community in which the young people are raised.

- The need for every young person, regardless of their particular needs, abilities and intellectual interests, to be catered for and included in the educational process. This might include children born with specific behavioural or physical needs that might be amplified in a space colony.

We can anticipate some predictable concerns. There is already a significant body of literature and empirical evidence on the medical and psychological impact of space travel on humans. We can predict with reasonable certainty that the children of space settlers would face the same issues of claustrophobia, isolation, lack of privacy, fear of equipment failure, and so on, that all settlers might be prone to.[185] Elsewhere in this volume Dunér describes a potentially new state of human consciousness that he terms "astrocognition";[186] the perceptual and emotional shift that might come about when humans commence long term habitation or exploration of space. The curriculum would almost certainly reflect this shift, and indeed in time might come to reinforce and reproduce it, further reinforcing a growing cognitive, as well as physical, distance between settlers and Earth-bound humans.

There are some specific concerns that might apply to young people born off-planet. One factor that could be easily overlooked is what might be termed "Earth envy", a potentially powerful source of alienation for young people who, as they mature, become more embittered that their parents have chosen to introduce them to such a harsh and potentially limited natural and social environment that forecloses any ambitions they might have for life and work beyond such confined circumstances. Unlike their parents, they have not volunteered for this life. For

Fig. 24. *The Children of Space Settlers have not volunteered for such a life and might develop a form of "Earth envy" (source: NASA).*

those children born on Earth but forced to leave their homes and friends behind to accompany their parents into Space there might be even greater negative reactions. Another legitimate concern would be the potentially disruptive impact of pubescent hormones in very enclosed spaces. Such alienation and associated developmental issues would have to be addressed through therapeutic and aesthetic educational approaches centred on the young people themselves.

Most contemporary analysts suggest that any serious attempt to colonise the Moon or one of the planets would require multinational collaboration in which the considerable expense and development would be shared by a coalition of different nations. This would entail a complex cultural dynamic that could be the source both of positive impulses around a shared transnational human enterprise, alongside potential fracturing of the community along national, religious or cultural lines. Such a dilemma is faced by many contemporary nations as they try to cope with increasingly complex societies through a "delicate balance of unity and diversity".[187]

Curriculum may hold the key to addressing all of these concerns in a sustainable space community. I would make a tentative proposal for three very broad "domains" of skills and knowledge that would be essential for children growing up in such a context: firstly *Technological and Practical*; secondly *Ethics, Values and Cognition*; and thirdly *Creative, Aesthetic and Well-being*. The first would address the obvious need for a strong technical and scientific skill-set within a very self-reliant community. The second would provide young people with the intellectual and ethical capacities and habits to cope with the complex and rarefied interpersonal dynamics of an enclosed community. The third would help them to find adequate means of artistic and emotional expression to ameliorate legitimate feelings of isolation and bitterness. Each domain could be addressed through coherent and progressive "spiral" programmes that take them from infant/kindergarten stage through primary to some form of qualification/graduation from the secondary stage. Thereafter their education would in all probability be required to be highly specialised, however with the caveat that a focus should remain on opportunities for choice, enjoyment, creativity, conflict resolution and well-being. At all stages I envisage not just a conventional model of nine-to-four learning in a "classroom". The proposed breadth in the curriculum would have to be delivered in partnership with parents and other members of the colony, and would benefit from active learning opportunities throughout the settler colony and its environment. The whole community would be involved in considering the shared values implicit in the curriculum, and the young people themselves would be consulted on educational issues that affect them; Article 12 of the United Nations Convention on the Rights of the Child already promotes such engagement, stating that "States Parties shall assure to the child who is capable of forming his or her own views the right to express those views freely in all matters affecting the child, the views of the child being given due weight in accordance with the age and maturity of the child".[188]

1.6.5.1 Technological and practical

Some of the most effective practice in present day education (measured by international comparisons in attainment[189]) suggests that until about the age of six or seven the bulk of "school time" is best spent in semi-structured play situations and informal learning which could nonetheless be related to the three proposed domains. Thereafter, the students could examine specific topics, sometimes in isolation but wherever possible in an interdisciplinary manner, thus promoting synthesised thinking and meta-cognitive development. At all stages a set of shared underpinning values would support the visible content. For example, they might learn the principles of terra forming in their particular lunar or Martian context through the integrated study of mathematics, physics, engineering, chemistry, microbiology, geology and architecture. Other subjects that might fall within the Technological and Practical domain could include: agriculture and biosphere maintenance; design; astronomy, including Earth observation and mapping through telescopes; and information technologies. Interdisciplinary studies across overlapping domains might also explore the ethics of scientific discovery and environmental sustainability, and, following Cockell's perspective,[190] there might be a need to confront such issues all the way down to the microbial level.

1.6.5.2 Ethics, values and cognition

Under Ethics, Values and Cognition the curriculum might concentrate on issues such as conflict resolution; intercultural communication, including languages (allowing for dialogue between the cultures represented in the colony as well as real time links with young people on Earth); the comparative study of world religions, as well as atheism and humanism; citizenship, politics and government; history of science; and philosophy, including epistemology and thinking skills. Such breadth of subject coverage may appear daunting and perhaps superfluous in what would be very much a technology and "hard science" focused community. However in order to build and maintain the cohesion of such a society I believe there would have to be a strong focus on these dimensions of the human experience, together with a shared understanding of the formal (legal) and informal (behavioural) parameters established within the community. The emphasis on understanding the nature of knowledge and learning would also build a flexible and critical dimension into the young people's intellectual abilities and growth, enabling them to react effectively to the unforeseen, and being able to learn new skills as the need arises while the colony develops and expands.[191]

1.6.5.3 Creative, aesthetic and well-being

Finally, the third domain, "Creative, aesthetic and well-being", would concentrate on developing the individual dispositions and creativity of each young colonist while taking care of his/her physical and emotional health. Topics might be explored in a more open-ended manner in this domain, with a significant element of personalisation and choice. These could include: habitat enhancement; understanding contemporary and historic human events and culture(s); appreciating a global canon of literature, art and music (as well as opportunities to develop these creative skills in their own right); physical education, including sports, games and special exercises to cope with the effects of low gravity; and emotional well-being: learning to recognise for themselves the signs of isolation, depression, interpersonal conflict and to act against them promptly. Sexual and reproductive health education would be a core necessity whatever the possible inhibitions of some parents. More generally, space within the curriculum (and the wider community) would also have to be set aside for the individual imagination to flourish in what might otherwise seem to be creatively impoverished surroundings. For example, there might be value in maintaining a culture of "story-telling" that brings memories of the Earth more vividly to life, such as the seasons, the grandeur of the natural landscape and the seas, and the planet's cultural diversity and biodiversity. Such an approach would resonate with "folkloric" traditions on Earth[192] and would provide opportunities for remembrance and continuity. Students who themselves were born on Earth but had "emigrated" at the behest of their parents to off-planet colonies might also bring fresh perspectives (as well as concerns) to the classroom, and considerable effort would have to be made to ensure their integration. The curriculum would also be enriched by some component of service to the settler community, promoting a sense of shared purpose, cultural identity and cohesion within the settler community.

The curriculum sketched out above could indeed be viewed as a "utopian text", seeking as it does to build an ideal society capable of transcending Earth-bound cultural differences (while acknowledging and promoting awareness of these) in order to create a new supra-national community. One can anticipate that such an aspiration would be vulnerable to a number of challenges and tensions that would have to be confronted in planning and delivering the educational programme.

Fig. 25. *An Illustration of a Space Settlement (source: ESA).*

Such tensions might include an emerging gap between the celebration of international cooperation in space exploration on the one hand, and the possible tendency of this educational programmes to foster a citizenry imbued with an "assimilationist ideology".[193] The probable multinational nature of the community would also suggest the need to reconcile different pedagogical traditions, as well as different versions of history, from the various nationalities and cultures.

Any unresolved cultural divergence might express itself in a gap between the educational values that underlie the proposed curriculum, and pedagogical instincts that might revert to traditional rote-learning under the pressure of time and the pragmatic need to focus on the short-term acquisition of key survival skills. There is also a danger that the desire to promote the significant curricular breadth proposed in my three domains might lead to lack of coherence and inevitably shallow learning. These concerns highlight the need for well-planned integrated curricular experiences and the input of a multinational panel of specialist educators at all stages in the process of curriculum development and delivery. Many educators around the world are already skilled in shaping educational experiences for a multinational and multilingual student body, whether in the various International Schools around the World, or in establishments that have a significant proportion of students from immigrant or second language communities. It is possible that teachers from such backgrounds would be natural candidates in any future recruitment drive.

1.6.6 The Apollo programme

The history of space flight has epitomised both the best and the worst of humankind. Despite the brief near-global euphoria of the first Moon landings and the Apollo programme they are often viewed in retrospect as a costly proxy of "ideological confrontation and nuclear war-often described by such euphemisms as world "leadership" and national "prestige".[194] The truly astronomical financial cost of Apollo, and the role of Wernher von Braun and his team in the development of rocket technology add to the ethical discomfort felt by many.[195] At the same time Apollo provided the "inescapable recognition of the unity and fragility of the Earth . . . (as) . . . its clear and luminous dividend.[196] Smith suggests that there were in fact two Apollos running in parallel: the "official one about engineering and flying and beating the Soviets" and another, "almost clandestine . . . about people and their place in the universe, about consciousness, God, mind, life".[197] The series of photographs of the Earth from space reinforced this impression, and according to some made a significant contribution to the emergence and growth of environmental movements.[198]

It was therefore easy to be inspired to research this field in preparing this chapter, given that it coincided with the 40th Anniversary of the Apollo 11 Moon landings. The ability of space and space exploration to inspire the imagination and ambition of humankind has been reasserted with serendipitous timing. A new generation of young people are learning about the inspirational exploits of the early pioneers at the same time as contemporary technologies are facilitating both greater visual understanding of space (notably through Hubble, and imminently, the European Space Agency (ESA) Planck and Herschel observatories) and the means to communicate and disseminate such understanding through the internet.

Education is one of the key routes towards public understanding of, and support for, space exploration, whether manned or robotic, and it has a central role in addressing the technological challenges and ethical complexities implied by the future of space exploration. At its best it has the capacity to engage young people through critical reflection and the fostering of alternative visions.[199] Inayatullah suggests that education can help in "decolonising the future",[200] in other words fostering the ability to imagine alternatives in the face of attempts by powerful forces to lay claim to the future in the same way that they dominate the present. In times of bleak assessments of future climate change, species extinction and habitat

Fig. 26. *Footprint from the Apollo 11 Moon Landing: Inspiring the Imagination (source: NASA).*

loss, the opportunity for young people to explore alternative, more optimistic futures, and the contemporary actions by which such alternatives may be realisable, is a fundamental ethical enterprise. For this reason alone, education ought to contribute to the interdisciplinary analysis of the future of humans on both Earth and in outer space.

[155] Addressing this question has been one of the more unusual tasks with which I have been charged in my life as an education academic. I was invited by the European Science Foundation (ESF) to contribute to the on-going "cross-disciplinary European dialogue" (Van Donzel, Monique and Jean-Claude Worms. "Preface: Humans and Space – Space and Humans." Humans In Outer Space – Interdisciplinary Odysseys. Eds. Luca Codignola and Kai-Uwe Schrogl. Vienna: SpringerWienNewYork, 2009: v–viii. v.) around the future of humans in outer space, by exemplifying and advocating a role for education as one discipline among many within that dialogue.

[156] Moore, Patrick. Philip's Atlas of the Universe. London: Chancellor Press, 2001: 27.

[157] Noted by Smith, Andrew. Moondust – In Search of the Men Who Fell to Earth. London: Bloomsbury Publishing Plc., 2005: 295.

[158] Brumbaugh, Robert S. "Applied Metaphysics: Truth and Passing Time." The Review of Metaphysics 19.4 (1966): 647–66. 649.

[159] Bernstein, Basil. "On the Classification and Framing of Educational Knowledge." Knowledge, Education and Cultural Change. Ed. Richard Brown. London: Tavistock Publications, 1970: 363–392. 363.

[160] Kerr, John F. Changing the Curriculum. Ed. London: University of London Press, 1968: 16.

[161] see Pinar, William F., William M. Reynolds, Patrick Slattery, and Peter M. Taubman. Understanding Curriculum: An Introduction to the Study of Historical and Contemporary Curriculum Discourses. New York: Peter Lang Publishing, 1995. for a comprehensive review of the different ideological and epistemological claims on the meaning(s) of curriculum.
[162] Bruner, Jerome S. The Process of Education. Cambridge, Massachusetts: Harvard University Press, 1966: 1.
[163] Pring, Richard. "Education as a Moral Practice." The RoutledgeFalmer Reader in Philosophy of Education. Ed. Wilfred Carr. London and New York: Routledge Falmer, 2005: 195–205, 204.
[164] See for example Pinar, William F., William M. Reynolds, Patrick Slattery, and Peter M. Taubman. op. cit.
[165] see Halpin, David. "Understanding Curriculum as Utopian Text." Schooling, Society and Curriculum. Ed. Alex Moore. Oxford, UK: Routledge, 2006: 147–57.
[166] Cuban, Larry. "Computers meet Classroom: Classroom Wins." Teachers' College Record 95.2 (1993): 185–209.
[167] Capella, Juan-Ramón. "Globalization, A Fading Citizenship." Globalization and Education: Critical Perspectives. Eds. Nicholas C. Burbules, and Carlos Alberto Torres. New York: Routledge, 2000.
[168] For full descriptors of the six scenarios, see: "Centre for Educational Research and Innovation (CERI) – The OECD Schooling Scenarios in Brief." Organisation for Economic Cooperation and Development. 6 Jan. 2010. http://www.oecd.org/document/10/0,3343,en_2649_39263231_2078922_1_1_1_37455,00.html.
[169] Illitch, Ivan. Deschooling Society. Calder and Boyars: London, 1971.
[170] Castells, Manuel. The Rise of the Network Society. Blackwell: Oxford, 1996.
[171] Kress, G. "A Curriculum for the Future." Cambridge Journal of Education 30.1 (2000): 133.
[172] One version of the fable can be found in Stuart Hall, and Paddy Whannel's book (Hall, Stuart, and Paddy Whannel. The Popular Arts. London : Hutchinson Educational, 1964.), which is in turn adapted from: Peddiwell, Abner J. The Sabre-Tooth Curriculum. New York, London: McGraw Hill, 1939.
[173] Hicks, David. "Re-Examining the Future: The Challenge for Citizenship Education." Educational Review 53.3 (2001): 229–40: 231.
[174] Inayatullah, Sohail. "From "Who am I?" to "When am I' – Framing the Shape and Time of the Future." Futures 25.3 (1993): 235–53: 236 Metafuture.org. 7 Jan. 2010. http://www.metafuture.org/Articles/framing–the%20future.pdf.
[175] see for example Remuss, Nina-Louisa. "Astronauts: From Envoys of Mankind to Combatants." (in this volume) on the role of the military (marines) as astronauts.
[176] Interestingly, Asimov's original satirical intent is often missed; the future school students *mistakenly* (in Asimov's view) imagine that learning was lots of fun in early 1950s classrooms.
[177] see for example Rogers, Thomas. "Creating the First City on the Moon." Beyond Earth – The Future of Humans in Space. Ed. Bob Krone. Ontario, Canada: Apogee Books, 2006: 53–63. 58–9; O'Neill, Gerard K. The High Frontier. Bantam Edition. New York: Bantam Books, 1978. 266.
[178] Schrunk, David G., Burton L. Sharpe, Bonnie L. Cooper, and Madhu Thangavelu. The Moon: Resources, Future Development, and Settlement, 2nd edn. Chichester, UK: Praxis Publishing, 2008.
[179] Crossman, Frank, and Robert Zubrin. On to Mars. Exploring and Settling a New World. Ontario Canada: Collector's Guide Publishing/Apogee Books, 2005.
[180] See NASA website for details of previous entries: "NASA Space Settlement Contest." Last updated 30 Oct. 2009. NASA. 7 Jan. 2010. http://www.nas.nasa.gov/About/Education/SpaceSettlement/Contest.
[181] "Space Settlement Design Contest 2009 Results." Last updated 30 Oct. 2009. NASA. 7 Jan. 2010. http://www.nas.nasa.gov/About/Education/SpaceSettlement/Contest/Results/2009/index.html.

[182] see for example, Schorer, Lonnie Jones. "Children's Visions of Our Future in Space." Beyond Earth – The Future of Humans in Space. Ed. Bob Krone. Ontario, Canada: Apogee Books, 2006: 127–34.
[183] O'Neill, Gerard K. op. cit. 295.
[184] Sagan, Carl. Pale Blue Dot: A Vision of the Human Future in Space. New York: Random House, 1994: 274.
[185] see for example Schrunk, David G., Burton L. Sharpe, Bonnie L. Cooper, and Madhu Thangavelu. op. cit. 393–5.
[186] Dunér, David. "Astrocognition: Prolegomena to a future cognitive history of exploration." (in this volume).
[187] Banks. James A. "Citizenship Education and Diversity." Global Citizenship Education: Philosophy, Theory and Pedagogy. Eds. Michael A. Peters, Alan Britton, and Harry Blee. Rotterdam, The Netherlands: Sense Publishers, 2008: 317–31. 318.
[188] Excerpt from Article 12. Full Text of Convention available from UNICEF. 7 Jan. 2010: http://www2.ohchr.org/english/law/crc.htm#art12.
[189] See for example the OECD PISA Assessment Framework: "OECD Programme for International Student Assessment (PISA)." Organisation for Economic Co-operation and Development. 7 Jan. 2010. http://www.pisa.oecd.org/pages/0,2987,en_32252351_32235731_1_1_1_1_1,00.html.
[190] Cockell, Charles. Ethics and Extraterrestrial Life. (in this volume)
[191] An example of new needs that arise is Mills elaboration on arising challenges for the international legal framework. See Mills, Kurt. "Who will own Outer Space? Governance over Space Resources in the Age of Human Space Exploration." (in this volume).
[192] A similar idea is explored in the 1993 Science Fiction novel, "The Giver" by Lois Lowry, published by Bantam Books. For a summary, see the Wikipedia article. 7 Jan. 2010. http://en.wikipedia.org/wiki/The_Giver.
[193] Banks, James A. op. cit. 318.
[194] Sagan, Carl. op. cit. 170.
[195] Smith, Andrew. op. cit. 336.
[196] Sagan, Carl. op. cit. 171.
[197] Smith, Andrew. op. cit. 287.
[198] Poole, Robert. Earthrise: How Man First Saw the Earth. New Haven and London: Yale University Press, 2008: 108–10.
[199] Schorer, Lonnie Jones. op. cit. 127.
[200] Inayatullah, Sohail. "Pedagogy, Culture, and Futures Studies." Advancing Futures – Future Studies in Higher Education. Ed. James A. Dator. Connecticut: Praeger, 2002: 109–22. 111.

1.7 Ethics and extraterrestrial life

Charles Cockell

1.7.1 Introduction

The study of other planets and moons in the Solar System has revealed the presence of environments that may be conducive to life. The discovery of sulphate-bearing rocks on Mars,[201] and the suggestion that they were formed in bodies of standing water, has invigorated the debate on the subject of the past, or even present, existence of life on Mars. In parallel, the investigation of Europa,[202] a satellite of the giant gas planet Jupiter, has shown the presence of a salty ocean beneath an icy crust.

Another rapidly developing area of study is investigations into distant planets orbiting other stars.[203] These extra-solar planets will eventually be examined for the composition of their atmospheres with the intention of finding biomarkers, such as the gases ozone and oxygen, out of equilibrium with purely chemical processes, suggesting the presence of oxygen-evolving photosynthetic life.

Somewhere within this milieu of exploration it is possible that life will be discovered, even if only microscopic. An essential challenge, which has to be addressed alongside scientific discoveries, is developing an environmental ethic for other life, and formulating a basis for understanding how we should treat it.

There are two important reasons for formulating such an ethic at this stage:

(1) Even prior to the discovery of extraterrestrial life, the construction of an ethical position for hypothetical extraterrestrial microscopic life is an important exercise since it ensures that the issues have been addressed and thoroughly debated in advance of any potential discovery.
(2) Regardless of the results of the search for extraterrestrial life, this debate remains an important exercise since it sheds light on our treatment of microscopic life forms on Earth.

It is the subject of this chapter to address the question of whether we can construct a consistent ethical framework for extraterrestrial microscopic organisms, prior to their detection, that is independent of their nature and origin and is consistent with our treatment of microorganisms on Earth.

In this chapter I will use the term "microscopic organism" to refer broadly to any life form analogous to single-celled microscopic organisms on Earth.

1.7.2 The instrumental value of extraterrestrial microscopic organisms

The utilitarian or instrumental value of extraterrestrial microscopic life is not a point of great ethical complexity or controversy. Most scientists and the public alike would view microscopic extraterrestrial organisms as having various forms of potential instrumental value, whether the organisms were related to life on Earth or not. Thus, valuing extraterrestrial microscopic organisms, and seeking to conserve them for their instrumental uses, is perhaps the easiest part of our perspective on the value of extraterrestrial life to understand. I will briefly discuss the two cases of Earth-related and non-Earth-related life and consider the instrumental value of this life, and the reasons why we might regard it as valuable.

The recovery of lunar and Martian meteorites on the Earth provides support for the theory that if life is found on Mars or on other planets or moons in our Solar System it could have been transferred from Earth to these places or *vice versa* on pieces of rock ejected by asteroid or comet impact events.[204] Thus, we should not be surprised to find that extraterrestrial life in our Solar System, if we find it at all, is related to life on Earth. While the discovery of life genetically related to life on Earth might be thought to be less interesting than a completely novel life form, we would view this life as valuable for a number of reasons. It would reveal how the process of microbial evolution occurs in organisms biogeographically separated from life on Earth for a long period of time, and so it could provide new insights into the evolution of the phylogenetic tree of microbial life on Earth. Although the life may be genetically related to life on Earth, an independent course of evolution (depending on when it was separated) might still have resulted in the evolution of novel taxa of microscopic organisms that might have biotechnological use. In the absence of higher animals and vegetation, microscopic organisms on other planets might provide insights into the characteristics and biogeochemical cycles of early life on Earth, when the planet was entirely dominated by microscopic organisms without the complicating contributions of higher life.

The discovery of life with a novel biochemical architecture, and with origins independent of life on Earth, would open up a much broader range of potential instrumental uses. Scientifically, understanding how this life worked could reveal information on entirely new enzymatic and genetic systems. Apart from biotechnological uses, this life might be used as a basis to understand how life on Earth originated. The bio-chemicals from which this novel life was made and the architecture of its construction might reveal principles about the origin of life; we would have a second data point to guide us in understanding how common life is in the Universe. The life forms would be ecologically significant because they would reveal how planets can be separated from one another as planetary bio-geographical islands and how, under such separations, they can give rise to independent origins of life. When one considers the disciplines from biochemistry to zoology that have arisen from the study of life on Earth, the potential instrumental value and interest of an entirely new type of life probably cannot be adequately estimated.[205]

From an instrumental point of view, then, there is little doubt that extraterrestrial microscopic organisms, whether related to life on Earth or not, would have substantial value and would be protected and studied so that this value could be realised.

1.7.3 The intrinsic value of extraterrestrial microscopic organisms

An ethical question with a less obvious answer is the matter of the intrinsic value of extraterrestrial microscopic organisms.

Fig. 27. *Microorganisms have many instrumental values on Earth, including in the production of wine (left) and fermented food such as Sauerkraut (right). But do microorganisms have intrinsic value? (source: "Wine may slow Dementia." Life without memory 1 June 2010. http://alzheimersadvocacy.com/2007/06/18/wine-may-slow-dementia/; "Ja its time for Oktoberfest!" spacecoastblog 1 June 2010. http://spacecoast.wordpress.com/2009/10/21/ja-its-time-for-oktoberfest/).*

"Intrinsic value" encompasses the notion that something has a value beyond its instrumental (utilitarian or practical) uses to people. Do microorganisms possess a worth independent of their utilitarian value to humans? The resolution of this question for any life form is important because it determines whether it will be treated as something with a greater worth than a mere resource. I have argued in previous papers for the intrinsic value of individual microscopic organisms on Earth.[206] The intrinsic value of individual microscopic organisms might rest on the basis that they exhibit the basic attributes of conation, at least within the definition of Feinberg:

> "mere things have no conative life: neither conscious wishes, desires and hopes; or urges and impulses; or unconscious drives, aims, and goals; or latent tendencies, direction of growth, and natural fulfilments. Interests must be compounded somehow out of conations; hence mere things have no interests."[207]

As microscopic organisms on Earth exhibit unconscious drives and aims, latent tendencies (genetically programmed) and directions of growth, we can consider them to be, even if only in the most limited sense, conative. These attributes of conation, by extension, give them "interests" and a "good of their own". We understand them to have interests because within the context of their conative qualities we can treat them "well" or "badly", depending on whether their latent tendencies are being realised or not. A microorganism adapted to life in the sea can be treated badly by placing it into fresh water, or well (it will grow and reproduce) if it is placed into salty water. It is, of course, not sufficiently conative to *know* that it is being treated badly or well, but as moral agents we know what is good and bad for it. If we were to equate this with having "interests", even if interpreted in the most biologically limited way, then we, as moral agents, are capable of recognising and responding to these interests. This notion of moral considerability for individual microscopic organisms is consistent with Goodpaster's view that "nothing short of being alive"[208] is sufficient for moral consideration.

This notion of moral considerability was, of course, constructed with Earth microscopic organisms in mind but, as Callicott observes, any life forms, even extraterrestrial microscopic organisms of unknown characteristics, probably exhibit the basic attributes of conation that we associate with life on Earth.[209] Indeed, it would be difficult to try and conceive of any life form that exhibited none of the attributes of conation defined by Feinberg and for it to still be recognisable as a life form. Directions of growth, latent tendencies and unconscious aims would seem to be universal qualities of what we expect from "life".

Thus, it would appear that extraterrestrial microscopic organisms can be considered morally relevant, even without a knowledge of their characteristics and origins, if we accept conations in Feinberg's meaning as equating to interests. If biological interests equate to the possession of rights, then we might even argue that individual extraterrestrial microscopic organisms possess rights: the rights to have their interests protected and respected by any moral agent that might come across them during its interplanetary or interstellar travels. However, the notion of "rights" is politically charged and it is not necessary to pursue this line of discussion in order to enquire as to whether extraterrestrial life is morally relevant and whether this places responsibilities on our treatment of it.

These arguments do not require us to define whether an organism is pleasant to humans or not. They apply equally to a highly pathogenic organism as they do to organisms that produce useful drugs. We may decide in some rare cases that our instrumental needs outweigh the protection of an extraterrestrial organism that happens to be capable of slaying billions of humans, but that is beyond the scope of the discussion here.[210]

The problem with recognising the interests of individual microscopic organisms on Earth is the impracticality of actually implementing such an ethic. It can only be regulative – an ethical position we would like to adopt but we cannot practically implement, as opposed to an "operative" ethic that can be implemented.[211] The reason for this impracticality lies in our inability to live our lives without killing microscopic organisms on a daily basis: whilst eating food, brushing our teeth, or cleaning our houses.

Given the impracticality of operatively recognising the interests of individual microscopic organisms, a more holistic ethic embedded in "bioempathy" has been considered to be a more practical basis to formulate an ethic for the treatment of microscopic organisms on Earth and ultimately their conservation.[212] The respect for the natural world that bioempathy embodies implies an approach of "minimum destruction", but it does not require us to protect all organisms, down to the individual organisms, particularly when that would place such severe constraints on our lives as to make daily living impossible. The ethic implies that on some occasions human instrumental goals must trump microbial intrinsic value, which is discussed in more detail later.

This dilemma would not exist for microscopic organisms on planets which we have not yet explored with humans, or that we have not contaminated with terrestrial microscopic organisms on robotic craft. In these cases we can respect the rights of individual microscopic organisms because it imposes no constraints on our lives. An example of a planet where respect for individual microscopic organisms might be a workable, operative ethic would be an Earth-like extrasolar planet with oxygen in its atmosphere suggesting the presence of life, including microscopic life.

But is it necessary for us to abandon space exploration to protect the rights of individual extraterrestrial microscopic organisms on planets nearer to Earth, such as Mars? The question that we must ask is – is there a similar holistic ethic to bioempathy that applies to extraterrestrial microscopic organisms, which allows us to respect them and encourages us to minimise our damage, but recognises that human exploration will unavoidably result in the killing of many individual microscopic organisms (caused, for example, by digging into soils and building research stations on other planets)? If bioempathy can lead to an environmentally practical treatment of microscopic organisms on Earth, can we find an identical ethical approach for extraterrestrial microscopic organisms?

1.7.4 Teloempathy

Bioempathy is the empathy we have with other living organisms on Earth and our sense of their intrinsic value as organisms with a good of their own. A limitation of bioempathy is that the prefix, "bio", is biased, as we understand "biological" to mean organisms that grow and reproduce using DNA and that are primarily constructed from carbon as the backbone of their molecules, among other characteristics. Even if some dictionaries define biology as "the science of life", "biological" is fundamentally a description of life on Earth or Earth-like life.

A sense of kindred attachment need not necessarily apply only to large animals that appear to be sentient. We can equally experience a sense of attachment to "lower" life forms. I have previously argued for a respect leading to the conservation of microorganisms because they are the base of our family tree – the ancestors of all life on Earth, including us.[213]

But a sense of the intrinsic value of organisms could extend beyond our terrestrial biological prejudices. Consider, for example, a hypothetical world colonised by life forms that are biochemically utterly novel. The life forms have no DNA or any other genetic information-reproducing machinery that is recognisable to us, but they seem to do all the things that we recognise as being "life": they reproduce, grow, live in particular environments, and seem to evolve and develop in response to their environment. Would we have any respect for these organisms, i.e. any sense that we should not arbitrarily destroy them? It seems certain that we would. These organisms are not "biological", in the sense that they show no biochemical similarities to life on Earth, so why is it that we respect them as having intrinsic value? The answer might plausibly lie in the fact that we recognise them as having a *telos*, or a "purpose". They do not necessarily have to exhibit all of the characteristics of "life", which itself is a difficult term to define.[214] However, it is their direction of development, their responses to their environment, and their exhibiting the properties of having a "good of their own" that we respect.

We might call this empathy that we feel with creatures, which have their own purpose "teloempathy", where *telos* is the Greek for "end"; as in English it means both "end" as in "final" and "end" as in "purpose".[215] I emphasise that when I talk of *telos* I am not implying that evolution has some predefined direction of progress or development, or that organisms "know" what their purpose or end is. When I speak of *telos,* I mean that organisms have pre-programmed latent tendencies, such as to reproduce and grow, which have evolved independent of external "design" and imbue life with behaviours and directions more complex than the purely physical laws operating around them. My definition of *telos* here comes closest to Mayr's concept of "teleonomic processes",[216] but I do not suppose that the programme in extraterrestrial organisms has any similarity to the "genomics" familiar to us. Seen in the context of Feinberg's definition, I am equating *telos* with basic attributes of conation. We can assume that any life on other planets will have latent tendencies of some form. Teloempathy is ethically based on a similar line of ethical reasoning to bioempathy, but it presupposes nothing about the characteristics of life from a biological or chemical point of view, or even about how it subdivides into "individuals", "species", and the other hierarchies of biological classification familiar to us.

Teloempathy has a number of important characteristics as an ethical construct that any position we adopt for extraterrestrial life should exhibit:

(1) It allows for a practical environmental ethic to be developed for microscopic life on a planet, even if we do not know what the characteristics of that life are;
(2) It is consistent with our treatment of terrestrial microscopic organisms.

Whatever ethical framework we adopt for extraterrestrial microscopic organisms, it must be normatively consistent with the way in which we treat microscopic organisms on Earth. Thus, it avoids the "ethical eclecticism" that Callicott warned us against.[217]

What practical constraints does a teloempathy impose upon us in our duties to extraterrestrial microscopic organisms? The practical constraints are identical to those that bioempathy or similar ethics impose on us in the treatment of Earth's microscopic organisms. These constraints can be generally identified as:

(1) No duty to preserve individual microscopic organisms when such individuals put constraints on human activities that are considered to be part of daily life, but a duty to preserve individual organisms when we reasonably can. This ethic implies a hierarchy whereby human instrumental needs must sometimes outweigh microbial intrinsic value.

(2) A duty to implement environmental policies that protect ecosystems and communities of microscopic organisms. The jump from individuals to collectives (i.e. ecosystems) is not an ethically simple one, but I make the assumption here that because microorganisms are part of communities or ecosystems we cannot actually conserve individual microorganisms without protecting the ecosystems to which they belong, so that an ethic directed towards individual microorganisms must necessarily also be directed at the larger scale of ecosystems.
(3) A general duty to show respect towards microscopic organisms in our activities.

Point 1 above raises a particularly difficult issue because it implies the need to develop a robust hierarchy as to when microbial interests outweigh human interests and *vice versa*. On Earth, is it acceptable to drain a pond in a person's back yard, which will kill many microbial individuals? It is unlikely we would prohibit such an activity but we would probably take offence at a systematic draining of all ponds in one country, for instance. Clearly, as Attfield recognises with trees,[218] there are many situations where the protection of individual microscopic organisms, and possibly even of some communities (such as the communities within adrained pond), is impossible. Broadly, teloempathy, like bioempathy on Earth, certainly requires us to be concerned when:

(1) We begin to systematically destroy communities of microscopic organisms of which individual microorganisms are a part: the localised killing of a community near a station on a planetary surface would be acceptable, provided that we did not systematically, for whatever reason, deliberately destroy these communities on a planetary scale.
(2) We cause the destruction of the biological diversity of microscopic organisms on other planets by systematic environmental change or destruction.

The points raised above highlight the need to develop a consistent and robust categorization of those situations where microbial killing is acceptable for the requirements of human life and those where it is not. As with the case for trees and other plant life, there is every reason to believe that some workable set of guidelines is possible. Such a set of priorities have been discussed for terrestrial microbial conservation and are applicable to the extraterrestrial case.[219]

Taking putative life on Mars as an example, the duties of a teloempathy are not incompatible with some writers' views on the ethical status of microscopic "Martians". Sagan's statement that if life is found on Mars, "Mars then belongs to the Martians, even if they are only microbes"[220] has an appealing rhetorical resonance, but on closer inspection it is more of a potential truism than an ethical guideline. For example, we could equally say that "Antarctica belongs to Antarctic life", in the sense that the Antarctic continent is almost entirely dominated by indigenous life, including microscopic life, and this life should be allowed to continue its existence. The agreement that a planet "belongs" to its indigenous microscopic biota does not necessarily prevent representatives of another biosphere, who hold a respect for that biota, from landing and establishing stations to understand that life, if they can demonstrate that their presence will not cause its widespread destruction.

This view is consistent with those enunciated by others. McKay suggests that Martian life does have rights and that "it has the right to continue its existence, even if its extinction would benefit the biota of Earth".[221] Lupisella urges similar caution: "first and foremost is a requirement that the rights of any indigenous biota be respected",[222] and Race and Randolph suggest that, if we find life on Mars, the organisms should be allowed to follow their "natural evolutionary or development trajectory".[223]

1.7.5 Planetary protection

Planetary protection deals with our concerns about the contamination of other solar system bodies with organisms from Earth (forward contamination) and the contamination of Earth with extraterrestrial microscopic organisms on board sample return missions (back contamination).[224]

Fig. 28. *Should the construction of a base on a planetary surface that harbour life (left) be forbidden if that planet has microbial life, when the construction of buildings, resulting in the destruction of many microorganism on the Earth (right), is acceptable? (source: NASA; "Legal Considerations." 1 June 2010. http://iuyopy.50webs.com).*

In principle, the contamination of the surface of another planet with microscopic organisms from Earth does not present an ethical problem within the framework just discussed, provided that the contamination does not cause widespread destruction of the indigenous life. The line of reasoning behind this view is partly embedded in practical microbiology and partly in ethics. Firstly, consider the living microscopic organisms. Marshall considers a number of ways in which non-indigenous microscopic organisms might threaten an indigenous biota.[225] He correctly points out that direct host-parasite interactions are unlikely, as on Earth these relationships are highly co-evolved. A much more likely route to destruction would be competition for habitat and nutrients. This competition would be caused by the fact that certain micro-habitats might have physical and chemical conditions conducive to the non-indigenous microscopic organisms. They would then grow and use habitat and nutrients of potential use to the indigenous biota. On a local level, this may not be a concern, provided that the indigenous life continues to co-exist and is not driven to extinction by the introduction of the non-indigenous life.

There are a number of reasons to find an extinction scenario unlikely:

(1) Almost all microscopic organisms released into the environment of other planets will be human contaminants that will die when exposed to extraterrestrial environmental conditions, just as they do on the surface on the Earth outside the human body. Similarly, even non-human Earth microbial contaminants are likely to find it difficult to persist in extraterrestrial environments.[226]
(2) Well-cleaned spacecraft apparently carry very few extremophile species.[227]

None of these points rules out the chance introduction into an extraterrestrial environment of an organism that happens to thrive on a planetary scale and cause widespread extinction in an indigenous biota unprepared for it. We can never eliminate this possibility with complete certainty, particularly if we explore using humans, with the inevitable shedding of many diverse microorganisms from their imported equipment and clothes.[228] Lupisella recognises this challenge in planetary protection and discusses in detail a decision-making tree for assessing the extent of the contamination of Mars by introduced organisms.[229] On Earth there are instances of non-indigenous multicellular organisms driving indigenous organisms to extinction, despite the high level of adaptation of the indigenous life to its environment.[230]

If microbial contaminants are localised, or would have only localised biological consequences, then with the careful monitoring of the shedding of microscopic organisms on other planetary surfaces, and by maintaining high levels of cleanliness, the exploration and even settlement of other life-bearing planets should be possible. As I have postulated that the destruction of individual microscopic organisms is regulatively undesirable, but operatively acceptable, even localised contamination and destruction of life with microscopic organisms from Earth might be acceptable, provided that the integrity of whole ecosystems and communities on the planetary scale is maintained and provided that we show an empathy towards that life and do our utmost to minimise its destruction. However, work would be needed to quantify the risk of their destruction from introduced non-indigenous microscopic organisms and to implement mission planning accordingly.

1.7.6 "Originism"

I will address in more detail the question of whether there is a basis for treating differently a life form, that evolved independently of life on Earth. I have assumed so far that there is not and that an identical ethical framework should apply to both types of life, but this position needs further clarification.

An unrelated life form would arise from a situation where life originates on a planetary body in isolation from life on Earth. Traditionally, life not related to life on Earth has been seen to be something that might be viewed very differently from life related to us. Callicott, in examining the relevance of the Land Ethic to extraterrestrial life, observes:

> "Extraterrestrial life forms, assuming that they were not of Earthly origin and inoculated somehow on some foreign body, or *vice versa*, would not be our kin .... Hence they would lie outside the scope of Leopold's land ethic."[231]

By contrast, Hargrove suggests that:

> "I suspect that the organisms would be considered more valuable because they were not part of our system or our history than if they were".[232]

Similarly, Randolph, Race, and McKay posit that:

> "From an ethical point of view, the need to preserve a life-form, however lowly, must be more compelling if that life form represents a unique life form with an evolutionary history and origin distinct from all other manifestations of life."[233]

And Lupisella, too, sees a potential difference between related and unrelated life:

> "If, however, that life occupies its own separate, unique phylogenetic tree, we may find ourselves asking fundamental questions regarding its status in our world views."[234]

These points suggest that life of independent origin might be treated either better or worse than terrestrial life. Callicott's observation suggests that if it is not kin and it does not fit within a Land Ethic, it might not command any claim to our respect. We might not even recognise this other life as falling within the purview of our ethics at all.

By contrast, as life derived from an independent origin would provide opportunities for novel scientific understandings of the nature of life, then it is likely to be treated as something very valuable, a point explicit in Randolph et al.'s comments.

In analogy to the word "speciesism", which denotes the treatment of life differently based on the species it belongs to,[235] we might refer to the treatment of life differently depending on its origin as "originism".[236] Therefore the question I wish to address is – *Is there a basis for an originist position to exist with respect to our treatment of extraterrestrial life?*

The passages just quoted show that originism could derive from several potential viewpoints. It could involve the treatment of extraterrestrial life as being inferior on account of a perceived inferiority of one or several of its characteristics (such as sentience). It could involve the treatment of Earth life as being inferior on account of the novelty of the extraterrestrial life and its perceived uniqueness – a sort of interplanetary and interstellar affirmative action. It could derive from the simple fact that extraterrestrial life is so different that we do not recognise it as fitting

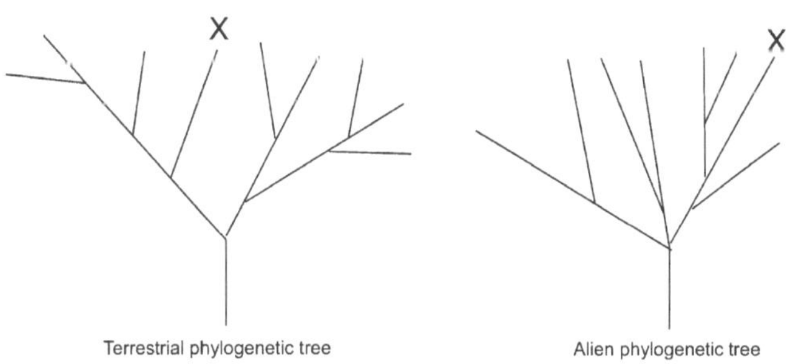

Fig. 29. *Two hypothetical microorganisms (denoted X) are identical in biochemical function and structure. However, one is terrestrial and fits into the terrestrial phylogenetic tree of life (left) and one is alien and comes from an entirely different tree of life (right). Do they have the same value in our system of ethics?*

within any ethical context at all – in this case originism is not a result of prejudice or even of ignorance, but just a difference so extreme it is outside our comprehension.

Throughout the history of ethics the division between organisms that are morally relevant and those that are not has been predicated on exactly where we draw the line in numerous different concepts thought to be important for assigning moral relevance. To dualists such as Descartes and Kant, humans are the only organisms worthy of moral consideration because we can "reason".[237] Other ethicists, such as Singer, Regan and Bentham, draw the line between organisms that feel "pleasure and pain", or that can suffer and those that do not.[238] For them, many non-human animals, such as dogs, merit moral consideration, but trees do not. For other ethicists, such as Taylor or Goodpaster, who assert that simply being "alive" is sufficient for moral consideration, then all life forms are morally relevant.[239]

Common to all these ethicists is the implicit assumption that being alive is a necessary prerequisite for having any moral relevance. Superposed on this requirement may be the further requirements of being sentient and being able to reason. This is not a wholly accurate characterisation. Stone, for example, has posited legal status for abiotic objects.[240] Significantly, in the context of space exploration, Rolston has considered abiotic extraterrestrial planetary bodies and suggests that anything with formed integrity deserves our respect and should be preserved.[241] The apparent arbitrariness of the biotic–abiotic division in ethics is explored by Hunt, who argues that drawing the line of moral consideration at "being alive" is as arbitrary as using "sentientism" or "reason" as the defining quality required for moral consideration, and he suggests that simply "being in existence" might be sufficient.[242,243]

One obvious approach to preventing prejudice is therefore simply to grant moral relevance to *everything*; however, this does not get us very far in understanding how to treat extraterrestrial life. However, the implicit assumption made by many ethicists that being alive is a necessary starting point for having moral relevance is also problematic because our definition of life remains in debate and obscure.

One aspect of life that is common to all life forms on Earth is the possession of a genetic code made from nucleic acid, and it is instructive to ask what impact its discovery has had on our approach to ethics and our knowledge of the definition of life. Despite the importance of this notion of "being alive" as the basis for being morally relevant, the biochemical architecture of life does not seem to have had a bearing on the our knowledge of the definition of life and thus the boundaries of the moral considerability of organisms or on arguments about where the line of moral relevance should be drawn. Most debates on moral relevance were developed long before the discovery of DNA, so one might view this pre-DNA-discovery world to be an age of pre-DNA ethics since debate occurred within the milieu of a lack of knowledge of the fundamental genetic structures that link all life on Earth.

For all of these early thinkers, the question of moral relevance revolved around "reason", "suffering" and "being alive".

Since the discovery of DNA, knowledge of the fundamental genetic structure of all living organisms has not informed ethical debate in any meaningful way. There has, apparently, been no post-1953 renaissance in ethical debate or understanding. The likely reason for this is that the discovery of DNA has not brought us any closer to a coherent definition of life; and without such a definition, we cannot determine what "being alive" means, and thus where the beginnings of moral relevance lie.

To avoid a diversion, I do not intend to review the numerous and lengthy attempts to define life. A review by Cleland and Chyba[244] draws attention to the fact that life may be a "non-natural kind", an old term used by philosophers to designate a class of objects that are bound together by an artificial definition. In other words, life is anything, that we choose to include within our definition of "life". By contrast, gold, for instance, is an example of a natural kind – something, which has a very specific definition based on its characteristics as an element.[245] Perhaps one day we will understand life sufficiently to be able to accurately quantify and define it and it will then, like gold, become a natural kind.

If life is a non-natural kind, then biochemical architecture would have an influence on moral relevance if the organism did not show the characteristics of "being alive" within the definition that prevailed at the time. What has this got to do with originism? To determine whether something is alive and whether the characteristics it possesses of being alive would be sufficient to grant it entry into that, often exclusive, club of having any moral relevance turns on our ability to define life. If we cannot adequately define life, then it is likely that we cannot develop an adequate and consistent morality to encompass an independent origin of life. Thus, biochemical architecture is not *in itself* a basis for treating extraterrestrial life differently, but it might be the *cause* of it, because one of its manifestations is to take an entity outside our definition of life. This is egregious for the other entity, because it invites the possibility of ad hoc definitions of life to include or exclude other entities at our own discretion – allowing for the possibility of deliberate originism.

With no other life with which to test our definitions of life, there is currently no resolution to this dilemma. There is a good chance that we would recognise other entities to be alive within our current definitions of life. If we do recognise the entities to be life because they reproduce, grow, evolve and have the collective attributes of life that fit within a definition, then it seems to me that they would fall within the purview of many of the moral debates with which we are familiar.

However, it is clearly not this simple. Viruses are based on DNA or RNA and yet we still debate whether they are alive or not. As they are responsible for some of the most influential diseases in human history and we are capable, in some cases, of driving them to extinction through concerted global efforts (the 1970s smallpox eradication programme being the exemplar, although it stopped short of driving smallpox to extinction), understanding whether their deliberate extinction is acceptable turns on our ability to apply an ethic to them. Our inability to decide whether viruses are alive or not may explain why, thirty years after the question was first raised,[246] we still have not resolved the matter of whether deliberate viral extinction is acceptable.

The inability to find a robust definition for life has important ethical implications, because it implies an inability to define moral considerability. The inability to define life probably lies at the heart of our inability to define "intrinsic value". The question of whether living things have intrinsic value has been one of the most pervasive and controversial debates in environmental ethics. The debate has been vigorous because its resolution defines the extent to which we should respect living things in themselves, or whether the biosphere and its components are just a resource to be used by humans for instrumental ends.

Callicott believes that any value in an object must derive from a *valuer*, although he suggests that things can still be valued for themselves, and not merely for their practical uses to people, leading to his notion of "truncated" intrinsic value (Callicott).[247] In contrast, Rolston has, for a long time, defended intrinsic value as an objective quality of an object, quite independent of a *valuer*, something that derives from its very being.[248] I suspect that the disagreement over the nature of intrinsic value lies in our inability to define "life",[249] a necessary precondition to really understanding what "intrinsic value" means in terms of being alive and what property of life might be examined to understand whether intrinsic value is a subjective or objective property.

A coherent and consistent definition of life has proven, at least so far, to be an insuperable problem. Unless such a definition is forthcoming, the fate of extraterrestrial life depends on its luck in fitting within our definition of life. If, for whatever biochemical reason, it does notmatch our definition of life, it will find itself at the receiving end of a most pernicious form of prejudice – an originism resulting from a failure to fit within a definition that lies at the heart of our deliberations concerning moral relevance.

Thus, in summary, there is no chemical reason for an originist position to exist with extraterrestrial life, but the effects of its chemical architecture on its characteristics might well cause it to lie inside or outside our definition of life, which would determine its moral relevance in our systems of ethics. However, there are instrumental reasons to invoke an originist view of extraterrestrial life. Extraterrestrial life forms of an independent origin would offer profound scientific insights into the origin of life. This instrumental view of the value of an independent origin of life probably lies at the heart of originist comments from previous writers that assert a potentially higher value for life unrelated to life on Earth.

1.7.7 Highest moral relevance

Despite the foregoing discussion, it may be a wise position to accept originism, i.e. to accept other life forms as very different from life on Earth and to treat them accordingly out of prudence. A life form of independent origin could exhibit characteristics that are not familiar to us. Its biochemical architecture and its behaviours may be so novel that they do not fit within any previously understood formal definitions of "conations" or "sentience", or at least their behaviours might require research to determine the extent to which they are aware of the things happening around them.

This argument bears similarities to those about the difficulty in defining "life", but it extends the problem to our difficulty in defining the other previous benchmarks of moral relevance, including "sentientism", "reason", and any other etymologically diffuse terms that lack a robust biological underpinning. These terms, furthermore, make fundamental assumptions about the nature of the evolutionary process on Earth and its universal application. As Goodpaster notes:

> "Nor is it absurd to imagine that evolution might have resulted (indeed might still result?) in beings whose capacities to maintain, protect, and advance their lives did not depend on mechanisms of pain and pleasure at all."[250]

In such a case, the lack of a capacity to feel pleasure and pain has the potential to put an otherwise reasoning being outside the ambit of any moral position of utilitarianism. Imagine a hypothetical situation, far in the future, where we find some plant-like forms on an extra-solar planet. They appear sessile and exhibit all the qualities of plants on Earth. We imagine them to have moral relevance but they do not appear sentient, so we kill many of them to make way for a research station, just as we might with plants on Earth. Later, we discover we have killed a population of intelligent organisms. They just happen to work on timescales far longer than ours and they do not feel pleasure and pain in our conventional sense of these words. We have, essentially, murdered them. Utilitarianism is a most unpleasant philosophy for a rational extraterrestrial that does not feel pleasure or pain in our neurological understanding of these terms.

To avoid such a situation we might take "an assumption of highest moral relevance". Within the purview of this axiom we assume that life forms are sentient, until proven otherwise. In other words, prior to a proper understanding of how extraterrestrial life forms fit into our existing ethical frameworks, we assume that they possess the attributes that give them the highest moral relevance in any ethical code available. This position minimises damage to extraterrestrial life and it minimises the chances of mistakes in the treatment of life. Most importantly, it allows us to *indefinitely* treat any living thing with respect (it may be that we *never* manage to resolve whether they are "sentient", a problem that even afflicts our assessment of many life forms on Earth and thus their position in our ethics).

Within this position is a form of originism – in that we are according special status to extraterrestrial life just because it is different. However, in this case it is not a systemic prejudice born of a flawed ethic, but rather a pragmatic approach to minimise the chances of the mistreatment of extraterrestrial life. The originism might be temporary, in that once we understand the new life better we might be able to place it into an ethical code.

"An assumption of highest moral relevance" is Kantian in its establishment of a rule or principle concerning extraterrestrial life. I regard this as reflecting a more fundamental ethical problem. A potential lack of knowledge about extraterrestrial life and its character and behaviours might frustrate any ability to make a robust moral calculation on how to behave towards it, which is required for a practical utilitarian ethic. If extraterrestrial life was discovered and its nature could be understood, then a *consequentialist* approach to dealing with this life would become realistic since the consequences of various moral actions could be calculated and the best course of treatment could be evaluated. I suspect that this would in fact be the case with most microscopic life as I have already implied in this essay. However, in some cases Kantian ethics is suited to extraterrestrial life – it puts in place a principle to follow that minimises the chance of mistreatment in *lieu* of an understanding of that life which allows for a moral calculation to be made on the consequences of our actions.

Thus, in summary, a space-faring civilisation would adopt a utilitarian ethic when it had sufficient information about extraterrestrial life to make a moral calculation, but it would fall back on a Kantian (deontological) ethic as a matter of caution when its knowledge was inadequate.

1.7.8 "Originism" as an obligation to extraterrestrial life

At least from a practical, operative, position we might even argue that originism is not merely prudential, but a moral *obligation*. The term "speciesism", brought to public attention by Singer (Animal Liberation) has been widely adopted by the animal liberation movement because Singer was interpreted by this community to have made the case that speciesism is simply the next tier beyond racism – the next level of prejudice to be confronted and removed. In fact, Singer called for equal *consideration*; he never claimed equal moral *significance* for all animals, something we would posit for all human beings. But as has been pointed out so many times, the analogy between racism and speciesism is specious anyway. There is no basis to treat humans differently from one another because of their race. There is no empirical evidence, for example, that one nationality is less intelligent than another, although there have been many execrable attempts in human history to prove otherwise. By contrast, to *not* recognise species differences would be absurd because there clearly are very big differences between species, from their physiological to their social needs. To treat an elephant like a microorganism would be most unpleasant for the elephant. Midgley describes the idea of the treatment of all species as being the same as a "supercilious insult".[251]

Perhaps the same argument would, *a fortiori*, apply to an alien species. Perhaps to treat all extraterrestrial life like terrestrial life would be an insult to the obvious differences in physiological needs. Originism would, like speciesism, not result from a flawed prejudice but, on the contrary, it would result from the recognition of the differences between terrestrial life and this "other" life, and the importance of understanding those differences.

1.7.9 Conclusion

These ethical discussions might appear indulgent, in that at the time of writing we have not discovered extraterrestrial life. However, the ethics of dealing with extraterrestrial life shed important light on our treatment of life on Earth and contribute to several important points of discussion in ethics in general. Thus, it is, at the very least, a useful device for understanding ethics on Earth. Some of the points relevant to terrestrial ethics that this discussion sheds light on are:

(1) The foundations of moral consideration rely on a definition of "life". Without a coherent definition of life, there can be no coherent ethic for dealing with all living things. The discussion is circular, because if something lies outside the prevailing definition of life, then it is not used, as a matter of definition, to expand the envelope of moral consideration. The problem is not merely one of

academic interest – it has direct practical implications, for instance, for the ethics we apply to viruses, which have been a point of discussion about whether they are alive or not. Without a coherent definition of "life", we are left in a position of being unable to determine the moral relevance of this important branch of biological entities, which are responsible for many influential diseases. The ethically problematic decision of whether to attempt to take the smallpox virus to extinction[252] depends on the value of viruses within environmental ethics and on the definition of life.

(2) The discussion brings us closer to an understanding of the nature of "intrinsic value", a ubiquitous point of discussion in environmental ethics. Do life forms have a value in themselves? By considering the importance of the definition of life within ethical debate and our inability to define the quality of "being alive", we start to arrive at a more fundamental understanding of the locus of intrinsic value and whether it is a property of valuers, or something actually manifest in the living thing itself. The problem of defining "intrinsic value" may itself rest in our ability to define "life". If "life" is a non-natural kind – a mere semantic construction – then it is likely that so too is intrinsic value.

(3) The inability to define "life", "sentientism", and "reason" cautions us to be ready to take a position of "highest moral relevance" with extraterrestrial life that we cannot fully understand to avoid erroneously assigning it lower moral relevance than it deserves, or maybe none at all (this position may not be required for microscopic organisms, which could be characterised and the consequences of our actions towards them properly evaluated). Such a position might apply to life on Earth that we have not fully understood and may lead to a more cautious and enlightened treatment of the biosphere.

(4) The discussion allows us to understand better the ethical arguments needed to construct a robust conservation policy for microorganisms that complements the existing focus on plants and animals in environmental conservation.[253]

(5) The discussion allows us to begin to draw up a practical code for the ethics we apply in space exploration in the same way as the ethics we can develop for microbial conservation on the Earth.

[201] Squyres, S. W., J. P. Grotzinger, R. E. Arvidson, J. F. Bell, III, W. Calvin, P. R. Christensen, B. C. Clark, J. A. Crisp, W. H. Farrand, K. E. Herkenhoff, J. R. Johnson, G. Klingelhöfer, A. H. Knoll, S. M. McLennan, H. Y. McSween, Jr., R. V. Morris, J. W. Rice, Jr., R. Rieder, and L. A. Soderblom. "In Situ Evidence for an Ancient Aqueous Environment at Meridiani Planum, Mars." Science 306 (2004): 1709–14. 1709.

[202] Chyba, Christopher F. and Cynthia B. Phillips. "Europa as an Abode of Life." Origins of Life and Evolution of the Biosphere 32 (2002): 47–67. 47.

[203] Mayor, Michel, Pierre Frei, and Boud Roukema. New Worlds in the Cosmos: The Discovery of Extrasolar Planets. Cambridge: Cambridge University Press, 2003.
[204] Mileikowsky, C., F. A. Cucinotta, J. W. Wilson, B. Gladman, G. Horneck, L. Lindegren, J. Melosh, H. Rickman, M. Valtonen, and J. Q. Zheng. "Natural Transfer of Viable Microbes in Space, Part 1: From Mars to Earth and Earth to Mars." Icarus 145 (2000): 391–427. 391.
[205] It thus seems obvious that evolving new disciplines also will need to be mirrored in school curricula of both children on Earth and particularly the children of space settlers, who will be directly dealing with many of the evolving aspects. For a sketch on how such a school curriculum for the children of space settlers could look like, see Britton, Alan. "A School Curriculum for the Children of Space Settlers." (in this volume).
[206] See for example Cockell, Charles S. "The Rights of Microbes." Interdisciplinary Science Reviews 29 (2004): 141–50. 141; Cockell, Charles S. "The Value of Microorganisms." Environmental Ethics 27 (2005): 375–90. 375.
[207] Feinberg, Joel. "The Rights of Animals and Unborn Generations." Philosophy and Environmental Crisis. Ed. William T. Blackstone. Athens: University of Georgia Press, 1974: 49–68. 49.
[208] Goodpaster, Kenneth E. "On Being Morally Considerable." Journal of Philosophy 22 (1978): 308–25. (Reprinted in Environmental Philosophy: From Animal Rights to Radical Ecology. Ed. Michael E. Zimmerman. Englewood Cliffs, New Jersey: Prentice Hall, 2001: 58–65. 308.
[209] Callicott, Baird. "Moral Considerability and Extraterrestrial Life." In Defense of the Land Ethic: Essays in Environmental Philosophy. Ed. Baird Callicott. Albany: State University of New York Press, 1989: 263–6. 263.
[210] although see Cockell, Charles S. "The Value of Microorganisms." Environmental Ethics 27 (2005): 375–90. 387.
[211] Goodpaster, Kenneth E. op. cit. 308.
[212] Cockell, Charles S. "The Value of Microorganisms." Environmental Ethics 27 (2005): 375–90. 387.
[213] Cockell, Charles S. "The Rights of Microbes." Interdisciplinary Science Reviews 29 (2004): 141–50. 141.
[214] Cleland, Carol E. and Christopher F. Chyba. "Defining Life." Origins of Life and Evolution of the Biosphere 32 (2002): 387–93. 387; Ruiz-Mirazo, Kepa, Juli Paretó, and Alvaro Moreno. "A Universal Definition of Life: Autonomy and Open-Ended Evolution." Origins of Life and Evolution of the Biosphere 34 (2004): 323–46. 323.
[215] Cockell, Charles S. "Duties to Extraterrestrial Microscopic Organisms." Journal of the British Interplanetary Society 58 (2005): 367–73.
[216] Mayr, Ernst. What Makes Biology Unique? Cambridge: Cambridge University Press, 2004: 51.
[217] Callicott, Baird. "Moral Considerability and Extraterrestrial Life." In Defense of the Land Ethic: Essays in Environmental Philosophy. Ed. Baird Callicott. Albany: State University of New York Press, 1989: 263–6. 264.
[218] Attfield, Robin. "The Good of Trees." Environmental Ethics. Eds. David Schmidtz, and Elizabeth Willott. Oxford: Oxford University Press, 2002: 58–70. 18.
[219] Cockell, Charles and Jones, Harriet. "Advancing the Case for Microbial Conservation." Oryx 43 (2009): 520–6.
[220] Sagan, Carl. Cosmos. New York: Random House, 1980: 130.
[221] McKay, Christopher P. "Does Mars Have Rights? An Approach to the Environmental Ethics of Planetary Engineering." Moral Expertise. Ed. D. MacNiven. New York: Routledge, 1990: 184–97. 184.
[222] Lupisella, Mark. "The Rights of Martians." Space Policy 13 (1997): 89–94. 89.
[223] Race, Margaret S. and Richard O. Randolph. "The Need for Operating Guidelines and a Decision Making Framework Applicable to the Discovery of Non-Intelligent Extraterrestrial Life." Advances of Space Research 30 (2002): 1583–91. 1583.
[224] Rummel, John D. "Planetary Exploration in the Time of Astrobiology: Protecting Against Biological Contamination." Proceedings of the National Academy of Sciences 98 (2001): 2128–31. 2128.

[225] Marshall, Alan. "Ethics and the Extraterrestrial Environment." Journal of Applied Philosophy 10 (1993): 227–36. 227.
[226] Schuerger, Andrew C., Rocco L. Mancinelli, Roger G. Kern, Lynn J. Rothschild, and Christopher P. McKay. "Survival of Bacillus Subtilis on Spacecraft Surfaces Under Simulated Martian Environments: Implications for the Forward Contamination of Mars." Icarus 165 (2003): 253–76. 253.
[227] Venkateswaran, K., M. Satomi, S. Chung, R. Kern, R. Koukol, C. Basic, and D. White. "Molecular Microbial Diversity of a Spacecraft Assembly Facility." Systematic and Applied Microbiology 24 (2001): 311–20. 311.
[228] McKay, Christopher P. and Wanda L. Davis. "Planetary Protection Issues in Advance of Human Exploration of Mars." Advances in Space Research 9 (1989): 197–202. 197.
[229] Lupisella, Mark. "Ensuring the Integrity of Possible Martian Life." International Academy of Astronautics (IAA) Paper (1999): IAA–99–IAA.13.1.08.
[230] Drake, J. A., H. A. Mooney, F. Di Castri, R. H. Groves, F. J. Kruger, M. Rejmanek, and M. Williamson. Biological Invasions: A Global Perspective. New York: John Wiley & Sons, 1989.
[231] Callicott, Baird. "Moral Considerability and Extraterrestrial Life." In Defense of the Land Ethic: Essays in Environmental Philosophy. Ed. Baird Callicott. Albany: State University of New York Press, 1989: 263–6. 262.
[232] Hargrove, Eugene C. Foundations of Environmental Ethics. Englewood Cliffs, New Jersey: Prentice-Hall, 1989. 130.
[233] Randolph, Richard O., Margaret S. Race, and Christopher P. McKay. "Reconsidering the Theological and Ethical Implications of Extraterrestrial Life." CTNS (Center for Theology and the Natural Sciences, Berkeley) Bulletin 17 (1997): 1–8. 1.
[234] Lupisella, Mark. "The Rights of Martians." Space Policy 13 (1997): 89–94. 89.
[235] Ryder, Richard. Speciesism: The Ethics of Vivisection. Edinburgh: Scottish Society of Vivisection, 1974.
[236] Cockell, Charles S. "Planetary Protection – A Microbial Ethics Approach." Space Policy 21 (2005): 287–92.
[237] Kant, Immanuel. The Metaphysics of Morals. Cambridge: Cambridge University Press, 2004.
[238] Singer, Peter. Animal Liberation. New York: Avon Books, 1975; Regan, Tom. The Case for Animal Rights. Berkeley: University of California Press, 1983; Bentham, Jereny. An Introduction to the Principles of Morals and Legislation. Oxford: Oxford University Press, 1823.
[239] Taylor, Paul W. Respect for Nature: A Theory of Environmental Ethics. Princeton: Princeton University Press, 1989; Goodpaster, Kenneth E. op. cit. 308.
[240] Stone, Christopher D. Should Trees Have Standing? Towards Legal Rights for Natural Objects. California: Los Altos, Kaufmann Press, 1974.
[241] Rolston, Holmes. "The Preservation of Natural Value in the Solar System." Spaceship Earth, Environmental Ethics and the Solar System. Ed. Eugene C. Hargrove. San Francisco: Sierra Club Books, 1986. 140–82. 140.
[242] Rolston, Holmes. Environmental Ethics: Duties to and Values in the Natural World. Philadelphia: Temple University Press, 1983: 59.
[243] Hunt, W. Murray. "Are Mere Things Morally Considerable?" Environmental Ethics 2 (1980): 59–65.
[244] Cleland, Carol E., and Christopher F. Chyba. op. cit. 387.
[245] Kripke, Saul A. Naming and Necessity. Cambridge, MA: Harvard University Press, 1980.
[246] Dixon, Bernard. "Smallpox, Imminent Extinction and an Unresolved Dilemma." New Scientist 26 (1976): 430–2. 430.
[247] Callicott, Baird. "On the Intrinsic Value of Non-Human Species." The Preservation of Species. Ed. B. Norton. Princeton: Princeton University Press, 1986: 138–72.
[248] Rolston, Holmes. Environmental Ethics: Duties to and Values in the Natural World. Philadelphia: Temple University Press, 1983: 186.
[249] Cleland, Carol E. and Christopher F. Chyba. "Defining Life." Origins of Life and Evolution of the Biosphere 32 (2002): 387–93. 387.

[250] Goodpaster, Kenneth E. op. cit. 221.
[251] Midgley, Mary. "The Significance of Species." The Animal Rights/Environmental Ethics Debate: The Environmental Perspective. Ed. E. Hargrove. New York: State University of New York Press, 1992: 122–36. 122.
[252] Dixon, Bernard. op. cit. 430.
[253] Cockell, Charles and Jones, Harriet, "Advancing the Case for Microbial Conservation." Oryx 43 (2009): 520–6.

1.8 Encounters among the stars – exosociological considerations

Michael T. Schetsche

1.8.1 Science rather than fiction

The encounter between humans and extraterrestrials has till today been one of the most important topics of science fiction. Narratives on alien beings and their relation to humans became increasingly differentiated in the course of the 20th century.[254] The millions of successful novels, films and TV series produced in this genre since the middle of the last century make it quite clear that the issue of the encounter with the "strangest stranger"[255] has a lasting effect on many people. Where and how could we meet them? What do they look like? And what are their interests and motives? Such issues not only structure the narrative patterns of science fiction, but also supply ideas for dealing with basic philosophical questions about the essence of human beings and their place in the cosmos.

At first glance it is all the more astonishing that the sciences take up this issue only very reluctantly. Whereas the scholars of the Renaissance period like Nikolaus von Kues, Johannes Kepler or even Giordano Bruno enquired after the *inhabitants of alien worlds*[256] as a matter of course in their astronomical and philosophical works, it is very difficult for many modern astronomers today to take this topic seriously. Academic monographs on real contact with extraterrestrials have been the exception.[257] This issue is mostly found only in the final chapters of astrobiological works – where the authors permit themselves witty, purely speculative or even critical-ironical views.

The reason for this reticence is not solely the extreme differentiation of the present system of science, which makes it almost impossible for astrophysicists or astrobiologists to develop a well-founded understanding of psychological or sociological problems. In line with this, perspectives dominate in cultural studies today, by which the basic ontological issue of the exceptional position of humans in the cosmos is epistemologically reduced to historic and psychological dimensions.

It is moreover a fact that – as the history of the Search for Extraterrestrial Intelligence (SETI) programmes[258] make it abundantly clear – research on "space aliens" is not funded by governments today but at best by private enterprise.[259] Thus, not only for this reason, enquiry into the "strangest stranger" is rather counter-productive for an academic career and this fact does not exactly strengthen the willingness of individual scientists to deal with it.

1.8.2 Good reasons for the scientific study of the topic "first contact"

Such deliberations can *explain* why the scientific study of the topic "extraterrestrial civilisations" has remained so marginal even in disciplines like astrobiology. They do not, however, supply any objective arguments for abandoning appropriate research. In fact, if one looks at the development of astronomical sciences in the last three decades, there is every reason for scientific attention to the issue of "*first contact*" between humans and extraterrestrials and its consequences.

First, according to the astronomical knowledge gained in recent years, it can be considered reasonably certain that the universe teems with planets, which in principle offer the possibility of the emergence of life. Simultaneously, biological basic research has shown that wherever there has been a possibility on this Earth for life to emerge, life actually exists. The deduction from these two findings that the emergence of life in our universe is not the exception but the rule is made not only by SETI enthusiasts. As we do not know today whether self-conscious intelligence is an inherent potential of biological evolution, we humans should reckon that we are not alone in the universe as an intelligent species.

Second, we not only passively collect more and more information about the universe with numerous instruments, but also actively advance more and more into the solar system with space probes. If the thesis of widespread life in the universe is true, the probability of finding evidence of extraterrestrial life grows with every step into the depths of the cosmos, with every new telescope and each new space probe. If in addition the assumption that biological evolution leads to self-conscious intelligence at least with a certain probability is correct, the probability of finding proof of existence of extraterrestrial civilisations also grows.

Third, above all, the confrontation with an extraterrestrial civilisation would undoubtedly be one of the most difficult, decisive turning points in the history of humanity so far –[260] the *certainty* that we are not alone in the universe as an intelligent species would not only revolutionise our scientific and philosophical thinking, but also *could* have a great number of serious consequences affecting even our everyday life.

1.8.3 Cultural consequences of the first contact

Is it possible at all to make scientifically founded statements on events that have not yet taken place? For this purpose, social sciences have developed the method of *scenario analysis.*[261] Exemplary scenarios are constructed on the basis of known or estimated data and a series of justified presumptions, which attempt to depict the essential features of an anticipated situation as realistically as possible, are formulated. Such a scenario analysis is on the one hand used to prevent all superfluous or unjustifiable presumptions, and on the other hand, used to disclose the necessary preconditions of its analysis. The scenario analysis that I have conducted[262] on the first contact has four such *preconditions*:

1. *There exists the possibility of the coexistence of ready-to-communicate civilisations in the cosmos.* A scientific study of the possible consequences of a first contact makes sense only if there is at least a definite probability that humankind is not alone in the universe at the time of its existence as a technical civilisation. As with the SETI programmes,[263] it should be assumed here as well that there *could* be extraterrestrial civilisations in the far reaches of outer space, with whom contact – no matter how – is possible in principle.
2. *The extraterrestrials must be recognisable as such.* A cultural contact takes place in our perception only if we are capable of identifying the others as intelligent extraterrestrial entities. This is not a matter of course; situations can be envisioned (science fiction is acquainted with corresponding thought experiments) where intelligent beings themselves do not mutually recognise each other during direct contact – perhaps because their spaces of perception are incompatible or their experiences of time diverge strongly. (*There is an apodictic facticity of nescience.* Before the actual contact, we cannot know anything about the physical and psychological nature, possible social structures or even the interests and motives of the aliens. This compels us to strictly ignore all "qualities" of the others when contemplating the consequences of a first contact. We can ponder on the consequences of a first contact only on the basis of our knowledge of human thought structures and behaviour patterns).
3. *Anthropocentric presumptions should be minimised.* As we can know literally nothing about the aliens before the first contact (see third precondition), we should dismiss all anthropocentric prejudices in particular, by which we regularly "humanise" extraterrestrials in our thoughts – in science fiction and in science.

The last point applies not only to the appearance, cognitive patterns or the motives of the extraterrestrials, but also the preceding question – the way in

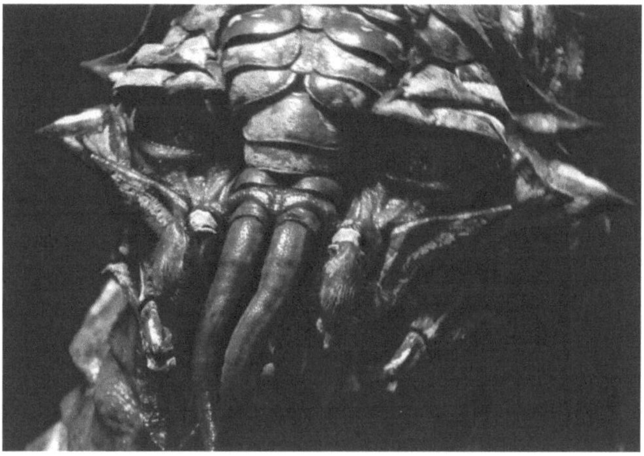

Fig. 30. *"Alien from "District-9"" (2009) (source: All Movie Photo. 31 May 2010. http://images.allmoviephoto.com/2009_District_9/2009_district_9_006.jpg).*

which the first contact could come about. Thus many SETI researchers still argue today that contact between civilisations could be realised exclusively by means of radio waves or the like.[264] The possibility of a direct physical encounter is however categorically ruled out in most cases. The main argument here is the great distances between the planetary systems. This entails travel times of centuries, if not millennia, which is too hard for humans to imagine.[265] This argument is consistent however only if an array of *invalid anthropocentric presumptions* are made: human travel technology and temporality of the travellers, subject-oriented travel planning or even the "biological quality" of potential visitors.[266] All these are implied, mostly unquestioned, in the debate on contact with extraterrestrial civilisations. However, aliens could possess a hundred times greater life expectancy than humans, use space ships to carry successive generations on board or send high-tech robots, etc. We do not know all that. For this reason we cannot say whether the first contact, if it at all happens, would actually take place through a radio signal or in a completely different way. A scenario analysis has to take this into account.

Based on these presumptions, three fundamental scenarios can be differentiated[267] from a sociological perspective, each with very specific cultural consequences from first contact: (1) the *long-distance contact scenario*, in which an exchange of information takes place over great distances, (2) the *artefact scenario*, in which the material legacies of an alien civilisation are found, and (3) the *direct contact scenario*, which examines a physical encounter with aliens on Earth, somewhere in the solar system or even "among the stars". Let us look at these scenarios one by one.

1.8.3.1 The long-distance contact scenario

This is the scenario on which the known SETI programmes are also based. Radio telescopes or other technical devices detect signals of artificial origin from the far reaches of space. From the technical parameters of the transmission, the origin coordinates, the distance and the relative velocity of the sender can be deduced, and perhaps something more can be found out about his technical possibilities.[268] What additional information can be extracted from the signals is still disputed in the SETI research.[269] Even though diverse deliberations on the decoding of radio and laser signals have taken place,[270] they are all based on human thought patterns and cannot be unconditionally transferred to aliens.[271]

Probably we will initially learn no more than the fact: we are here. More appropriate here would be the statement: we *were* there. The detection of a signal travelling at the speed of light from 1000 light years away also means that the signal was sent 1000 years ago. On the one hand this aspect is of significance in terms of the idea of a *cosmic dialogue*. On the other hand the extent of this distance will probably be crucial in both senses of the term for concrete responses on Earth: the larger the space and time the message had to travel, the more indifferent might

Fig. 31. *Very Large Array Radio Telescope, New Mexico (SETI) (source: Website of the SETI Institute: www.seti.org).*

be the psycho-social responses on Earth: "Distance is critical because it structures the nature of the contact (. . .) the closer the contacting civilisation, the greater the impact".[272] A signal from those 1000 (light) years distance would, for instance, primarily affect the scientific, philosophical and religious subsystems of the Earth but would be rather irrelevant for the life of humans and their perception of everyday life. Initially it would certainly create great public interest and also discussions in the various mass media. Given the low amount of information received (assumed here) and due to the impossibility of direct dialogue, the topic would quickly vanish again from public focus. The reception of the signal therefore would have consequences primarily for religious-philosophical systems and the sciences.[273]

Things would be different if the signal came from Earth's immediate *cosmic neighbourhood*. The proximity threshold to be assumed here socio-psychologically might be in a range that corresponds to the lifetime horizon of the average member of our society, that is thirty or maximum 50 (light) years. Alien civilisations would be perceived cognitively and emotionally as "neighbours within reach" only if they were within this time frame and this distance – within reach both in terms of the possibility of a civilisational dialogue by means of radio waves or other signals and in terms of the theoretically conceivable possibility of direct contact (today on Earth) in the foreseeable future. The consequences of such a (however statistically improbable)[274] "close long-distance contact" could – besides the religious and philosophical change in worldview which is rather independent of the distance from the alien civilisation – on the one hand, consist of a massive change in the national and international strategies of research policy to achieve a corresponding dialogue or even direct contact. On the other hand, the *apprehensions* of the earthly population with regard to the consequences of the contact might increase with the decrease in the distance (see below).

1.8.3.2 The artefact scenario

The artefact scenario[275] depicts the situation where humans in their advance into the cosmos come upon the material legacy of an alien civilisation sometime in the near or distant neighbourhood of the Earth. Compared to the distant contact, this scenario leads to an "increased explosiveness" only if the corresponding discovery is made close to the Earth, let us say, somewhere in our solar system. In this case, the "we are here" message of the distant contact would not only be superimposed by the "we were here" message, but also be scientifically and psychosocially dominated.[276] With such a finding, at the least, all theses (hitherto dominating the area of traditional SETI research) on the limitations of spaceflight engineering in

Fig. 32. *"2001 – A Space Odyssey": A Monolith on the Moon (source: http://millenniumppl.blogspot.com/2009/07/monumenta.html).*

covering interstellar distances would be debunked in one swoop as anthropocentric prejudice. It would be proven that other civilisations are very capable and willing to bridge the relevant distances at least with automatic space probes.

This fact alone would compel human societies to rethink in scientific and ideological terms. How immense the cultural impacts of such a discovery would be beyond this depends primarily on two factors: (1) A possible determination of age would shift the found objects into the human time horizon, or on the contrary, out of it. An estimated or calculated age of one hundred years would have a completely different significance here than one of ten million years. In the former case, we would be confronted with immediate "neighbours in time", who, if the find was close to the Earth, were probably informed about the existence of a civilisation on Earth. In the latter case, on the contrary, all such deliberations would be unnecessary.[277] (2) If one of the found objects could be interpreted as being a material base for some kind of technical functionality, this would immediately lead to speculations about the type of that/those function/s and certainly about the current operability of the concerned object. The question then would not only be what the artefact can do, but particularly what consequences its doing could have for its near or far environment.

Such and similar factors would influence not only the scientific handling of the found object, but also the opinion of the public and of policy makers. Should the object remain untouched as far as possible, or should it be scientifically examined? Can or should it be transported to another location, or even brought from space to the Earth, as the case may be? Should it – if technically possible – be in some way manipulated or even taken apart?[278] All these questions are extraordinarily explosive. The third scenario should reveal what the cultural and mass-psychological context would look like, in which these issues might be discussed with the discovery of the artefact.

1.8.3.3 The direct contact scenario

A *direct contact* takes place if an extraterrestrial object, which due to its flight manoeuvres or the like is presumably controlled by an intelligence or at least by technically advanced artificial agents, appears in space close to Earth. The object could also transmit signals in the direction of the Earth, one or more of such objects could enter the Earth's orbit, or even land on Earth. In the extreme case, even alien entities could be put down on the surface of our planet. The genre of science fiction features numerous variants of this scenario.

In the cultural consequences of such an epochal event, in the truest sense of the term, we should differentiate between short-term and long-term consequences.

Fig. 33. *"District 9": The Aliens Arrive (source: Movie Impulse. 31 May 2010. http://www.movie-impulse.co.cc/?p=756).*

With respect to the short-term consequences, social psychology has at least some empirical data: the findings of quasi-natural experiments conducted occasionally by mass media by disseminating *fictitious reports* in a pseudo-documentary format about the landing of extraterrestrials on Earth. Perhaps the first and most spectacular case of this type till today was the broadcasting of the radio play "The War of the Worlds" (based on the novel by H. G. Wells) in 1938. Even though this case continues to be controversially discussed in media history,[279] it reveals at least two things: first, willingness among the media recipients to believe in the reality of such contact events is quite high; and second, the human emotions associated with such an event need not necessarily be of a positive nature. Whether this depends solely on the concrete behaviour of the "visitors", though empirical data are wanting here, should be doubted for theoretical reasons.

A meaningful theoretical background for prognosis here is the familiar psychological mechanisms of an unanticipated confrontation with a completely unknown (and for this reason at least potentially dangerous) counterpart. The fear impulses that are triggered in humans in this situation are even stronger, the more it affects one's own *psycho-social environment*. If one transfers this to the collective responses to a cosmic first contact, at least the Earth itself and its orbit have to be considered as an exceptionally problematic area. In the event of an unexpected encounter with extraterrestrials in this area, the development of a *collective existential shock* entailing severe psychosocial consequences for the individual and society is absolutely possible.

But what does it mean in the mid- and long-term? If our experience with contacts between *human* cultures in the past centuries is taken as the basis, the crucial issue here too would be whether the encounter takes place in a territory (understood here in the social-psychological sense) claimed by one of the parties, or some other space interpreted as "neutral". In the former case, the terrestrial players at least will carry out a role assignment of the participants on the basis of "discoverers vs. the discovered": For the "discoverers", the discovery far away from their home proves their own superiority, and the cultural and technical inferiority for the "discovered" due to the fact that they are confronted with the aliens in their own territory.[280]

The systematic examination of such *asymmetric cultural contacts* on Earth[281] shows that these encounters have oftentimes threatened the cultural existence of the "discovered" – largely independent of the concrete event of the first contact.[282] The devastation of the culture *that viewed itself as inferior* was often not the outcome of the evil motives of the "conquerors", but the consequence of the mass-psychological impact of "being discovered".[283] In a theoretically oriented combined analysis of the most varied empirical findings, Groh (1999) concludes that an encounter of cultures with widely varying degrees of technical development always leads to an economic and cultural dominance gradient, which directly threatens the existence of the inferior culture.[284]

Pertinent to the case of the human-alien contact, Shostak[285] also argues that the confrontation with a technically superior extraterrestrial civilisation *would* in all probability *signify the end of the human culture*. The verity of this prognosis will become evident if at all the event takes place. All experiences that we have had on Earth with such asymmetrical cultural contacts speak for the probability of a global culture shock[286] and the endangerment of our cultural and social systems. This is valid at least if the encounter took place close to Earth, which represents the "living room of humanity" in mass-psychological terms.

1.8.4 Conclusions

If one proceeds from the current status of the scientific presumptions regarding the prevalence of life in the universe and associates it with knowledge in the social sciences regarding the responses of individuals, groups and entire societies to a possible confrontation with the "strangest stranger", various necessities of action arise in more than the scientific realm.

The Brookings Report,[287] commissioned by NASA as early as 1960, recommended the conduct of systematic scientific studies on the fallouts of a human encounter with extraterrestrials. To my knowledge, such studies have rarely been undertaken so far. I consider them, however, as urgently necessary. In this context cooperation between the disciplines of the sciences and the humanities would have a distinctly stronger significance than what we observe at present in the SETI programmes (which are unilaterally dominated by astronomy). The contribution of the cultural studies would consist in supplying data on historic cases of cultural contacts on Earth, enabling the avoidance of factually unfounded anthropocentric presumptions or, as has been the case here, providing methods for their prediction.

Moreover, an open-ended debate on the special consequences of space research and space travel discussed here would be required. In this context, for instance, one might deliberate whether the current and future SETI programmes should not be subjected to a technical impact assessment as is common in other technologies. Contrary to the assumption of a broad group of the public, the search for extraterrestrial intelligence in particular is not a "harmless hobby of eccentric space researchers", but is high-tech research in scientific terms and *high risk research* in cultural terms. This applies in particular to the "active SETI research" that is not only proposed but also practised by several research groups, where – without any social debate – contact signals are emitted into the far reaches of space.

In catastrophe research, the extent of the *risk of an event* (e.g. an earthquake) is determined by two factors: its probability of occurrence and the magnitude of its negative consequences. If contact with an extraterrestrial civilisation were not only one of the most dramatic events in the history of humanity, but – at least under certain conditions – also could have devastating cultural and social impacts, the probability of the first contact event could be almost arbitrarily minor, without the overall risk being negligible. For, if the negative consequences of an event tend to infinity, only they, but not the probability of occurrence,[288] determine the risk-related relevance of the event. This would here mean that, in accordance with the current state of knowledge, it is not acceptable either scientifically or politically to *not* deal with the consequences of a possible first contact of humans with an extraterrestrial civilisation.

The ignorance of this issue displayed today by broad groups of the scientific community, and even more by the political-administrative system, is a viable option for action only because and as long as there are no public indices of the existence of extraterrestrial intelligence. However this strategy becomes precarious at that instant when indications of extraterrestrial life accumulate or the first contact event obviously occurs.

To prevent a global existential shock, the development of a *strategy of a global preparation* prior to this moment is essential. On the part of science, this includes the systematic exploration of the potential impacts of this event on psychosocial and religio-economic areas. On the part of the public, the dissemination of information to the population on these possible consequences of space research is necessary. Finally, nation States and international organisations would be required to develop administrative codes of conduct and appropriate emergency plans. Another meaningful strategy could ultimately be the promotion of manned and unmanned space travel with the aim of being permanently present even far outside the Earth orbit. This might not exclude the risk of a global culture shock, but would distinctly minimise it because a direct contact would lose some of its asymmetry. In terms of mass psychology it would be distinctly more favourable to meet the aliens, if at all it happens, at the edge of the solar system than to encounter them on Earth. The "being discovered" has never gone down well with human cultures.

---

[254] Hurst, Matthias. "Dialektik der Aliens. Darstellungen und Interpretationen von Außerirdischen in Film und Fernsehen." Von Menschen und Außerirdischen. Transterrestrische Begegnungen im Spiegel der Kulturwissenschaft. Eds. Michael Schetsche and Martin Engelbrecht. Bielefeld: transcript, 2008: 31–53; Engelbrecht, Martin. "Von Aliens erzählen." Von Menschen und Außerirdischen. Transterrestrische Begegnungen im Spiegel der Kulturwissenschaft. Eds. Michael Schetsche and Martin Engelbrecht. Bielefeld: transcript, 2008: 13–29.

[255] Schetsche, Michael, René Gründer, Gerhard Mayer, and Ina Schmied-Knittel. "Der maximal Fremde. Überlegungen zu einer transhumanen Handlungstheorie." Berliner Journal für Soziologie 19.3 (2009): 469–91.
[256] Heuser, Marie-Luise. "Transterrestrik in der Renaissance. Nikolaus von Kues, Giordano Bruno, Johannes Kepler." Von Menschen und Außerirdischen. Transterrestrische Begegnungen im Spiegel der Kulturwissenschaft. Eds. Michael Schetsche and Martin Engelbrecht. Bielefeld: transcript, 2008: 55–79.
[257] to name a few: Connolly, Bob and Robin Anderson. First Contact. New York: Viking Penguin, 1987; Harrison, Albert A. After Contact. The Human Response to Extraterrestial Life. New York, London: Plenum Trade, 1997; Michaud, Michael A. G. Contact with Alien Civilizations. Our Hopes and Fears about Encountering Extraterrestrials. New York: Springer, 2007.
[258] Website of the SETI Institute http://www.seti.org.
[259] Shostak, Seth. Nachbarn im All. Auf der Suche nach Leben im Kosmos. München: Herbig, 1999: 198–201; Engelbrecht, Martin. op. cit.
[260] Heidmann, Jean. Bioastronomie. Über irdisches Leben und außerirdische Intelligenz. Berlin: Springer, 1994: 195; Davies, Paul. Are We Alone?: Philosophical Implications of the Discovery of Extraterrestrial Life. New York: Basic Books, 1995: 14.
[261] Fink, Alexander and Andreas Siebe. Handbuch Zukunftsmanagement. Werkzeuge für strategische Planung und Früherkennung. Frankfurt am Main: Campus, 2006: 15–56; Wilms, Falko E. P. Szenariotechnik. Vom Umgang mit der Zukunft. Bern: Haupt, 2006.
[262] Schetsche, Michael. "Auge in Auge mit dem maximal Fremden? Kontaktszenarien aus soziologischer Sicht." Von Menschen und Außerirdischen. Transterrestrische Begegnungen im Spiegel der Kulturwissenschaft. Eds. Michael Schetsche, and Martin Engelbrecht. Bielefeld: transcript, 2008: 227–53.
[263] Engelbrecht, Martin. op. cit.
[264] Michaud, Michael A. G. Contact with Alien Civilizations. Our Hopes and Fears about Encountering Extraterrestrials. New York: Springer, 2007: 123.
[265] See for example Hoerner, Sebastian von. Sind wir allein? SETI und das Leben im All. München: C. H. Beck, 2003: 112–9.
[266] Kuiper, T. B. H. and M. Morris. "Searching for Extraterrestrial Civilizations." Science 196.4290 (1977): 616–21; Michaud, Michael A. G. "Ten Decisions that could Shake the World." 17 Dec. 2007. Setileague 7 Jan. 2010. http://www.setileague.org/iaaseti/decision.pdf. 2.
[267] A possible fourth scenario (Schetsche, Michael. "Auge in Auge mit dem maximal Fremden? Kontaktszenarien aus soziologischer Sicht." Von Menschen und Außerirdischen. Transterrestrische Begegnungen im Spiegel der Kulturwissenschaft. Eds. Michael Schetsche and Martin Engelbrecht. Bielefeld: transcript, 2008: 227–53. 244–5), which assumes that the first contact has actually already taken place without the knowledge of the public, is not discussed here.
[268] Harrison, Albert A. After Contact. The Human Response to Extraterrestial Life. New York, London: Plenum Trade, 1997: 199–200; Shostak, Seth. op. cit. 231–2; Harrison, Albert A. and Joel T. Johnson. "Leben mit Außerirdischen." S.E.T.I. Die Suche nach dem Außerirdischen. Ed. Tobias Daniel Wabbel. München: Beust, 2002: 95–116. 100; Hoerner, Sebastian von. op. cit. 133.
[269] Schmitz, Michael. Kommunikation und Außerirdisches. Überlegungen zur wissenschaftlichen Frage nach Verständigung mit außerirdischer Intelligenz. Unveröffentlichte Magisterarbeit. Essen: Universität-Gesamthochschule, 1997.
[270] See for instance, Freudenthal, Hans. LINCOS. Design of a Language for Cosmic Intercourse. Amsterdam: North-Holland Publishing, 1960; Fuchs, Walter R. Leben unter fernen Sonnen? Wissenschaft und Spekulation. München: Knaur, 1973: 47–93; McConnell, Brian. Beyond Contact. A Guide to SETI and Communicating with Alien Civilisation. Sebastopol/CA: O'Reilly, 2001: 181–346.
[271] Shostak, Seth. op. cit. 233; Finney, Ben. "The Impact of Contact." Acta Astronautica 21 (1990): 117–21.

[272] Michaud, Michael A. G. Contact with Alien Civilizations. Our Hopes and Fears about Encountering Extraterrestrials. New York: Springer, 2007: 211; Shostak, Seth. op. cit. 235.
[273] Shostak, Seth. op. cit. 228–34.
[274] The number of the neighbouring stars of our sun tends to increase with its distance by the third power: with approximately 150 stars in up to thirty light years distance, statistically there are 150,000 stars in up to a distance of 300 light years, etc.
[275] Harrison, Albert A. and Joel T. Johnson. "Leben mit Außerirdischen." S.E.T.I. Die Suche nach dem Außerirdischen. Ed. Tobias Daniel Wabbel. München: Beust, 2002: 95–116. 113; Michaud, Michael A. G. Contact with Alien Civilizations. Our Hopes and Fears about Encountering Extraterrestrials. New York: Springer, 2007: 135–40.
[276] Michaud, Michael A. G. op. cit. 211.
[277] Ibid. 212.
[278] For ethic considerations on how to treat extraterrestrial life see Cockell, Charles. "Ethics and Microscopic Extraterrestrial Life." (in this volume).
[279] Harrison, Albert A. and Joel T. Johnson. "Leben mit Außerirdischen." S.E.T.I. Die Suche nach dem Außerirdischen. Ed. Tobias Daniel Wabbel. München: Beust, 2002: 95–116; Bartholomew, Robert E. and Hillary Evansk. Panic Attacks. Media Manipulation and Mass Delusion. Stroud: Sutton Publishing, 2004: 40–55.
[280] Shostak, Seth. op. cit. 121.
[281] Bitterli, Urs. Die "Wilden" und die "Zivilisierten": Grundzüge einer Geistes- und Kulturgeschichte der europäisch-überseeischen Begegnung. München: Beck, 1991; Bitterli, Urs. Alte Welt – neue Welt. Formen des europäisch-überseeischen Kulturkontaktes vom 15. bis zum 18. Jahrhundert. München: Beck, 1986.
[282] Rausch, Renate. "Der Kulturschock der Indios." 1492 und die Folgen: Beiträge zur interdisziplinären Ringvorlesung an der Philipps-Universität Marburg. Ed. Hans-Jürgen Prien. Münster, Hamburg: LIT, 1992: 18–32. 19; Finney, Ben. op. cit.
[283] Michaud, Michael A. G. "A unique Moment in Human History." Are we alone in the Cosmos? The Search for Alien Contact in the New Millennium. Eds. Byron Preiss and Ben Bova. New York: ibooks, 1999: 265–84. 272. Numerous peoples of America and Oceania suffered a long-lasting culture shock after the arrival of "white people", which caused the collapse of their religious and cultural conceptual system and led to the complete disintegration of economic and social systems in the medium term. In some cases, entire population groups committed suicide as a response to the culture contact (Müller, Klaus E. "Tod und Auferstehung. Heilserwartungsbewegungen in traditionellen Gesellschaften." Historische Wendeprozesse. Ideen, die Geschichte machten. Ed. Klaus E. Müller. Freiburg: Herder, 2003: 256–87. 270–1; Müller, Klaus E. "Einfälle aus einer anderen Welt." Der maximal Fremde. Begegnungen mit dem Nichtmenschlichen und die Grenzen des Verstehens. Ed. Michael Schetsche. Würzburg: Ergon, 2004: 191–204. 196.
[284] Groh, Arnold. "Globalisierung und kulturelle Information." Die Zukunft des Wissens. Workshop-Beiträge, XVIII. Deutscher Kongreß für Philosophie. Ed. Jürgen Mittelstraß. Konstanz: UVK, 1999. 1076–84.
[285] Shostak, Seth. op. cit. 236.
[286] Kuiper, T. B. H. and M. Morris. op. cit. 620.
[287] Michael, Donald N. (Ed.). Proposed Studies on the Implications of Peaceful Space Activities for Human Affairs. (Brookings – Report. A report Prepared for the Committee on Long-Range Studies of the National Aeronautics and Space Administration by The Brookings Institution). Washington D.C.: Brookings Institution, 1960. (source: http://www.nicap.org/papers/brookings.pdf).
[288] At least so long as this is not *equal to zero*, which however hardly any space researcher would argue today.

CHAPTER 2

HISTORY AND RELIGION

2.1 Astrocognition: Prolegomena to a future cognitive history of exploration

David Dunér

2.1.1 The astrocognitive question

The human desire for exploration and man's encounters with the unknown are a fundamental part of the cultural history of mankind, from the first stumbling steps on the African plains to the recent explorations of our globalised and urbanised world. From the dawn of the hominids to the days of the modern man, this ever changing terrestrial being has expanded in ever increasing circles of spatial consciousness, in an endeavour to climb over mountains to the next valley, transcend vast oceans and fly through the air. The next small step, or giant leap for mankind, that of going far beyond the atmosphere and gravitation of the Earth to the unknown outer space, is decisive, but that, too, is part of the long history of mankind.

A true universal history includes space, not only the human history on this Earth, not only the short history of human civilisation: it must include the immense space, and, as we know it, the long history of matter from the big bang, through the formation of our solar system and the evolutionary history on Earth, to the future speciation of man and the final big crunch or the ever-expanding universe and eternity of time. History is crucial for an understanding of the perennial enigmas of who we are, where we come from, and where we are going. Historical narratives are essential, and the only way for an understanding of the big questions of science: the origins of the universe and the formation of galaxies, stars and planets, the evolution of life on Earth, the human species and the human mind.[289] Cognitive neuroscience can offer, as Mark Turner says, a wider conception of human history, not just brief temporal spans of decades or centuries we usually find in cultural and sociological history, but a phylogenetic history that runs over thousands and millions of years.[290] Here, I will radically extend the notion of human history to also include our evolutionary past and situate our thoughts in humankind's deep history.

In the course of everyday events and encounters the human mind has been enabled, through an evolutionary process, to understand, interact, deal with and adapt to the environment of this particular planet and to the minds of other human beings. The brain is thus well adapted to the biological, ecological, and physical characteristics of our planet, and the cultural, social and cognitive characteristics of the species *Homo sapiens*. The question is then:

> *What will happen to human cognition when humans encounter a totally different environment, physically, biologically and culturally; an extraterrestrial environment that the human brain is not accustomed to and developed for?*

That is what I will call *the astrocognitive question*. In short, what will happen to our minds in an unknown extraterrestrial environment?

In order to try to answer this question I will propose a new multidisciplinary field that I call astrocognition; that is literally "acquaintance of the stars", from the Greek *astron*, star, and Latin *cognitio*, knowing or acquaintance. Astrocognition could be defined as: "The study of human cognitive processes in extraterrestrial environments". This is the main issue here. But in a wider sense, it can also be defined as: "The study of the thinking universe".

This as an allusion to the definition of astrobiology in the 1996 NASA Strategic Plan: "the study of the living universe".[291] Astrobiology tries to find the necessary conditions for life in the universe and to develop theories of plausible features of extraterrestrial life. Astrocognition goes even further. The universe is not only living, it contains not just self-reproductive entities but it has also apparently led to self-reproductive organisms that are able to reflect on their own existence and on the universe they live in. We humans are the only species here on Earth that is to a greater extent able to reflect upon the universe. Through the self-conscious human being, the universe is, one might say, also self-conscious and able to reflect upon itself. It seems, then, that the universe is not only bio-friendly,[292] but also "cogito-friendly". Astrocognition deals exactly with this: the thinking universe and issues concerning cognitive aspects in a universal perspective.

The astrocognitive paradigm states that space and extraterrestrial environments, as well as contacts with other forms of life and societies, will change our thinking and belief systems. This leads to a discussion of what we can know and not know about the extraterrestrial; or, put another way, the limits of our human-based epistemology and the constraints of our evolutionary history. And further, what cognitive challenges we are likely going to face when we encounter the unknown; this is in line with the "phenomenology of space" proposed by Gísli Pálsson – the challenges our Earth-bound perceptual, cognitive, and psychological capacities face in a space context.[293] Philosophical, ethical and theological implications of extraterrestrial discovery have been discussed to only a limited extent.[294] But the cognitive implications have not been discussed at all. Thus, the purpose of this article is to show the possibilities of future space research for the humanities. Astrocognition is the science of the future firmly tied with the past and present.

I will give some examples of the possibilities for future astrocognition. Astrocognition studies for example:

(i) *universal astrocognition*, the mind in a universal context;
(ii) *incognita cognita*, the human mind's encounter with the unknown;
(iii) *situated astrocognition*, what will happen to the human mind situated in and in interaction with an extraterrestrial environment;
(iv) *cosmocognition*, the human mind's search for order in the universe, i.e. kosmos in its original Greek sense;
(v) *xenocognition*, the human mind's encounters and interactions with foreign beings, and other forms of non-human cognition;
(vi) *astrocognitive epistemology*, what we can learn through extraterrestrial explorations, interactions and encounters;
(vii) *interstellar communication*, how we can communicate and understand extraterrestrial messages;[295]
(viii) *extraterrestrial imagination*, how we think about and imagine the extraterrestrial.

Astrocognition is a multidisciplinary approach consisting of a wide range of perspectives from cognitive science, philosophy of mind, the history of science, theology, ethics, cultural studies, semiotics, computer science, neurology, evolutionary theory, and space technology. The only field dealing with the human being in itself and what it is to be human – especially in its relation to the non-human – is the humanities, and thus the humanities must be the core of any future research on humans in outer space.

2.1.2 The astrocognitive premise

The astrocognitive paradigm rests on a supposition that man is a historical creature. If we want to understand how space travel and encounters will change human culture, we need first to understand our prehistory; the limits and possibilities it has given us. Thus astrocognition starts from a fundamental, basic premise, the *astrocognitive premise*: "The human brain has developed through an evolutionary process here on Earth, with its specific environment."

Our thinking has an evolutionary origin.[296] The human mind has to a large extent evolved as an adaptation to certain problems that our ancestors have faced during the evolutionary development of our species. That is, the human brain is adapted to, firstly, the physical and biological environment of the Earth – to understand and interpret, interact and deal with, and orientate itself in the Earth's physical and biological environment, in relation to its specific conditions, such as planetary orbit, gravitation, light conditions, atmosphere, radiation, temperature, chemistry, geology, ecology, fauna and flora. Secondly, the human brain is also adapted to the mind and culture of the tellurian species *Homo sapiens*, to understand and interact with other beings of our species, to understand human feelings, thoughts, motives, etc., in a psychological and sociological interplay that forms our human culture. Culture can here be defined, in accordance with Chris Sinha, as the existence of intra-species group differences in behavioural patterns and repertoires, which are not directly determined by ecological circumstances, and which are learned and transmitted across generations.[297]

Thus, *Homo sapiens* could only have developed on this particular planet on which we now stand, with its specific conditions and evolutionary history. That is why it is nearly impossible for our earthly minds to really grasp for example:

(i) *the infinite universe*. We tend to think that, if the universe is finite, then it must have limits, but what in that case is beyond the limits of the universe? Everything seems to have a border, and there is always something on the other side of the border;

(ii) *the creation or beginning of life on Earth*. How could the universe be created from nothing? Everything must have a beginning, we assume, nothing can be created out of nothing, without a purpose or a creator;

(iii) *meaningless things*, the meaninglessness of existence, that something does not have any meaning or purpose;

(iv) *space*, such as the relative space of Einstein's universe, the multidimensionality of space, and long distances like light-years;

(v) *time*, such as the relative time of Einstein's universe. Even our common objective, uniform clock time is sometimes hard to grasp – our subjective time could go slower and faster – and time spans longer than some few generations, or speeds faster than a running horse;

(vi) *the modern world*, the artificial environment of the city and the virtual reality of cyber space. There are arguments based on evolutionary psychology that claim that our brains are actually maladapted to our modern society.

Our brains are not developed for situations and phenomena – as those mentioned above – far from ordinary life. Sometimes it is said that the world is inherently understandable and that we can understand it by observing how it works. This conception is based on an idea that our brains are developed for understanding such things, but that is unlikely from an evolutionary perspective. The brain has developed for handling ordinary daily life experiences in a rather narrow space and time scale. Human understanding is constrained by its evolution and history; it has limits. The different worlds out there are likely to be far beyond our most vivid imaginations. We will be surprised.

2.1.3 Astrocognitive theory

Cognitive science can give clues to how we understand and think about the universe and reveal new perspectives on human encounters with the unknown. In short, cognitive science studies how the external world is represented, how we use cognitive tools for our thinking – such as language, image schemas, mental maps, metaphors, and categories, but also how we use and interpret, for example, drawings and images to enhance communication. It is about perception, attention, memory, learning, consciousness, reasoning and other things that we include in what is called "thinking". Recent theories about human cognition could open up new perspectives on space encounters. The following does not aim to be a full, complete list of relevant cognitive theories, just a few examples from cognitive science that I believe could be useful in our task. First of all is the theory of the *embodied mind*.[298] We think with the body. The mind is not detached from the body. Thus bodies of other kinds and evolutionary backgrounds will have other minds and ways of thinking. This opens another question of astrocognition: how aliens from other evolutionary trails understand their world – or our world.

The theory of *situated cognition* claims that our cognitive processes are not just inside our brains; we also use our environment for thinking.[299] The brain not only needs the body but also the surrounding world in order to function efficiently. The environment has an active role in driving the cognitive processes of what Andy Clark and David Chalmers called the *extended mind*.[300] Thus, cognition emerges in the interaction between the brain, the body and the world. There exists no sharp line between the brain and the world. We cannot be isolated observers. In other words, cognitive activity cannot be separated from the situations in which it occurs. There is a dynamic interplay between mind and environment, making a cognitive system of agent and environment. The agent both adapts to the world and changes the world. This change is not just pragmatic, but also epistemic, so the world becomes easier to adapt to.[301]

But during the evolution of human thinking, as Peter Gärdenfors argues, it seems that the mind became less dependent on the situation here and now, and more detached from the current environment.[302] Human cognition depends on constraints in the surrounding culture and evolves in a dynamic interaction with the technological environment. To this we can add what is called *distributed cognition*: that we are using our environment and tools for enhancing thinking, and we place our ideas and memories in things – books, computers, etc. So, in conclusion, where we are in time and space is totally fundamental to cognition. Extraterrestrial environments will probably act on and change our thinking. Our

Fig. 1. *Interpretations of perceptions. Unusual celestial occurrences, like comets, are interpreted as terrifying omens in the shape of a blazing cross (source: Lycosthenes, Conrad. Die Wunder Gottes in der Natur (1744)).*

knowledge of extremophiles shows also that life can thrive in very different conditions than those we are used to. Perhaps the same applies to intelligent life. Aliens, developed in and adapted to an entirely different physical and cultural environment would probably have very different modes of thinking. Thinking in the universe could be very different from what we are used to in our westernised, anthropocentric and earthbound human culture.

A central cognitive process is the *perception* of patterns. We do not passively receive images and sounds from the surrounding world. Instead the brain actively searches for patterns in what it receives from the senses, and interprets them. This process is determined by both biological and cultural factors. We are especially efficient at seeing faces everywhere, for example the man in the Moon, or the face on Mars that Viking 1 captured with its camera in 1976. Other examples of our ability to see patterns are the star constellations, or how Giovanni Schiaparelli and Percival Lowell in the nineteenth century saw canali on the surface of Mars, which were interpreted as artificial waterways built by a Martian civilisation.

We also think with *metaphors*.[303] Spatial experience is fundamental for cognition, as mentioned above, which leads to the fact that many basic words relate to bodily experiences. Our cognitive capacities, especially concerning concept formation, can be explained as a kind of metaphorical extension of spatial reasoning.[304] Abstract concepts relate to concrete, basic human experiences. In order to be useful our concepts must not only be applicable to known cases, they should also be able to be generalised to new situations as well. The unknown is then understood, described, and grasped through known experiences. Thus, our thoughts and descriptions of the unknown often tell more about ourselves than the things unknown. Metaphors are like tools for dealing with the unknown. Actually, astrobiology rests on metaphors, on a great analogy. From just one known case of a life-bearing planet we infer with an inductive mode what potential extraterrestrial *loci* must be like in order to sustain life. When we think about and try to grasp the universe and the unknown, we have to use experiences from the known ordinary life on Earth that we are familiar with.

Another cognitive ability is *categorisation*.[305] All living creatures seem to categorise the environment in terms of edible versus inedible, benign versus harmful, etc. Categorisation becomes more complex in humans. The human mind tends to categorise, seeing hierarchies and similarities between things, for example stellar spectra or species and genera in taxonomy. But do they really exist outside our minds? The classification of species – whether it is artificial or natural – has been debated ever since Linnaeus and Buffon.[306] Behind this is our ability to see sameness in diversity. Our categories are based on experiences and a sort of inductive reasoning that, according to Eleanor Rosch, give rise to prototypes from which we construct our categories.[307] These prototypes, I would underline, are developed through experiences of this actual world we are living in and will be unfit for use in another world where we have never set foot. New encounters will force us to change those of our categories that have ceased to work.

Intersubjectivity, the sharing and representing of others' mentality is an important part of our inner worlds. Empathy, the representing of other human beings' emotions, motives, intentions and desires, bodily expressions of emotions, beliefs and knowledge, are impossible without a rich inner world. Cooperation on detached goals requires advanced coordination of the inner worlds of the individuals.[308] Future encounters with aliens will face severe problems concerning intersubjectivity, coordination of inner worlds, and empathy etc., due to totally different biological and cultural conditions. A human and an extraterrestrial will probably have trouble even in perceiving the same target, in aligning their attention, adjusting their actions, and imitating each other.

Communication is situated in and constrained by its surroundings.[309] Language has an evolutionary history and has evolved due to its enhancement of communication between humans, used for describing the world around us, but perhaps more importantly as a social interplay: to express feelings, for socialising and creating bonds, etc. Cognitive linguistics aims to situate language within more general cognitive capacities. According to John Taylor, language can be understood as a set of resources that are available to the language user for the symbolisation of thought, and for the communication of these symbolisations.[310] The reference of a symbol is a detached representation; the symbol refers to the inner world, in contrast to the signal that refers to something in the outer environment. Symbols are conventional signs, or arbitrary as Ferdinand de Saussure called them, dependent on culture.[311] We may figure out the reference of the signal, but will probably have severe problems understanding extraterrestrial symbols. Cognitive semiotics, which deals with our use of signs, symbols, etc. for communication between humans, will I believe have great importance for interstellar communication.

Can we understand a message from outer space? Are we able to recognise it at all as an intelligent message? The first problem that arises in this situation of interstellar communication is realising that it really is a message at all, as Göran Sonesson has pointed out.[312] Some regularity and order, finding a repetition in the pattern, is not enough. We have to understand that someone has an intention that we should understand it as a message. Next comes the problem of deciphering what the message means. Cultural semiotics, developed among others by Yuri Lotman, studies sign systems and the correlations between different systems.[313] In order to understand a message the receiver must be able to fill in the gaps between the receiver's perception of the message and the sender's intention with it.[314] The problem is that the creator of the message and the receiver of it are situated in different and specific cultural and social contexts. Relating to interstellar communication, this gap will be huge, with totally different ecological and cultural contexts. As Douglas A. Vakoch clearly states: "In the absence of knowledge of physical and cultural clues, communication between two species can be almost impossible."[315] If we want to formulate a message comprehensible to an intelligent extraterrestrial we must, as Sonesson has also stated, go beyond our specific human biology, ecology, and culture. Designing a language for cosmic intercourse, like Hans Freudenthal's *lingua cosmica*, will probably be in vain.[316] The famous Pioneer plaque now traversing deep space is also too firmly restricted by human culture and cognition, and will most likely be incomprehensible to an extraterrestrial. Even though the aliens would be familiar with the content of the same kind of mathematics, chemistry, or physics as we are, their expressions of it would probably not be the same as ours when our life-worlds are different.[317] The aliens and ourselves live in different cognitive or, if you wish, semiotic worlds with ways of thinking and signification that are not in agreement with each other.

Finally, man is constantly searching for *meaning*. When we find that something is caused by an agent we presume that there is a purpose behind it, that the act is goal-directed. We tend to search for meaning in everything. That is a driving force for culture, theology and sciences. We search for a meaning in the universe and creation. These cognitive abilities mentioned above – and many others – are important and will be tested in future encounters with the unknown.

2.1.4 How can we get empirical data?

As yet we have no or very little experience of extraterrestrial environments. What, then, should we do in the meantime, if we want to study human encounters with the unknown? How could we increase our understanding of astrocognitive processes? To begin with, we actually have experience of unknown environments – here on Earth. Of course, we nearly every day encounter greater and lesser unknown things in unfamiliar situations: when we meet foreigners, read new books, enter new buildings, perform new tasks, and so on. Human encounters with the unknown seem to be a human universal. These encounters with the unknown can here and now be studied in at least three different empirical ways.

We can get further empirical data from:

(i) *Observations* of authentic cognitive processes in practical action. This can be done from anthropological studies of cultural encounters, when mutually unknown cultures meet and interact. This includes trying to learn other ways of understanding reality and living it in a more basic physical sense – going native, so to speak. The anthropologists' encounters with unknown cultures and their attempts to find meaning in foreign rituals and ways of life, is a study in itself on the meta-level. Further information can be gained from psychological and sociological studies of our modern life: how our palaeolithic hunter brain works in the postmodern city jungle. Even zoological and ethological studies of animal communication, and intraspecies communication between man and animal, can give some clues to future communication with non-human entities.

(ii) *Neurological experiments*, such as experiments that study perception of the self and the body. For example, experiments have been carried out by Henrik Ehrsson that test reactions to changes in the normal sensory feedback, in order to imitate clinical conditions that disturb normal brain functioning – for instance illusions of out-of-body experiences.[318] Simulations, or manipulations, of this and other kinds may reveal reactions to astrocognitive possibilities.

(iii) *Computer simulations*, such as virtual reality experiments with parameter variation. I propose that three-dimensional virtual reality environments should be constructed in a virtual reality lab where the test parameters can be slightly altered, such as the speed and echo of sounds, light conditions, humidity, temperature, distances, transparency, air, etc. Then the test subjects can be observed: what they do and how they act in these environments. For example, the attention of the test subject in an unknown environment can be observed with eye-tracking equipment. Even haptic devices (the virtual sense of textures, etc.) may be used in order to act on a complete set of sensory input.

These empirical methods have limits, though. The first (i) is restricted by the fact that it is all but impossible to find isolated situations without previous knowledge and preconceptions. The other two alternatives (ii, iii) are limited by our putting into the experiment factors that we already know of, and elaborating by using known parameters? But we also have another option for studying authentic cognitive processes in practical action: (iv) *historical encounters* with the unknown found in travel diaries, scientific reports and other historical documents. In history we can find isolated cultures and environments with little or no previous knowledge of each other. Throughout history, man has always ruminated and thought about the unknown. Science deals actually with the unknown; about how we can grasp it and make it into something known and possible to understand. In all these cases of confrontation with the unknown, man has used his earthbound cognitive abilities.

2.1.5 Historical questions of astrocognition

In order to understand how we think as humans, I would say it is necessary to put the human being in a historical time setting and study the changes of cultures, environments, modes of thoughts, etc., both diachronically and synchronically; that is to say, phylogenetic, ontogenetic and historical studies of change – for example the development of species, individual organisms and cultures –, but also of fixed points in time for comparative and multidisciplinary perspectives. Human distinctiveness, as Merlin Donald puts it, is not only a function of biology, but also owes much to the specificities of human history.[319] The human being is an historical animal; it has an evolutionary history but can also reflect on its history and is more or less conscious of it. In one respect, past historical cultures are alien, unfamiliar and unknown environments. The quest for understanding these distant pasts is not unlike the understanding of contemporary distant extraterrestrial cultures and environments. But there is one big difference: in the case of the distant human past on Earth we can use our cognitive abilities developed through our evolutionary history that we have in common. Two fields have special relevance to the astrocognitive paradigm: the history of imagination and the history of exploration.

To begin with, the *history of imagination* has deeper implications than we can readily imagine. Throughout history, man has always imagined other worlds – the nonexistent and the utopian; in science fiction man has imagined extraterrestrial worlds and life forms; and in fantasy societies outside our time and space have been conceived of. But all these speculations are actually about our own existent world and say more about ourselves than the outer world itself. Human imagination always comes back to actual sensory experience and is based, among other things, on metaphorical processes of the mind. You might say, according to a Lockean tabula rasa: all ideas that come to be in the mind must have first been in the senses. With the help of our cognitive abilities we put together known things, actual sensory experiences, to create new ideas. Narrative constructions of reality, as Jerome Bruner has explained, are also about how we organise our experiences and memories in the form of stories and myths.[320] Our occupation with science fiction is a way of envisioning various actions and their consequences, which can have further use as a tool for planning and for readiness for the future. Our thoughts about science fiction are also mixed with our ideas of utopias, of the perfect society. A future Mars colony is a utopia and raises the classic utopian question: what is needed to build a future society from scratch? So in all, how we talk and write about space is not in the least irrelevant, something that has been underlined by Ulrike Landfester.[321] This leads to the need for historical analyses of ideas and speculations concerning the unknown, extraterrestrial life, space travels and astrobiology.[322]

The other field, the *history of exploration*, itself needs to be further explored. This approach leads to studies of the early history of the exploration of the Earth, from a new angle. In these early travels, man encountered the unknown: new continents, life forms, cultures, and other unfamiliar things. How did these travellers actually handle the unknown?

2.1.6 The history of exploration

Exploration appears to be inherent to humans, as it says in the rationale of the European Science Foundation's Standing Committee for the Humanities and the European Space Sciences Committee position paper of Humans in Outer Space. There seems to be an "inherent human quest for odysseys beyond the atmosphere". Throughout the ages explorers have searched for

> "fire, fresh water, food, milder weather, new hunting grounds, stone, minerals, spices, terra incognita, gold, precious stones, other life forms, rare animals, high mountains to climb, mysterious places to reach, and in the process bringing back answers, novel things to study, theories, and many more questions asked".[323]

Several drives and behaviours such as curiosity, the acquisition of new territories and riches, national prestige, etc. lay behind this. It is also an intellectual endeavour. Looking around us, in the world, in space, we may understand a little bit more of what it means to be human and how we as an earthly species fit into the universe.[324] In a study of eighteenth-century exploration of the Earth, Mary Louise Pratt calls this endeavour a movement towards a "planetary consciousness", in reference to circumnavigation and mapping of the world's coastlines, the Linnaean taxonomy project and the French expedition of Charles Marie de La Condamine and its determination of the shape of the Earth.[325] Exploration of the universe extends this quest to a deeper cosmic consciousness.

History as a guide to frame our thinking about the implications of encounters with extraterrestrial life has been proposed by, for example, Steven Dick and Luca Codignola.[326] As an attempt to look forward, we look back. By analysing historical responses to past discoveries we can gain perspectives on how worldviews have changed in connection with encounters. In line with this, I would like to complement it with an evolutionary and cognitive perspective on explorations: the zest for exploration has an evolutionary origin. Explorations to a great extent have to do with emotions as a vital part of the drive, such as curiosity and admiration. In the "Vienna Vision on Humans in Outer Space" it is emphasised that the inherent human curiosity for exploring the unknown is at the heart of the dialogue.[327] To this I would add that the emotion of curiosity has cognitive, evolutionary benefits. Without curiosity and a drive for exploration we would not have found food, water, shelter, mates, etc.

For the most part, the history of exploration has dealt with what travellers actually saw, whom they met, where they travelled, what islands and geographical features they observed. A great part concerns logistic problems and how they travelled from spot to spot, for example navigation on the seas and how they followed unbeaten tracks, coastlines, rivers, mountain ranges, and valleys. Another interest has been the sociological and political aspects of travelling, such as networks, power relations, contacts, collaboration, crewmembers' careers, and how the encounters have resulted in political and economical change.[328] But if we want to understand how encounters with the unknown in a more fundamental way change our thinking and belief systems we have to rewrite the history of exploration and study it from a new angle: from a cognitive perspective. A cognitive history of exploration could go further and, in fact, study the lacunae of the mind – what they did not see because of a selective mind directed to particular things. It could also study how our minds try to grasp and interpret what the senses offer us and how our preconceptions and earlier experiences direct us to conclusions about what we encounter. The narrations follow also genre specific norms, and the "real" travels merge with the fictive travels. The history of polar expeditions has already been useful for studying technical and logistic problems with isolated expeditions in unfriendly environments. I think that travels in tropical and subtropical environments can also help us to understand the cognitive processes of encounters with unknown organic environments.

As a consequence of this approach, we have to go far back in time, before our globalised world; to a time when there were still entirely unknown white spots on the Earth, cultures that had no deliberate external contact whatsoever, and having no knowledge or awareness of each other. The pre-globalised world can be dated to a time in human history when differences between foreign cultures and environments seemed to be greater, or more accurately: the gaps between their life-worlds were greater; their cultures were more isolated, independent, and self-sufficient; mutual contacts were fewer. There was a wide gulf between cultures in regard to culturally constrained preconceptions, knowledge, and ways of life, etc. Yet they had the same human cognitive abilities and an evolutionary history in common.

With respect to scientific expeditions of the early-modern period, we can find that a cognitive approach reveals and sustains a wide range of arguments for what happens to the human mind in unknown environments. First and foremost, there is an interaction with the environment leading to a wider *spatial conception* or

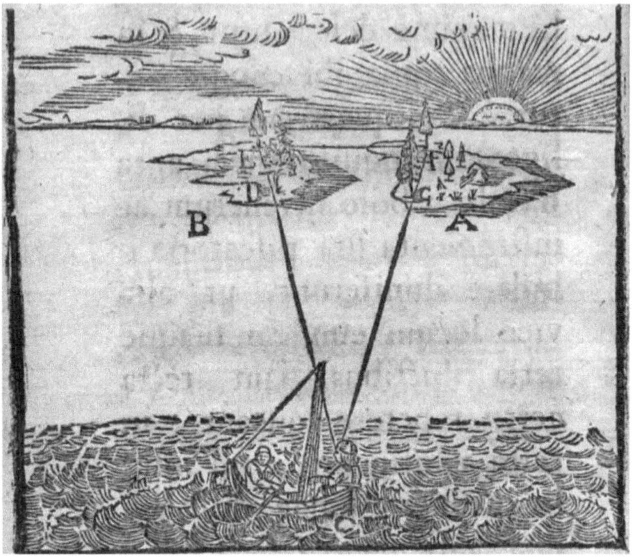

Fig. 2. *Spatial consciousness. Two cod fisherman who orientate themselves with reference to the angles between the boat and the trees on two small islands. Harald Vallerius describes the spatial significance of geometry, how right angles can be used in geodesy, in a dissertation on angles, De angulo, from 1698 (source: Lund University Library).*

consciousness. Not just texts, ideas, traditions and cultures lead to historical change, but also the body, the environment, space and matters around us, dimensions, topography, logistics, buildings, communication routes, etc.[329] We relate to home and away, inside and outside, existing and non-existing. Travels into unknown geographies change our spatial conceptions and develop the relationship between mental space and real space.[330]

Encounters lead to *cultural change* and change our thinking, categories, and belief systems. While many theories and disciplines – such as anthropology, cultural studies, post-colonialism – claim to study cultural encounters, they have focused more on political, ideological and sociological issues. A cognitive history, but also cultural semiotics, can in a more fundamental way study communication and what is going on in the individual mind's encounter with the unknown.

We are often not able to fully understand other intelligent life; often we actually cannot even see that other lives are intelligent lives. It is difficult to see the complexity of cultures and the differences between the tribes or groups we encounter. We tend to *dehumanise* the others and see complex cultures as primitive cultures. This is due to cognitive challenges for the human mind concerning, for example, empathy, intersubjectivity and the coordination of the inner worlds. One way to cope with this – if it concerns other members of our own species – is to go native, like for example the German explorer Carsten Niebuhr, who in the 1760s travelled through Arabia Felix and tried to fit into the Arabian culture using his cognitive abilities, learning and imitating in order to acquire language skills and customs.[331]

We often have severe *communication problems*. Misconceptions are frequent in cultural encounters due to different life-worlds, cultural backgrounds and experiences. Culturally- and environmentally-bounded concepts, symbols, categories, etc. differ and complicate communication.

The unknown is a *reflection of the known*. Travelogues always tend to refer back home, to the familiar and known, to former experiences and knowledge; traditionally to western culture and its preconceptions and ideologies. In descriptions of

Fig. 3. *A cognitive history of exploration. Our minds try to grasp and interpret what the senses offer us, and our preconceptions and earlier experiences direct us to conclusions about what we encounter (source: Sigfridus Aronus Forsius's Physica. Sea monsters, fishes, whales, sirens, crocodiles, etc. (1611)).*

unknown new things, we always use known things that we are used to. When the American settlers saw the red-breasted *Turdus migratorius*, it reminded them of the European Robin (*Erithacus rubecula*) and they called it *Turdus migratorius* for American Robin. It is actually more closely related to the Common Blackbird (*Turdus merula*) than the Robin. There are lots of such examples from biological taxonomy.

The interpretation of the seen, the environment and nature, goes back to cognitive abilities like memory, categorisation, perception of patterns, and so on. This acquired pre-knowledge, conscious or unconscious, our *preconceptions*, forces us to interpret what we see in one way or another. We use previous knowledge, narratives, myths, and religious conceptions in order to make sense of what we encounter. A famous example is when Christopher Columbus encountered America. He did not see a new continent and he did not encounter a new human race. He came to India, and met Indians. Columbus' account shows also, like many other descriptions from unknown environments, how difficult it is to describe the seen. Concepts and words are lacking. He then saw high and beautiful mountains, wide fields, and forests, fertile plains that, as he says in his letter from America immediately after the first voyage, surpassed your conceptions if you had not previously seen them.[332] The new things seen are impossible to imagine or describe – you have to see them with your own eyes.

This is also true for interpretations of foreign cultures. The traveller takes concepts from his own known culture and superimposes them on the different, unknown culture, projecting himself onto the unknown other. We see the other as a mirror image of ourselves. The very term "discovery" unveils an egocentric perspective in the modelling of cultures.[333] Discoveries and encounters are seen from a local, personal point of view, not from a global perspective, and we have no other option. The ego is in dialogue with the other, alter and alius, in Sonesson's words: our own, the ego-culture, stands in relation to the extra-culture and the non-culture.[334] The ego-culture deforms texts from other cultures and times in line with its own needs. Our own culture stands for order in contrast to the other's chaos and disorder. Anna Cabak Rédei gives striking examples of this from the travel accounts of Germaine de Staël from her journey to Germany and Russia during her exile from 1803–1812.[335] Most travelogues appoint western culture to the ego and the visited culture to the other. In some few cases we have the opposite – for example, the Japanese Hoshu Katsuragawa's description of Russia in the eighteenth century.[336]

We need the other in order to create an image of ourselves, not just in the Saidian collective sense, but also from the perspective of the individual. This ego-perspective of the human mind leads to *subjectivity*. Thus travelogues and

Fig. 4. *Cultural encounter with an Indian family in North America (source: Woodcut by Thomas Campanius Holm from 1702 after a sketch by Per Lindeström from 1654. Thomas Campanius Holm, Kort beskrifning om provincien Nya Sverige uti America (1702)).*

descriptions often say more about the traveller than about the actual objective world. According to Edward Said, ideas about the other say more about western culture than the Orient itself.[337] From a cognitive perspective, what we always have are human internal conceptions of the world, not the real objective, external world. The world is filtered through our minds.

Finally a kind of *nostalgic isolation* during expeditions is a frequent theme. Emotions of isolation, that there is no turning back, the loss of contact with relatives and friends, evolve in the mind's journey through territories detached from the familiar environment. An unknown environment – like distant planets in outer space – is frightening, likewise unknown creatures never seen before. This anxiety is due to of a lack of orientation. Our ordinary cognitive tools for handling our known environment are put aside and do not work anymore.

2.1.7 Astrocognitive hypotheses

Something similar to this that took place in the history of exploration of the Earth will happen again when we encounter an entirely new and unknown extraterrestrial world. But the human mind will then face even harder cognitive challenges, when the gap between the worlds is likely to be huge, far greater than the gaps upon our Earth.

Based on historical research on the history of exploration of our Earth, I suppose that we – when we in the future tread a new world – will face similar cognitive problems. Thus we can assume at least nine *astrocognitive hypotheses*. Encounters with the unknown outer space will:

(i) change our spatial consciousness;

(ii) change our thinking – conceptions, categories, belief systems, culture, and the meaning of things. What we so far have come to believe through science and human cogitation will face anomalies. The old categories, systems, and beliefs will fall short in the understanding of new unfamiliar things. Our thinking, science and belief systems will then have to be revised, which will lead to adjustments, adaptations, and compromises.

The children of space settlers will use same cognitive abilities as their parents and they will belong to the same historical tradition. The first generations of space settlers will have nearly exactly the same cognitive tools as we have. But in the long run after many generations in space the genetic-cultural co-evolution of the human descendents might have resulted in slightly different or new cognitive tools that enhanced their ability to live efficiently in their new extraterrestrial environment. The conceptual organisation of space and time would probably change for example.

We will:

(iii) not be fully aware of whether that which we encounter is intelligent life, perhaps not even that it is something living;

(iv) not be able to understand their feelings, motives, and desires, i.e. their inner worlds. To feel empathy for an alien will be problematic.[338] Different evolutionary and cultural history makes different ethics and social behaviours;

(v) probably not be able to or at least have severe problems in communicating with "it";

(vi) describe the extraterrestrials with human, earthly conceptions, and anthropocentric terms and qualities. We will start from our known culture, biology and science and force it onto the unknown. But why does intelligent life have to look like us? We need an awareness, as Landfester brings to mind, that what we may encounter out there cannot be presumed to be like us, neither in knowledge or learning, in body or spirit.[339] The way we think about the universe is anthropocentric and geocentric, firmly caught in the evolutionary history of our Earth as a part of the anthropocentric history of mankind. A truly cosmocentric view will be hard to achieve.[340]

And furthermore:

(i) it will be difficult, if not impossible, not to force our preconceptions onto the unknown. We cannot escape from a non-preconceptualised understanding;

(ii) the travelogues and descriptions of these new worlds will for the most part say more about ourselves than about the real, or so to speak objective,

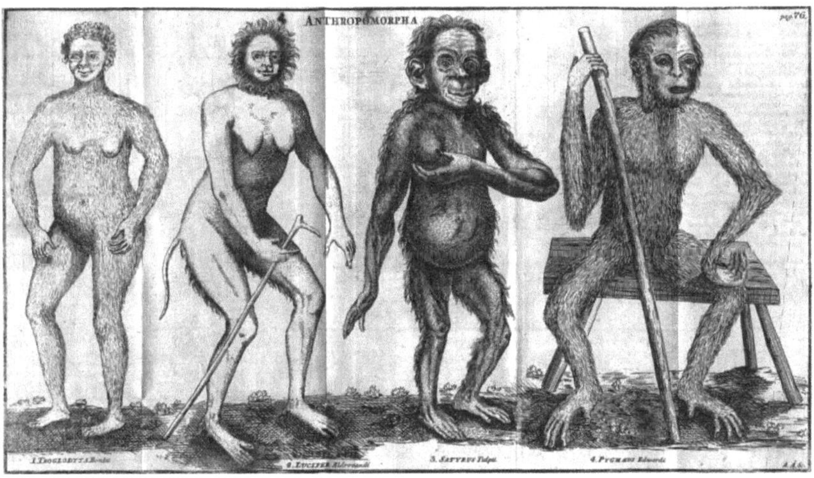

Fig. 5. *We describe the aliens with human conceptions, and anthropocentric terms and qualities. Anthropomorpha or manlike creatures, the missing links between man and animal: Homo troglodytes, Lucifer or Homo caudatus, Simia saturus and Simia pygmaeus, Linnaeus, Carl. Anthropomorpha (1760) (source: Lund University Library).*

extraterrestrial world. An independent world outside us might exist, but we will never be able to reach it;

(iii) crewmembers and settlers in outer space will feel emotions of isolation.

2.1.8 Conclusions

My aim was to search into the limits of our earthly brains that human evolution and culture have given us, to try to find out what we can know and what we will likely encounter in the future. I have tried to show how earthbound and historically constrained we are as thinking beings. Our intellects do not transcend space, instead, our cognition is situated in space.

To summarise: what will be achieved with astrocognition? We will get further theoretical, scientific knowledge of:

(i) how we encounter the unknown;
(ii) how the human mind interacts with space and the environment around us.

These two achievements will be valuable even though we might never go beyond our solar system. From this we would learn more about human cognition and how

Fig. 6. *Cultural life on the mountains of the Moon (source: Dunér, David. "Centro Cultural de Monte de Luna" La Palma, Canary Islands, Spain: 2009).*

it has been developed here on Earth. But we will also get practical knowledge. Astrocognitive research will prepare us for:

(iii) future space expeditions, thus preparing a new kind of explorer astronaut and how we behave in a new unknown world;
(iv) terra-forming and how to live in an extraterrestrial environment;
(v) communication with aliens.

And furthermore:

(vi) future comparative research on cognitive processes of extraterrestrial minds can reveal new knowledge of how humans think. We will then know more about the specifically human ways of thinking and meaning-making, specific or typical characteristics of our species. Are we unique in the universe? This approach can lead to the development of not just an anthropocentric, but also a cosmocentric perspective on cognition.
(vii) Finally, an astrocognitive inquiry will give indications as to what a human being is from a truly universal point of view. Encounters with other minds here on Earth or on extra-solar foreign planets out there in deep space among billions of stars and galaxies can give a more universal answer to the question: "What is thinking?"

Acknowledgment

Special thanks to Erik Persson, Per Lind, Lars Hall and Göran Sonesson for valuable comments during the preparation of this article.

---

[289] Jakosky, Bruce M. "Philosophical Aspects of Astrobiology." A New Era in Bioastronomy, Vol. 213. Eds. Guillermo A. Lemarchand and Karen J. Meech. San Francisco, CA: ASP Conference Series, 2000: 666.
[290] Turner, Mark. "The Cognitive Study of Art, Language, and Literature." Poetics Today 23.1 (2002): 18; See also Richardson, Alan and Francis F. Steen. "Literature and the Cognitive Revolution. An Introduction." Poetics Today 23.1 (2002): 1–8. 3; Dick, Steven J. and Mark L. Lupisella, Eds. Cosmos and Culture: Cultural Evolution in a Cosmic Context. Washington, DC: NASA History Series, 2009.
[291] NASA. "Strategic Plan 1998: With 1999 Interim Adjustments." NASA Policy Directive (NPD)-1000.1a of 1998/1999. NASA. 10 Jan. 2010. "http://www.hq.nasa.gov/office/codez/plans/NSP99.pdf; Chyba, Christopher F. and Kevin P. Hand. "Astrobiology: The Study of the Living Universe." Annual Review of Astronomy and Astrophysics 43 (2005): 31–74.
[292] Davies, Paul. Cosmic Jackpot: Why Our Universe Is Just Right for Life. Boston: Houghton Mifflin, 2007.

[293] Pálsson, Gísli. "Celestial Bodies: Lucy in the Sky." Humans in Outer Space – Interdisciplinary Odysseys. Eds. Luca Codignola and Kai-Uwe Schrogl. Vienna: SpringerWienNewYork, 2009: 69–81. 79.

[294] Davies, Paul. Are We Alone? Philosophical Implications of the Discovery of Extraterrestrial Life. New York: Basic Books, 1998; Race, Margaret S. "Societal and Ethical Concerns." Planets and Life: The Emerging Science of Astrobiology. Eds. Woodruff T. Sullivan and John A. Baross. Cambridge: Cambridge University Press, 2007: 483–97; Bertka, Constance M., ed. Exploring the Origin, Extant, and Future of Life, Philosophical, Ethical and Theological Perspectives. Cambridge: Cambridge University Press, 2009.

[295] Schetsche, Michael. "Encounters Among the Stars. Exosociological Considerations." (in this volume).

[296] Gärdenfors, Peter. How Homo Became Sapiens. On the Evolution of Thinking. Oxford: Oxford University Press, 2006.

[297] Sinha, Chris. "Language as a Biocultural Niche and Social Institution." New Directions in Cognitive Linguistics. Eds. Vyvyan Evans and Stephanie Pourcel. Amsterdam: Benjamins, 2009: 289–309. 292; Cf. Tomasello, Michael. The Cultural Origins of Human Cognition. Cambridge, MA: Harvard University Press, 1999.

[298] Varela, Francisco J., Evan Thompson, and Eleanor Rosch. The Embodied Mind. Cognitive Science and Human Experience. Cambridge, MA: MIT Press, 1991; Lakoff, George and Mark Johnson. Philosophy in the Flesh. The Embodied Mind and its Challenge to Western Thought. New York, NY: Basic Books, 1999; Krois, John, Mats Rosengren, Angela Steidele, and Dirk Westerkamp, eds. Embodiment in Cognition and Culture. Amsterdam: Benjamins, 2007.

[299] Clark, Andy. Being There. Cambridge, MA: MIT Press, 1997; Brinck, Ingar. "Situated Cognition, Dynamic Systems, and Art. On Artistic Creativity and Aesthetic Experience." Janus Head 9.2 (2007): 407–31.

[300] Clark, Andy, and David Chalmers. "The Extended Mind." Analysis 58.1 (1998): 7–19.

[301] Gärdenfors, Peter. "Cognitive Science: from Computers to Ant Hills as Models of Human Thought." A Smorgasbord of Cognitive Science. Eds. Peter Gärdenfors and Annika Wallin. Nora: Nya Doxa, 2008: 11–34. 28; Wallin, Annika, and David de Léon. "How Should We Study Interaction?" A Smorgasbord of Cognitive Science. Eds. Peter Gärdenfors, and Annika Wallin. Nora: Nya Doxa, 2008: 133–48. 139.

[302] Gärdenfors, Peter. "The Evolution of Thought." A Smorgasbord of Cognitive Science. Eds. Peter Gärdenfors and Annika Wallin. Nora: Nya Doxa, 2008: 81–97. 81.

[303] Lakoff, George and Mark Johnson. Metaphors We Live by. Chicago, IL: University of Chicago Press, 1980.

[304] Gärdenfors, Peter. "Concept Learning." A Smorgasbord of Cognitive Science. Eds. Peter Gärdenfors and Annika Wallin. Nora: Nya Doxa, 2008: 165–81. 167.

[305] Lakoff, George. Women, Fire, and Dangerous Things. What Categories Reveal about the Mind. Chicago, IL: University of Chicago Press, 1990; Taylor, John R. Linguistic Categorization. Oxford: Oxford University Press, 2003.

[306] Mayr, Ernst. The Growth of Biological Thought. Diversity, Evolution, and Inheritance. Cambridge, MA: Harvard University Press, 1982; Atran, Scott. Cognitive Foundations of Natural History. Towards an Anthropology of Science. Cambridge: Cambridge University Press, 1990; Berlin, Brent. Ethnobiological Classification. Principles of Categorization of Plants and Animals in Traditional Societies. Princeton, NJ: Princeton University Press, 1992; Wilson, Robert A., ed. Species. New Interdisciplinary Essays. Cambridge, MA: MIT, 1999; Coyne, Jerry A. and H. Allen Orr. Speciation. Sunderland, MA: Sinauer, 2004; Atran, Scott and Douglas Medin. The Native Mind and the Cultural Construction of Nature. Cambridge, MA: MIT, 2008: 9, 23.

[307] Rosch, Eleanor. "Principles of Categorization." Cognition and Categorization. Eds. Eleanor Rosch and Barbara B. Lloyd. Hillsdale, NJ: Erlbaum, 1978.

[308] Concerning intersubjectivity, see Thompson, Evan. Ed. Between Ourselves: Second-Person Issues in the Study of Consciousness. Thorverton: Imprint Academic, 2001.

[309] Christiansen, Morten H. and Simon Kirby, eds. Language Evolution. Oxford: Oxford University Press, 1997; Deacon, Terry. The Symbolic Species: The Co-Evolution of Language and the Brain. New York: Norton, 1997; Tomasello, Michael. Origins of Human Communication. Cambridge, MA: MIT, 2008.
[310] Taylor, John R. Cognitive Grammar. Oxford: Oxford University Press, 2002: 30.
[311] Saussure, Ferdinand de. Cours de linguistique générale. Lausanne: Payot, 1916.
[312] Sonesson, Göran. "Preparation for Discussing Constructivism with a Martian." Les Signes du Monde: Interculturalité & Globalisation: Proceedings of the Eighth Congrès de l'Association International de Sémiotique, 7–12 July 2004, Lyon, France. Lyon: Université de Lyon, 2007. http://jgalith.univ-lyon2.fr/Actes/articleAsPDF/SONESSON_(martiens)_pdf_20061114153124.
[313] Lotman, Yuri. Universe of the Mind. A Semiotic Theory of Culture. London: Tauris, 1990.
[314] Cabak Rédei, Anna. An Inquiry into Cultural Semiotics: Germaine de Staël's Autobiographical Travel Accounts. Lund: Lund University, 2007: 43, 45f.
[315] Vakoch, Douglas A. "The View from a Distant Star: Challenges of Interstellar Message-Making." Mercury 28.2 (1999): 26–30; See also Arbib, Michael A. "Minds and Millenia: The Psychology of Interstellar Communication." Cosmic Search 3.1 (1979): 21ff; Vakoch, Douglas A. "Constructing Messages to Extraterrestrials: An Exosemiotic Perspective." Acta Astronautica 42.10–12 (2000): 697–704; McConnell, Brian. Beyond Contact: A Guide to SETI and Communicating with Alien Civilizations. Sebastopol, CA: O'Reilly, 2001; Dunér, David. "Cognitive Foundations of Interstellar Communication." Communication with Extraterrestrial Intelligence. Ed. Douglas A. Vakoch. Albany, NY: State University of New York Press (in press).
[316] Freudenthal, Hans. Lincos: Design of a Language for Cosmic Intercourse. Amsterdam: North-Holland, 1960.
[317] Sonesson, Göran. op. cit.
[318] Ehrsson, H. Henrik. "The Experimental Induction of Out-of-Body Experiences." Science 317 (2007): 1048, also see 1020f; Henrik Ehrsson's group – "Brain, Body & Self Laboratory." 28 Dec. 2009. Henrik Ehrensson's Group Website 10 Jan. 2010 www.neuro.ki.se/ehrsson/index.html.
[319] Donald, Merlin. Origins of the Modern Mind: Three Stages in the Evolution of Culture and Cognition. Cambridge, MA: Harvard University Press, 1991; Donald, Merlin. A Mind so Rare: The Evolution of Human Consciousness. New York: Norton, 2001.
[320] Bruner, Jerome. "The Narrative Construction of Reality." Critical Inquiry 18.1 (1991): 1–21. 4.
[321] Landfester, Ulrike. "Missing the Impossible: How We Talk and Write about Space." Humans in Outer Space – Interdisciplinary Odysseys. Eds. Luca Codignola and Kai-Uwe Schrogl. Vienna: SpringerWienNewYork, 2009. 94–106.
[322] Dick, Steven J. Plurality of Worlds: The Origins of the Extraterrestrial Life Debate from Democritus to Kant. Cambridge: Cambridge University Press, 1982; Guthke, Karl S. Der Mythos der Neuzeit: Das Thema der Mehrheit der Welten in der Literatur- und Geistesgeschichte von der kopernikanischen Wende bis zur Science Fiction. Bern: Francke, 1983; Crowe, Michael J. The Extraterrestrial Life Debate 1750–1900: The Idea of a Plurality of Worlds from Kant to Lowell, Cambridge: Cambridge University Press, 1986; Dick, Steven J. The Biological Universe: The Twentieth-Century Extraterrestrial Life Debate and the Limits of Science. Cambridge: Cambridge University Press, 1996; Dick, Steven J. and James E. Strick. The Living Universe: NASA and the Development of Astrobiology. New Brunswick, NJ: Rutgers University Press, 2004; Sullivan, Woodruff T. and John A. Baross, eds. Planets and Life: The Emerging Science of Astrobiology. Cambridge: Cambridge University Press, 2007.
[323] Worms, Jean-Claude. "Space and Humans." Humans in Outer Space – Interdisciplinary Odysseys. Eds. Luca Codignola and Kai-Uwe Schrogl. Vienna: SpringerWienNewYork, 2009: v–xiv. vii.
[324] Jakosky, Bruce M. op. cit. 663.
[325] Pratt, Mary Louise. Imperial Eyes: Travel Writing and Transculturation. London, and New York: Routledge, 1992: 15.
[326] Dick, Steven. "Humanity and Extraterrestrial Life." Proceedings of Life on Mars. What Are the Implications? Eds. John M. Logsdon, Bridget R. Ziegelaar, and Anne Marie Burns. 26 Nov. 1996,

Washington D.C.: Space Policy Institute George Washington University, 1997; Codignola, Luca. "Future Encounters: Learning from the Past?" Humans in Outer Space – Interdisciplinary Odysseys. Eds. Luca Codignola, and Kai-Uwe Schrogl. Vienna: SpringerWienNewYork, 2009: 14–21.

[327] "The Vienna Vision on Humans in Outer Space." Humans in Outer Space – Interdisciplinary Odysseys. Eds. Luca Codignola and Kai-Uwe Schrogl. Vienna: SpringerWienNewYork, 2009: 227–9. 229.

[328] Concerning cultural encounters and travel literature: Campbell, Mary B. The Witness and the Other World. Exotic European Travel Writing, 400–1600. Ithaca: Cornell University Press, 1988; Porter, Dennis. Haunted Journeys: Desire and Transgression in European Travel Writing. Princeton, NJ: Princeton University Press, 1991; Pratt, Mary Louise. op. cit.; Pagden, Anthony. European Encounters with the New World. From Renaissance to Romanticism. New Haven, CT: Yale University Press, 1993; Elsner, Jas and Joan-Pau Rubiés. Voyages and Visions: Towards a Cultural History of Travel. London: Reaktion Books, 1999; Bridges, Roy. "Exploration and Travel Outside Europe (1720–1914)." The Cambridge Companion to Travel Writing. Eds. Peter Hulme and Tim Youngs. Cambridge: Cambridge University Press, 2002; Sell, Jonathan P. A. Rhetoric and Wonder in English Travel Writing, 1560–1613. Aldershot: Ashgate, 2006; Abulafia, David. The Discovery of Mankind: Atlantic Encounters in the Age of Columbus. New Haven, CT: Yale University Press, 2008.

[329] We can find a new interest in spatiality, the spatial turn. See for example Crang, Mike and Nigel Thrift, Eds. Thinking Space. London: Routledge, 2000; Schlögel, Karl. Im Raume lesen wir die Zeit: Über Zivilisationsgeschichte und Geopolitik. München, and Vienna: Carl Hanser Verlag, 2003.

[330] Lefebvre, Henri. La production de l'espace. Paris: Anthropos, 1974.

[331] Niebuhr, Carsten. Reisebeschreibung nach Arabien und andern umliegenden Ländern. Copenhagen: Nicolaus Møller, 1774–837.

[332] Columbus, Christopher. Epistola de Insulis nuper Inventis. Rom: Eucharius Silber, 1493.

[333] Todorov, Tzvetan. La Conquête de l'Amérique. La Question de l'Autre. Paris: Seuil, 1982.

[334] Sonesson, Göran. "Ego Meets Alter. The Meaning of Otherness in Cultural Semiotics." Semiotica 128.3/4 (2000): 537–59; Sonesson, Göran. "The Globalisation of Ego and Alter. An Essay in Cultural Semiotics." Semiotica 148.1/4 (2004): 153–73; Cabak Rédei, Anna. op. cit. 2, 7, 70.

[335] Cabak Rédei, Anna. op. cit.

[336] Katsuragawa, Hoshu. Kratkie vesti o skitanijach v severnych vodach. Moscow: Nauka, 1978.

[337] Said, Edward. Orientalism. London: Routledge & Kegan Paul, 1978.

[338] We will thus not know how to treat them. For an analysis of the moral implications when encountering extra-terrestrial life see Cockell, Charles. "Ethics and Microscopic Extra-Terrestrial Life" (in this volume).

[339] Landfester, Ulrike. op. cit. 105.

[340] For the term "cosmocentric", see Lupisella, Mark. "Astrobiology and Cosmocentrism." Bioastronomy News, IAU Commission 51.10(1) (1998): 1–2, 8.

2.2 Looking back to Earth

S. J. Gustav Schörghofer

2.2.1 The Jesuit church in Vienna

The Jesuit Church in the centre of Vienna (Austria) was built in the beginning of the 17th century, consecrated in 1631. One hundred years later, from 1703 on, Andrea Pozzo created the interior decoration. Pozzo, a Jesuit brother and famous artist, reinterpreted the whole interior space. His idea was to centralise the space of the audience below a painted cupola. The high altar was built like a stage. The altar piece shows the assumption of Mary. The daylight illumination in the morning makes the whole scene brilliant like a movie.

The interior space with its warm colours and rich gold-platted decoration welcomes the visitors as a world closed unto itself, visualising non-visual truths. It shows a heaven full of angels, it shows saints and allegorical figures, it shows even God, the Holy Trinity. All this was made to attract the visitor through a highly dramatic presentation of religious truths.

Nowadays this attraction is still present. It works. Visitors come to admire the fake-cupola, the harmony of this unique interior. Many of them might not have any idea of all those religious truths, those baroque religious truths surrounding them inside the church. Who might be able to understand the language of baroque pictures? Who will know the saints? Who will recognise God on the stage of the high altar? So we stand in the midst of a strange world, a universe passed away a long time ago. For all of us that is gone. We have left the interior space of baroque. We have left the interior space of any bygone epoch looking back at it like travellers in outer space.

Till now the Jesuits take care of their church in the centre of Vienna. We celebrate mass together with a large community from all over the town and all over the world. The visitors are Catholics and non-Catholics, Christians and non-Christians. However they all are travellers – travellers in search of essential truths.

2.2.2 The "Jesuitenkosmos" – Cosmos of the Jesuits

Years ago we asked artists to create some contribution to the baroque interior of our church. These works of art remained for some months. We shared

Fig. 7. *The Nave of the Jesuit Church in Vienna (Austria) (source: "Fresco with Trompe l'oeuil – Andrea Pozzo – Jesuit Church Vienna." 9 Oct. 2006. Wikimedia 8 Jan. 2010. http://commons. wikimedia.org/wiki/ File:Fresco_with_Trompe_l%27oeuil_-_Andrea_Pozzo_-Jesuit_Church_Vienna.jpg).*

them in our daily life, in our services and masses. So did "Jesuitenkosmos" ("Cosmos of the Jesuits" in English) from November 2008 to May 2009. The huge reproduction of a NASA photograph from the internet was spanned all over the nave. The photograph showed the Earth, a view from the International Space Station (ISS). The blue surface of our planet, clouds, a part of the space station itself, an astronaut's legs, a narrow strip of black space.

Looking upward on a bright day visitors could still get some impression of the vaults' decoration glimmering through. After sunset, illumination of the pictures created another reality. There was no vault anymore. The whole interior of the church seemed to be opened. And we looked down or up

Fig. 8. *"Jesuitenkosmos", a huge display of the ISS in the Jesuit Church in Vienna (Austria) in 2009, designed by Christoph Steinbrener and Rainer Dempf (source: Steinbrener, Christoph and Rainer Dempf. http://www.steinbrener-dempf.com).*

to Earth as if the church itself would be in outer space gliding in weightlessness.

This is a strange situation for somebody unsuspectingly entering the church. Visitors expect the harmonious interior described in their guide: angels and saints, views from Earth to heaven. To find oneself confronted with that space shuttle view was astonishing. Evidently. Some artists rejected the work of their colleagues, Christoph Steinbrener and Rainer Dempf, as a shallow decoration, some kind of banal advertising without subject. Others complained about the destruction of the interiors' unity, criticising that these things should be done within the context of modern architecture and not within the context of a high quality baroque interior. Others again expected the end of "Jesuitenkosmos" in silence. They knew: there would be an end. Others again thought, the huge picture would hide on-going restoration works. But others – many young visitors and artists too – appreciated "Jesuitenkosmos". They came repeatedly and brought friends with them. They were able to play the game in another sense. They too, of course, found the ancient unity of the church's interior, the harmony of the space closed in itself, destroyed. But they did not find an end at the point of just complaining about the loss. They went further, they saw more, something new. An old structure opened to other and new experiences. What kind of experiences?

2.2.3 Ignatius of Loyola and his "spiritual exercises"

The Society of Jesus (shortly called Jesuits) was founded nearly 500 years ago in a stirring time of religious renewal. The primary objective and aim of Ignatius of Loyola and his friends, the nucleus of the quickly growing order, was to strengthen the Catholic Church and to help people to find their personal way to God, their personal vocation. Favoured means were education, sermons and spiritual exercises.

This basic intention has not changed throughout the centuries. From the beginning on, Jesuits had a global vision. Nowadays the 18,000 members are engaged all over the world. We focus on the poor and stay side by side with them. Our central engagements are to serve people so that they might increase their faith, to support justice and structures of justice, to cultivate interreligious dialogue and the dialogue with local and global cultures. In short: to get involved

Fig. 9. *Saint Ignatius of Loyola (source: Heilige Gebete und Andachten – Eine Sammlung von Gnadenschätzen der katholischen Kirche in Text und Bild. 10 Jan. 2010. http://immaculata.files.wordpress.com/2006/12/ignatius_loyola_1.jpg).*

in that world, to live a faith and a service directed to the world, to the people, to our whole planet.

Much inspiration comes from the "Spiritual Exercises" composed by Ignatius of Loyola. One of the most important and most impressive meditations of these exercises is devoted to the incarnation. It starts with a prelude to survey the history of the matter.

> "Here it is how the three Divine Persons gazed on the whole surface or circuit of the world, full of people; and how, seeing that they were all going down into hell, they decide in their eternity that the Second Person should become a human being, in order to save the human race."[341]

The artists were not aware, but their "Jesuitenkosmos" showed precisely this imagined view of the Divine Persons. Their work covered the painted Trinity of our church and hid it. But it showed our planet seen with the eyes of God. God gazes on the surface of the world from above and perceives not only the blue atmosphere but the people. Looking back to Earth there is much more to discover than a beautiful blue planet.

Michael Collins, a member of the Apollo 11 crew, suggested after his return from the Moon, that on a future flight to the Moon a poet, a priest and a philosopher should be on board: "Then we might get a much better idea of what we saw." Looking back to Earth from outer space, a view never experienced before, has changed our minds. It might not be by chance that the sense of responsibility for the environment developed together with the astronautics. We are able to see our beautiful and endangered planet with new eyes.

This vision is not so different from the one imagined by Ignatius of Loyola nearly 500 years ago. We are able to see the world with the eyes of God. But this requires our taking responsibility for what we see.

[341] The Spiritual Exercises of Saint Ignatius, 102, translated by Georg E. Ganss, S. J.

2.3 Alien life: Remarks on the exobiological perspective in recent terrestrial biology

Thomas Brandstetter

At the beginning of the 20th century, extraterrestrial beings started to appear in biological thought. In literature, of course, they have a long history, and astronomers have speculated about life on other planets since the 17th century. However, these creatures were not situated in the framework of biological discourse: they were mirrors for criticising the social or political shortcomings of earthly life, they were arguments invoked to corroborate theological arguments about the creation, or they served as a means to reflect on the imperialistic aspiration of European powers.[342]

In the last third of the 19th century, the discussions of extraterrestrial life gained increasing scientific credentials through the Darwinian theory of evolution and the application of spectrography to the investigation of celestial orbs.[343] Biologists like Alfred Russell Wallace started to discuss the possibility of extraterrestrial beings, and while Russell denied it, others started to speculate on the forms such alien creatures could have.[344]

In the following paper, I will concentrate on some late 20th and early 21st century examples of the appearance of extraterrestrials in the discourse of mainstream biology.[345] Extraterrestrial life is of course a phenomenon whose occurrence until now has been restricted to scientists' imaginations. The problems this simple fact poses for the methodology of a science that deals exclusively with such an object, i.e. astrobiology, is a question that has to be dealt with separately.[346] For the disciplines that concern themselves mainly with life on Earth, it is clear that arguments about the possibility and the nature of extraterrestrial life and intelligence are based on thought experiments. As the evolutionary biologist Ernst Mayr stated, the claim that such forms of life exist can neither be corroborated nor falsified and is therefore "outside the bounds of science".[347] If this is true, one can ask what function such thought experiments have. The physicist Ernst Mach once argued that thought experiments serve to test the consistency of arguments and ideas.[348] They provide an imaginary site where the consequences of certain assumptions can be examined, the coherence of deductions can be judged and the limits of theories can be probed.

Biologists' use of thought experiments therefore is to be distinguished from astrobiology proper, even if these discourses often refer to each other. As James Struck and Steven Dick have shown, astrobiology developed into an independent and distinct research programme in the early 1960s.[349] The principal reasons for this differentiation were: progress made in the planetary sciences, the "origin of life" research, and the first steps towards a search for extraterrestrial intelligence. Most important, however, was the funding by NASA, which provided the financial means to establish research projects and hire scientists. Especially in the beginning, astrobiology was closely linked to the space programme. Besides discussing questions such as the possible contamination of Earth by extraterrestrial microbes and vice versa, research concentrated on the development of experiments and devices for the detection of life on Mars. This was done in anticipation of unmanned missions, planning for which had been ongoing since 1959. I will come back to that later; for the moment, let it be stressed that the aim of astrobiology is the empirical examination of actual extraterrestrial life. Its methods are based on real experiments, not thought experiments. Of course thought experiments are conducted, but they serve to develop empirically verifiable hypotheses and to prepare real experiments. The development of technical means for the testing of these hypotheses is a constitutive part of astrobiological research. This is even clear to critics, who deny the reasonableness of such an endeavour. For Ernst Mayr, the problematic issues are not the hypothesis about extraterrestrial life or intelligence; these are for him, as already stated, "outside the bounds of science". However, according to Mayr it is highly problematic to try to corroborate these hypotheses because this requires a technical and economic effort, which is out of proportion to the prospects of success. Therefore his conclusion takes the form of a direct political statement: "In my view, SETI is a deplorable waste of taxpayers' money, money that could be spent more usefully for other purposes".[350]

Astrobiology presupposes the existence of life out there; otherwise the discipline would make no sense, and looks for ways of finding this life. If it looks back upon Earth in this process, it is to develop tools and theories that might help with such a search (e.g. in exploring the conditions of life in extreme terrestrial environments). But in principle, astrobiology's gaze is directed outward, the discourses I want to deal with in the following pages are characterised by an inward gaze: they refer to the question of the nature of life on Earth and, as we will see, to the question of the meaning of our existence.

2.3.1 Aliens in the fossil record

Sometimes, the work of palaeontology can turn up creatures whose strangeness is on *a par* with the constructs of literary writers. The animals depicted in Figures 10 and 11 are reconstructions from fossils found at the so-called Burgess Shale, a

Fig. 10. *Desert varnish seen through an electron microscope (source: "Desert Varish - In The Rough UNM Scanning Electron Microscope Images by M. Spilde and P. Boston" The Caves of Mars. 6 Sept. 2009. http://www.highmars.org/niac/niac06a.html).*

geological formation of the Canadian Rocky Mountains. The deposit dates back to the Middle Cambrian, around 505 million years ago. Its significance stems from the fact that it contains a great number of remarkably well preserved specimens from a time shortly after the Cambrian Explosion, a period when most major groups of complex animals first appeared, ancestors of today's fauna as well as some other groups that have become extinct meanwhile. The Burgess Shale fossils and their interpretation have led to a dispute between two famous palaeontologists, Stephen Jay Gould and Simon Conway Morris.[351] Interestingly, while arriving at contrary conclusions, both use the figure of the alien as a means to probe the consequences of their arguments.

In 1989, Stephen Jay Gould published his book "*Wonderful Life*".[352] He argued that based on a new description and classification of the Burgess Shale fossils by a group around Harry Whittington, a silent revolution has taken place in the disciplines of palaeontology and geology. When the deposit was originally discovered at the beginning of the 20th century, the entire specimens there were classified as ancestors of still existing animals. Since the 1970s, however, experts have realised that the organisms found represent totally unknown phyla, which are not related to our modern fauna. Gould took efforts to stress the strangeness of these beings. The pictures he included in the book, which were drawn specially for this purpose, as well as his descriptions convey the image of extraterrestrials: "Magnify some of them beyond the few centimetres of their actual size, and you are on the set of a science-fiction film".[353] He calls one of the creatures, a species named Opabinia (Figure 11)

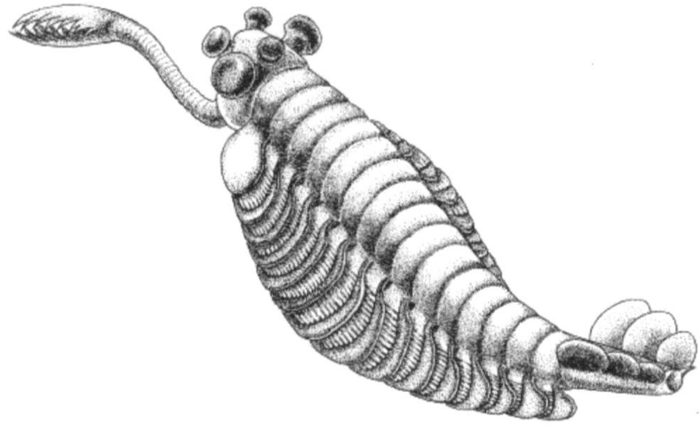

Fig. 11. *Reconstruction of Opabinia (source: Gould, Stephen Jay. Wonderful Life. The Burgess Shale and the Nature of History. London: Hutchinson Radius, 1990: 126).*

"an animal that might grace the set of a science-fiction film",[354] and a world where the Ediacara fauna, the ancestors to the Burgess Shale, survived "would have produced a world worthy of science fiction at its best".[355]

For Gould, this utter strangeness was an inducement to call for a fundamental rethinking of the mechanics of the evolutionary process. He argued that it should no longer be framed in terms of a linear narrative of continuous progress, which necessarily arrives at ever more complex organisms and culminates in the appearance of Man. Instead, evolution should be accepted as implying radical contingencies. For Gould, this means that at each moment in the course of evolution, a great number of divergent ways are possible, and it is mere chance as to which one is taken and proves successful. To illustrate his argument, he invoked a thought experiment: If the "tape of life" would be rewound to the time of the Burgess Shale fauna and then rerun, life on Earth would look completely different today.[356] That, in turn, also means that life on other planets would have no similarity to terrestrial life. As he argues in an article in defence of SETI, intelligence could develop in any form: "blobs, films, spheres of pulsating energy, or diffuse and unimagined forms far beyond the limited visions of most science fiction writers".[357] For Gould, the emergence of life was a simple and almost necessary process, its further development, however, was highly contingent and capable of taking many different ways. From these arguments, he drew a conclusion that can be called metaphysical insofar as it addresses the significance of human existence: "Homo sapiens, I fear, is a "thing so small" in a vast universe, a wildly improbable evolutionary event well within the realm of contingency".[358]

Of all people, it was a member of the team around Whittington and one of the heroes of "*Wonderful Life*" who brought forward a sharp rebuttal of Gould's theory. In 1998, Simon Conway Morris published "*The Crucible of Creation*", where he argued that the strangeness of the Burgess Shale fossils is really an artefact of human imagination.[359] His key concept is convergence: the independent but analogous development of organs and features in different life forms. The most famous example is the camera eye, which was developed by cephalopods (e.g. sepia) as well as by vertebrates (e.g. mammals). Conway Morris argued that the reason for such parallel developments is that while in theory, biological problems can be solved in an almost infinite variety of ways, in practice the constraints of physics, chemistry and the environment leave only a very limited number of solutions. This means that the thought experiment of re-running the tape of evolution leads to a result that is like the fauna we know: "the world, perhaps even any world, would have to look broadly similar".[360] In a later book, "*Life's Solution*", Conway Morris elaborated on these arguments and gave a central place to speculations about extraterrestrial forms of life.[361] His conclusions were almost symmetrically opposed to those of Gould: he considered the emergence of life as a very unlikely process, its evolution, however, as predictable. If there were life on other planets it would look very much like terrestrial life: "On any suitable planet there will I suggest be animals very like mammals, and mammals much like apes. Not identical, but similar, perhaps surprisingly similar".[362] Conway Morris also drew a metaphysical conclusion: the fact of the inevitability of Man's evolution for him hinted at the existence of God as creator.[363]

For the purpose of this article, the positions of these two prominent representatives of evolutionary biology are interesting because both use thought experiments about extraterrestrial life to frame their theory of evolution of life on Earth. Although they disagree fundamentally, both argumentations resemble each other insofar as they postulate evolution as a universal law and as the framework not only for the development of life, but also for the development of the universe itself. They share this assumption with exobiology, which also embraces the concept of a cosmic evolution that is the interdependent development of all parts of the cosmos from its origins to today.[364] Furthermore, both approach their object from a distance and as sovereign subjects. One could say that Gould represents a certain exoticism that indulges in the strangeness of the hypothetical aliens without, however, imagining them as too strange to communicate with (as is shown in his defence of SETI). Conway Morris, on the other hand, represents an anthropocentrism, which acts on the assumption that an encounter with extraterrestrials would be an encounter with ourselves – not because we are alien to ourselves but because the aliens are fundamentally akin to us. At the end of "*Life's Solution*" he closes an episode about an encounter with extraterrestrials with the remark: "We were, of course, looking at ourselves".[365] In spite of their differences, however, both positions leave our categories intact and do not unsettle their epistemological grounds.

2.3.2 Looking back, seeing things

A completely different stance is at work in speculations about alien life on Earth. The motif that on Earth itself something akin to "alien" life forms may be present can be found in several texts since the 1970s. Although they each stem from different contexts, they have one thing in common: they use the figure of the extraterrestrial as an incentive for fundamental epistemological considerations.

The 1970 book by French molecular biologist and Nobel laureate Jacques Monod, "*Le hasard et la nécessité*", begins with a thought experiment: The Mars equivalent to NASA sends a probe to Earth to look for signs of life; how would this probe have to be programmed to discover life on Earth? Monod discusses different criteria for life but arrives at no definitive result, as most properties would also cover one or another clearly inorganic thing, e.g. a technical artefact. In the end, he proposes a definition based on the structural properties of regularity and repeatability, combined with the specific conditions of formation of living beings, i.e. that fact that organisms owe their development to inner morphogenetic processes that act autonomously and spontaneously. Even with these criteria, however, it would still not be possible to distinguish organisms from crystals.[366]

Monod's thought experiment was clearly influenced by the discussions surrounding unmanned missions to Mars since the 1960s, which took place not only in specialised academic circles but also in journals read beyond the confines of disciplinary borders, such as "*Science*" and "*Nature*". In this context, the question of a definition of life took on an increasingly pragmatic and technical meaning.[367] The problem was exactly the one Monod described: How can a machine be programmed to recognise signs of life? From the beginning, the development of such devices by NASA scientists was accompanied by discussions about robust criteria for the detection of life.[368] Most biologists agreed that no definite and universally valid definition of life was available. As microbiologist Joshua Lederberg, who coined the term "exobiology", remarked: "Our only consensus so far is that such a definition must be arbitrary".[369] Therefore, it was deemed more sensible to look for pragmatic definitions i.e. definitions that could be implemented in an automatic probe. The question "What is life?" should be solved experimentally, while it was at the same time clear that the design and setup of the experiment provided the frame for what would be recognised as life and what would escape attention and fall, so to speak, through the cracks. In addition to the constraints imposed by the presuppositions of the experiments' designs (their theory-ladden-ness), there were also severe constraints imposed by the technical and economic necessities of space travel. A probe could only carry a limited number of instruments, and these had to be constructed so as to withstand considerable strain and to be controllable from a distance. Scientists had to accept that not every theoretically possible form of life could be searched for. As James Lovelock put it: "it would seem pointless and very uneconomic to send a space probe to detect a speculative life-form".[370]

As soon as biologists got involved with the down-to-earth labours of space missions they were confronted with the practical constraints of science. As a result, they became aware of their own blind spots: each question of the nature of life, each definition implies a certain perspective and therefore not only renders visible but also blots out aspects. The thought experiment of Monod took up this insight and applied it to mainstream biology itself: even there, "life" is a reflexive concept that carries the index of its context and marks the pragmatic point of origin of its generalisations.[371] That means that, the question of "life" is always accompanied by certain interests and is dependant on cultural, social, political and technical conditions of possibility. I want to argue that in the 20th century, this insight is inextricably linked to the figure of the extraterrestrial or alien, which provides the ground for such a reflective perspective.

Fig. 12. *Cover of Gold, Thomas. The Deep Hot Biosphere. New York: Springer, 1999.*

Apart from Monod, two other examples are apposite. One is a hypothesis brought forward by the astronomer and acknowledged gadfly Thomas Gold. In his 1999 book "*The Deep Hot Biosphere*" (Figure 12), he argues that on Earth as well as on other planets, like Mars, there exists a rich biosphere of microorganisms, up to a depth of 10 km, that feed on hydrocarbons rising from deep within the planets core.[372] The point is that Gold also stresses the perspectivity of definitions of life, arguing that we have not yet found proof of such a deep hot biosphere because our methods and instruments are inappropriate for the task: they unintentionally destroy the microorganisms which are adapted to survive at extreme conditions but which immediately die when exposed to the Earth's surface.[373] He therefore demands an epistemological "shift in perspective" that takes into account the possibility of other, hitherto unlikely forms of life.[374]

The topos of a hidden alien life on Earth can also be found in the work of the philosopher Carol Cleland, who in 1997 published an article where she speculated about the possibility of an "alternative terran biosphere".[375] Such organisms would per definition escape our attention as our methods and instruments are based on different assumptions. Cleland argues that we are bound to a specific perspective because of our concepts and practices and therefore are blind to other forms of life. However, according to her, candidates for traces of such alternative biospheres do exist on Earth, and we should direct our attention towards phenomena like desert varnish (Figure 10), a mineral coating found on rocks in deserts whose origin is as yet unknown.[376]

2.3.3 Conclusions

The extraterrestrial appears in biology in the late 20th and early 21st centuries in two ways. First, in the context of evolutionary biology as a thought experiment which is used to clarify the workings of evolution itself. Extraterrestrials represent a possible world, a possible outcome of the evolutionary process. The similarity or difference to our own world depends on whether contingency (Stephen Jay Gould) or convergence (Simon Conway Morris) is regarded as the most important mechanism. Both versions of the extraterrestrial, however, represent possibilities in an *a priori* defined framework, which is not challenged from an epistemological standpoint: "[T]he reality of evolution as a historical process is not in question".[377] This discourse operates on the same level as its empirical objects and phenomena. The epistemic ability and the fundamental epistemological presuppositions of the scientist are not questioned. Only his anthropological or existential feeling is affected, depending on whether he perceives himself as a product of chance or of necessity.

The second way of bringing extraterrestrials into biology is in a self-reflexive turn, which understands the alien to be a marker for a lacuna in our knowledge. This discourse operates at the level of the constitution of objects and the generation of concepts. It accepts the perspectivity of each definition of "life" by stressing the pragmatic and tentative nature of definitions and acknowledging their technical constraints. The effect of such an introduction of the alien at the very heart of scientific reasoning is that it makes us strangers to our own home: we can no longer be sure if deep beneath our feet or just around the corner, invisible to our complacent gaze, alien life forms proliferate.

---

[342] For a history of the extraterrestrial, cf. Dick, Steven J. The Biological Universe. The Twentieth-Century Extraterrestrial Life Debate and the Limits of Science. Cambridge: Cambridge University Press, 1996; Weber, Thomas P. (ed.) Science & Fiction II. Leben auf anderen Sternen. Frankfurt/Main: Fischer, 2004; Crowe, Michael J. The Extraterrestrial Life Debate, 1750–1900: The Idea of a Plurality of Worlds from Kant to Lowell. Cambridge: Cambridge University Press, 1986.

[343] Dick, Steven J. op. cit. 29.

[344] Cf. for example Wallace, Alfred Russell. Man's Place in the Universe: a study of the results of scientific research in relation to the unity or plurality of worlds. London: Chapmann & Hall, 1903.

[345] By mainstream biology, I understand a discourse that is framed by the Evolutionary Synthesis and has as its objects terrestrial organisms as they are empirically accessible, either via fossils or via living specimens. This concept, however, should not be taken too strictly as its borders are always in motion.

[346] Evolutionary biologist George G. Simpson voiced the accusation that exobiology is a "'science' [which] has yet to prove that its subject matter exists", cf. Simpson, George Gaylord. "The Non-prevalence of Humanoids." Science 143 (1964): 769–75. 769.

[347] Mayr, Ernst. Toward a New Philosophy of Biology. Observations of an Evolutionist. Cambridge, Mass.: Harvard University Press, 1988: 68.

[348] Mach, Ernst. Erkenntnis und Irrtum. Leipzig: Barth, 1917. 188.

[349] Dick, Steven J. and James Strick. The Living Universe. NASA and the Development of Astrobiology. New Brunswick, New Jersey, London: Rutgers University Press, 2005: 14.

[350] Mayr, Ernst. op. cit. 73. For an elaboration on the "good reasons" for the scientific study of the topic 'first contact' cf. Schetsche, Michael T. "Encounters among the stars – exosociological considerations." (in this volume).

[351] For the epistemological assumptions underlying this debate, cf. Baron, Christian. "Epistemic values in the Burgess Shale debate." Studies in the History and Philosophy of Biological and Biomedical Sciences 40 (2009): 286–95.

[352] Gould, Stephen Jay. Wonderful Life. The Burgess Shale and the Nature of History. London: Hutchinson Radius, 1990.

[353] Ibid. 25.

[354] Ibid. 132.

[355] Ibid. 314.

[356] Ibid. 48.

[357] Gould, Stephen Jay. The Flamingo's Smile. Reflections in Natural History. New York, London: W. W. Norton & Company, 1985: 406. Gould defends SETI on the basis that, at the moment, it represents the only available way to leave the realms of speculation.

[358] Gould, Stephen Jay. Wonderful Life. The Burgess Shale and the Nature of History. London: Hutchinson Radius, 1990: 291.
[359] Conwy Morris, Simon. The Crucible of Creation. The Burgess Shale and the Rise of Animals. Oxford et.al.: Oxford University Press, 1998: 170.
[360] Ibid. 205.
[361] Conway Morris, Simon. Life's Solution. Inevitable Humans in a Lonely Universe. Cambridge: Cambridge University Press, 2003.
[362] Ibid. 308.
[363] Conway Morris is a declared Christian who, however, strongly opposes Intelligent Design.
[364] Dick, Steven J. and James Strick. op. cit. 9.
[365] Conway Morris, Simon. Life's Solution. Inevitable Humans in a Lonely Universe. Cambridge: Cambridge University Press, 2003: 332.
[366] Monod, Jacques. Le hasard et la Nécessité. Paris: Éditions du Seuil, 1970: 25. For the motive of crystalline life, cf. Brandstetter, Thomas. "Imagining Inorganic Life: Crystalline Aliens in Science and Fiction." Imagining Outer Space: European Astroculture in the Twentieth Century. Ed. Alexander C. T. Geppert. New York: Palgrave, 2010 (forthcoming).
[367] NASA was actively engaged in establishing theoretical biology; however, I would like to argue that despite that effort, which did produce remarkable work, the discussion about criteria for life signs was driven by a pragmatic and self-reflecting approach. For NASA and theoretical biology, cf. Dick, Steven J. and James Strick. op. cit. 63.
[368] Dick, Steven J. and James Strick. op. cit. 30–5. 57.
[369] Lederberg, Joshua. "Exobiology: Approaches to Life beyond the Earth." Science 132 (1960): 393–400. 394.
[370] Lovelock, James E. "A Physical Basis for Life Detection Experiments." Science 207 (1965): 568–70. 568.
[371] For the term 'reflexive concept', cf. Grundwald, Armin and Yannick Julliard. "Technik als Reflexionsbegriff. Zur semantischen Struktur des Redens über Technik." Philosophia Naturalis 42 (2005): 127–57.
[372] Gold, Thomas. The Deep Hot Biosphere. New York: Springer, 1999.
[373] For an analysis of ethic considerations in the encounter of extraterrestrials cf. Cockell, Charles. "Ethics and extraterrestrial life." (in this volume).
[374] Ibid. 165; 191.
[375] Cleland, Carol E. "Epistemological issues in the study of microbial life: alternative terran biospheres?" Studies in the History and Philosophy of Biological and Biomedical Sciences 38 (2007): 847–61.
[376] Ibid. 859.
[377] Conway Morris, Simon. Life's Solution. Inevitable Humans in a Lonely Universe. Cambridge: Cambridge University Press, 2003: 5.

CHAPTER 3

CULTURE AND PSYCHOLOGY

3.1 Laokoon in Outer Space? Towards a transformative hermeneutics of Art

Ulrike Landfester

When Alan Sokal, professor of physics in New York, in 1996 published an article entitled "Transgressing the Boundaries: Towards a Transformative Hermeneutics of Quantum Gravity"[378] in the influential cultural theory journal *Social Text*, the journal's editors promoted its publication enthusiastically, taking it as a serious attempt by a representative of the so-called "hard sciences" to enter into a dialogue with the sciences on the other side of the gap between the two cultures of knowledge which Charles P. Snow had so paradigmatically diagnosed in his famous 1959 "Rede Lecture".[379] Instead of aiming at bridging the gap, Sokal's article was meant as a parody. It sharply denounced the – in the author's view – infuriatingly incompetent efforts of humanities' scholars to colonialise the natural sciences while only succeeding to produce an incoherent mishmash of overblown pseudo-theoretical postmodernist terminology and misunderstood miscellaneous fragments of scientific facts. After the prompt unveiling of the article's parodist intention, a bitter battle ensued. On the one hand some earnestly advocated the necessity of enabling and using interdisciplinary synergies between "soft" and "hard sciences"; on the other side some, like Sokal, viewed the delimitation of disciplinary work as an imposition on the purity of science on the other, as *Impostures intellectuelles*, as Sokal and his co-author Jean Bricmont put it in their subsequent book publication.[380]

By alluding to Sokal's 1996 parody in the subtitle of my own article, I take neither side in this quarrel but rather will try to follow up on the disciplinary transgression that Sokal performed: That his parody was not recognised as such by the editors of *Social Text* does not necessarily discredit their intellectual powers but rather points out that the inter- or trans-disciplinary work, that is already in progress has, in spite of such violent opposition, left its mark on contemporary discourse both in the humanities and social sciences and in the natural sciences. There is no doubt that Sokal is right when he attacks the promiscuous dilettantism that the post-humanist inter-disciplinarity movement has in some instances engendered; there are indeed some scholars who mistakenly believe their disciplinary training sufficient to cross over into any other scientific realms simply on the strength of their persuasion of the all-encompassing heuristic powers of cultural or social theory. At the same time, there is also no doubt whatsoever that scholars of disciplines even so far removed from each other like, for example, quantum physics and classical philology, can mutually greatly benefit from the other's expertise as long as they respect precisely those disciplinary differences which, in the end, must always remain the foundation of any productive dialogue.

Along those lines, the following article will apply the framework unwillingly suggested by Sokal in his use of the formula of "transformative hermeneutics" to an issue which, on first sight, is squarely based in a humanities' discipline – the theory of art – but which, on second sight, might very well at some stage in the future become relevant for the further improvement of space exploration technology. This issue is the possible development of the concept of Art under the conditions of human space travel and extraterrestrial settlements, that is to say, under conditions when humans leave their natural habitat on Earth either for a certain time span or even for good.

After a short introduction to the history of the concept of Art since the 18th century, three leading questions will be raised: Can Art be used to communicate with alien life forms? What, if any, impact might the physical differences between the environment on Earth and extraterrestrial environments, either space stations or other planets, have on human understanding and practicing of Art? Will Art and space technology remain separated by the gap between Snow's "Two Cultures" or is it conceivable that they might enter into a closer, mutually beneficial relationship?

3.1.1 The space of Art: The Laokoon paradigm

Before the imaging technologies of the 20th century started to delimit Art – with a capital A, to mark the term's conceptual dimension – towards the currently rather all-encompassing concept of Media, the space of Art in social and cultural theory was defined by the emergence of aesthetics out of two major historical paradigm changes.

The first of these changes took place in the early modernity during the 15th and 16th century, when the improvement in reproduction techniques both for images and texts – effective woodcut and copperplate technology and the newly-invented printing press – on the one hand, and the Reformation's iconoclast movement on the other, converged towards setting Art free from the fetters of its hitherto mainly religious function. From then on, the use of both images and texts was no longer confined within the boundaries of affirmative action towards Christianity as propagated by the Church. Instead, apart from becoming an important instrument in the distribution of scientific and other secular forms of knowledge. Art began to gain a certain independence, documented by and promoted through an increased theoretical interest in its specific merits, which had been in abeyance since the works of Aristoteles had first drawn attention to them.[381]

The second change came about not long after, during the 18th century, when philosophers like Alexander Baumgarten in his "*Aesthetica in nuce*" (1750/59) began to draw the conceptual consequences already implicitly laid out in the Renaissance theory of Art: Baumgarten was among the first who argued that beauty, in the sense of the sensual satisfaction a work of art gave to its recipient, must be conceived of as a value of its own, independent from any instrumental use for societal, political, religious or economic purposes, a value, moreover, which in itself constituted a sufficient reason both for the creation and the reception of Art.

Shortly after Baumgarten's "*Aesthetica in nuce*" had appeared, Gotthold Ephraim Lessing published a treatise which subsequently became one of the founding documents of the modern concept of Art: "*Laokoon: oder über die Grenzen der Mahlerey und Poesie*" (translated: Laokoon: or on the limits of poetry and painting).[382] Through the choice of his main title Lessing already alluded to an important part of the argument he was going to present in his treatise, i.e. that as Christianity in its use of both textual and iconic artefacts had not developed any pertinent concept of Art, the foundations of such a concept must be drawn from the artwork of Greek antiquity to be built upon according to the needs of modernity.

The name of Laokoon means both the mythical protagonist featured in antique renderings of the history of the siege of Troy and a famous marble statue representing him and his sons. According to the myth, the priest Laokoon had warned the people of Troy not to take the wooden horse, which the Greeks had offered to them as a present, into the city. Laokoon, it turned out, was right, as the wooden horse contained armed soldiers who the night after the gift had been taken in proceeded to vanquish the city from within. The Trojans, however, would not listen to his warnings, especially as Athene, seeing her scheme for the conquest of Troy endangered by him, sent out two huge sea snakes to kill the priest's two sons and subsequently also the father when he tried to rescue them. Taking up this myth, around 200 B.C. three stone masons from Rhodos created a marble statue depicting Laokoon and his sons desperately fighting against those snakes; the original is lost, but it had been copied several times around 100 b.C., and in the 16th century one of those copies turned up in Rome. Housed today in the Vatican Museum (cf. Figure 1), during the 18th century plaster copies of this statue were circulated through Europe, inspiring enormous interest and thus, among several others, Lessing's treatise on the nature of Art.

In this treatise, Lessing took Baumgarten's idea on the independent value of aesthetics a significant step further by applying it to create an epistemological

Fig. 1. *Laokoon, anon. copy, ca. 100 B.C., Rome, Vatican Museum (source: "Vatikan." Web Es. 16 Aug. 2010. http://www.web-es.eu/rom/kirchen-vatikan/sankt-peter-petersdom).*

framework, that allowed distinctions to be made between different types of Art, giving a set of criteria which, on the whole, served to systematically establish the aesthetic autonomy of Art. By contrasting painting – as a label for iconic representation including statues like the one he uses as his crucial point of reference – and poetry, he stipulates that while poetry may describe Laokoon and his sons screaming aloud with pain and fear, any iconic representation of the same subject had to abstain from showing the three figures with open mouths, because this would create a visual effect both ridiculous and disgusting. The stonemason therefore has to translate their pain and fear into an appropriately expressive body language with twisted limbs and bulging muscles, which in turn cannot be reproduced by poetry convincingly.

While Lessing's interpretation of the different impressions produced by painting and poetry in themselves of course are influenced strongly by 18th century ideas on what is or is not appropriate and/or beautiful and thus, from today's point of view, certainly may seem rather questionable, there still remains the fact that he for the first time diagnosed a feature which until the 20th century has proved a key element in any conceptual debates on Art: aesthetic value is created neither by the choice of subject nor by the choice of technique alone, but by the dialectical adaptation of one to the other. There is no privileged dominion of the instrument of expression over the issue expressed by it or vice versa, as is the case in any communication process outside the realm of Art. Instead, the aesthetic value of a piece of Art is constituted by the relation between the techniques used and its subject.

3.1.2 Anthropomorphism revisited: The alienness of Art

The question whether Art could be used to communicate with alien life forms, i.e. whether the concept of Art is bound so closely to humanity that it cannot be supposed to be comprehended by any other species on however abstract a level,[383] or whether there are any indicators that exactly because of its intrinsically human quality Art might yield some potential in that area, seems fairly self-answering: as we know nothing about what to expect, should we ever indeed meet with any extraterrestrial life-forms, except the fact that they are unlikely to exhibit any similarity to our species at all, we must presume that our concept of Art will be as alien to them as any of their modes of expression will be to us.[384] Still, the question is not quite as superfluous as it may seem at the first glance, as there are three aspects to it, which might still bear some consideration:

(1) Even if other life forms might not be able to relate to any kind or form or content in examples of our kind of Art, they might still recognise it on the whole as a sign of an advanced species;
(2) History has proven that in the process of exchange following an encounter between strange civilizations, artefacts circulating from one civilisation into the other often gain Art status through their sheer alterity, thus promoting a basic exchange of however superficial elements of knowledge about the other;
(3) Art with its experimental and speculative propensities has already been and will in the future be a useful heuristic instrument for developing an awareness of the probable non-anthropomorphous "alterity", i.e. otherness, of alien life forms.

The first aspect pertains to a structurally economics point of view. Since Art has been released into aesthetic autonomy through the paradigm changes described above, its very existence denotes a sphere of cultural activity which until today, spoken utilitaristically, is usually taken as superfluous: music and paintings, statues and ornamental gardens, literary texts and theatre performances neither feed nor clothe nor heal, nor can they be used as weapons or agricultural instruments. Their function is "merely" one of cultural self-identification and has been so even before the age of aesthetics, given that any pre-aesthetic religious artwork might seem to an alien life form equally unnecessary in the strict pragmatic sense of instruments to be used for the preservation of life. While the meaning of such luxuries can certainly not be expected to be evident to extraterrestrial species, there might still be the slight chance that the mere fact that humanity is able to afford them, could act as an indicator of a high civilisationally developed stage, thereby identifying the species which inhabits Earth as one which just might be worth to be taken into account. This argument is, of course, extremely speculative and, moreover, in a way guilty of exactly that kind of anthropomorphism which my former remarks have denounced as futile – but nevertheless it derives a certain kind of value *ex negativo*: Assumed that it is quite sure that Art as such is non-communicable to non-human species, the luxury meta-angle might well be the only conceivable way of using it in communication at all.

The second aspect of the question has no direct relevance for active communication purposes but instead concerns the possibilities of what might happen in the realm of Art after an inter-species encounter. As Luca Codignola has pointed out, the most significant of encounters in the history of humankind, the discovery of America with its subsequent process of trafficking between Europeans and Indians, has clearly shown both the dangers to be avoided and the strengths to be relied on for future encounters.[385] While Codignola mainly stressed the ideological and biological lessons to be learnt from the American encounter, it also had the by-effect of introducing artefacts to Europe. While these usually fulfilled religious or pragmatic functions in their culture of provenance, once in Europe they were incorporated into the field of Art, losing their dependence on a functional context and instead gaining the status of aesthetic autonomy. In the 18th and 19th century such items not only became proudly displayed collectors' items but also inspired European Art into assimilating characteristic forms as well as strange techniques. On that level, at least on the human side of any possible future encounter, the concept of Art might again well prove a transformational platform for integrating knowledge about the non-human other into the human self.

The third aspect of the use of Art in dealing with extraterrestrial life forms is probably the most fruitful for space exploration, seeing that its application has quite a long and productive history. Cardinal Nikolaus Cusanus for example, a 15th century philosopher and advisor to the pope, in his treatise "*De docta ignorantia*" (1440) came to the conclusion that not only was it highly probable that the planet Earth was not, in fact, the center of the universe but also that it was equally probable the other planets might be inhabited by "other species"[386] who were not remotely human. Johannes Kepler in his playfully fictional dream-narrative "*Somnium*" (1634) already imagined a whole lunar civilisation of non-human creatures. Even if Cusanus, Kepler and others of their time who thought along similar lines[387] did not see themselves as artists but rather as scientists, they still employed the freedom of speculative fiction. In the 18th century this freedom of speculative fiction was turned over into the realm of Art as a means for creating scenarios in which humans had to come to terms with categorically non-human *alterity*, scenarios which in their basic assumption of such *alterity* anticipated much of what Stanislaw Lem 1961 in his novel "*Solaris*" established as the leading topic of this most classic of science fiction classics. It is anything but a coincidence that Lem in his novel gave the spaceship that first discovered the planet Solaris the name of Laokoon – which served not only to emphasise the fact that Lem counted his novel as a work of Art, but also pointed to the paradigm of alterity inherent in Lessing's modern conceptualisation of Art: According to Lem's reading of Lessing's treatise, being constitutionally structured by insurmountable differences between poetic and iconic techniques of presentation, Art is uniquely equipped to make insurmountable differences between life forms a subject of heuristic exploration.

3.1.3 Unearthing Art: The "environment factor"

Art is not merely a conceptual entity. To the contrary, its conceptualisation is a meta-construction built on and derived from its material manifestations, which in turn are evidently hugely influenced by the concrete environment from which and in which they are brought forth. This begs the question of how far and in what ways the concept of Art might change in reaction to environments which are physically different from the planet Earth, i.e. the environments of space ships, space stations or even settlements on other planets.

Looking at what is currently available on the internet in terms of Art related to outer space, it quickly becomes clear that of course visual Art has long since opened up towards inspirations from that particular source – more so perhaps even than 20th century literature, seeing that space exploration from the beginning has communicated its findings mostly in images to the broader public. A typical example of the spin-offs of space exploration in that respect is "*The Cosmic Café*" with its "*Outer Space Art Gallery*", the latter featuring mainly the works of spray painter g.HARLAN whose fantastic renderings of cosmic landscapes are wonderfully colourful and nicely executed – but remain safely in the realm of earthly materiality in terms of format and technique.

Quite another and much more interesting example, however, is a project that is embedded in something of a work of art itself: "*Spacearts*".[388] The framework is a huge endeavour, an internet archive aiming to comprise all works of art concerned with outer space since the middle of the 19th century carried by the International Academy of Astronautics (IAA), the European Space Agency (ESA), the International Association of Astronomical Artists (IAAA), the London-based art and science agency Arts Catalyst, the Institute for Unstable Media, Rotterdam (V2), MIR, an international consortium of institutions having space art activities

Fig. 2. *Reddish Moon Rising (source: g.HARLAN. Cosmic Café and Outer Space Art Gallery. 27 Jan. 2010. http://www.outer-space-art-gallery.com/space-artist.html).*

including Leonardo/Olats, Projekt Atol (Slovenia) and the TV Gallery (Russia) and the Maison d'Ailleurs – Museum of Science Fiction, Yverdon. *Spacearts'* self-description puts forward precisely the heuristical argument, that I have made above.

Artists have been at the forefront of space exploration since its very beginning. Their works of imagination have stimulated and catalysed a new human endeavour. Works of art and literature about space have both anticipated and stimulated space development, while exploring destinations and technological concepts that were often too dangerous, too distant or too advanced for the science and technology of the moment.[389]

Spacearts currently displays 23 projects, among them "*The Mars Patent*", an interactive web project run from Hamburg in Germany by Claudia Reiche and Helene von Oldenburg. *The Mars Patent* self-describes itself as "[t]he first interplanetarian Exhibition Space on Mars":

> "THE MARS PATENT is a place for art and theory and sensible to its various concepts. THE MARS PATENT will exhibit projects, considering the special situation of the Mars Exhibition Site, e.g. geographic, geologic and meteorological aspects. THE MARS PATENT offers a suitable positioning for sculptures, internet relay chats, kinetic objects, art- and mediatheories, science fiction literature, videos, sound installations, manifestos, web-art etc. (You could even try it with paintings, but please send digitized formats. Remember: There are no walls on Mars!)"[390]

The introductory page of the exhibition space shows a map of a part of the surface of Mars (cf. Figure 3), where the exhibition is supposed to be located between "the

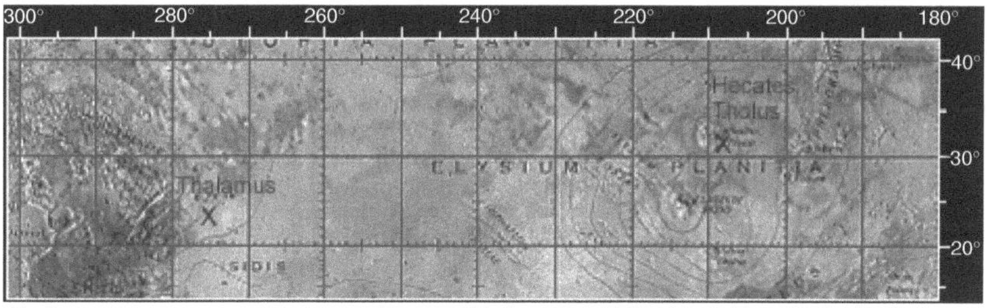

Fig. 3. *The Mars Patent (source: Reiche, Claudia, and Helene von Oldenburg. "The Mars Patent." The Mars Patent. 27 Jan. 2010 http://www.mars-patent.org/mes/mes.htm).*

steep volcanic mountain Hecates Tholus and the bizarre Thalamus region".[391] The map, together with information about the weather and chemical and geological data, offers a sense of material accessibility, which in the light of the factual inaccessibility of the area in question, underlines and heightens the purpose of the project. The conceptual impetus of Reiche, and von Oldenburg aims at forcing the spectator to accept certain conditions of perception: it is solely the spectator's eye which provides the interface between Space and Art, an interface connecting two things which are virtually not there – or rather are there only virtually, as the works of art which are to be exhibited. Thus, Reiche, and von Oldenburg simulate a future of Art where the materiality of artefacts will progressively lose – in fact, is already losing – its ground to the immateriality of digitally generated and transmitted images. Works of Art will then no longer be things from canvas and colour, stone or any other earthly substance but will consist instead of pixels and electronic impulses, gaining their reality solely through the visual perception of the spectator without any need for them to have a physically tangible existence – a state of Art which, however counterintuitive this may seem at first glance, will certainly reaffirm Lessing's notion of the necessary dialectical interplay between technique and subject matter under the conditions of the space age.

3.1.4 "Theatre level": Art and space technology

Thus, already through this "environment factor", Art is closely linked to space technology, both in terms of the inspiration derived from and the instrumental use of its achievements. There is, however, yet another angle to their relationship which even more directly pertains to the transformative hermeneutics paradigm. The first half of the title for this last part of my article is taken from an article by

Xavier Pasco on "Controlling the freedom of using space: The White House Policy dilemma" which dedicates a paragraph to 'theatre level' of 'force multiplier' space (1991–2007): adapting the space systems to the new strategic environment".[392] Pasco maintains that during this phase in the emergence of American space policy, space was conceived, on the one hand, as "the cornerstone of future defence architectures around which forces and doctrines would have to organise",[393] focussing, theatre-like, the perspective of the spectator solely on the military potential while, on the other hand,

> "space should also appear as one of the most flexible sectors of activity given its recent history, an a priori less heavily entrenched area than other military domains supposedly more heavily structured",[394]

thus open for combined or force-multiplying use by military and civilian stakeholders alike.

Though not spelled out by Pasco, and quite probably not even intended that way, his use of the term "theatre level" implies the awareness that the human use of space has a heavily theatrical dimension. On the face of it, this dimension is constituted by the fact that space exploration has always, at least in the public eye, manifested itself in the shape of avidly watched transmissions of images – which is one of the poetological foundations of the above-mentioned "*Mars Patent*" –, and it is to be supposed that the "theatre-level" of American defence activities in space at least in part means a pronounced display of military power towards potentially threatening other nations. More important, however, is the theatricality of the transmitted images themselves which is, of course, inherent in any image but which gains in significance when the images in question show things which in reality cannot – at least not yet – be seen by human eyes like, for example, the surface of Mars. Stimulated by the knowledge that for several decades perfectly convincing photo-like images have been generated digitally, so that what is sold to the spectator as an iconic re-presentation may in fact be an artefact with no relation whatsoever to any given reality. Doubts as to the validity of images from outer space have begun to undermine the average spectator's faith in their documentary character ever since, after the first landing of humans on the Moon in 1969, rumours began to circulate that the whole event had been an elaborately staged visual hoax.

As the fabrication of images without a direct object of mimetic reference is probably as old as mankind itself, there is no need whatsoever to broadly denounce the whole technology of visual communication as untrustworthy devil's work. Seeing, however, that especially in the area of space exploration, in both military and civilian operations in space and at some none-too-far future stage also in interplanetary exchange, this technology plays a crucial enabling role, it may be as well to meet head on possible problems in connection with the theatricality aspect instead of ignoring them altogether on the strength of the assumption that any truly scientific mind will always know exactly what can and what cannot be seen. Here, Art may serve the purpose of acting as a catalyst for critical reflection on both the dangers and the possibilities inherent in the circulation of images,[395] developing strategies for coping with the one and optimising the other – for example by experimentally opening up avenues towards getting rid of the moralist notion of simulacra being a priori lies, and instead propagating the sheer rationality of transmitting information encoded in images without any need for those images' mimetic anchorage in reality. Thus, interpreting space technology through Art might at least partly succeed where the mythical Laokoon fatally failed, warning humankind against the dangers lurking on the "theatre-level" of the Trojan horse of visual communication in outer space.

[378] Sokal, Alan. "Transgressing the Boundaries: Towards a Transformative Hermeneutics of Quantum Gravity." Social Text 46/47 (1996): 217–52. University of New York, Faculty of Physics 21 Sep. 2010. http://www.physics.nyu.edu/faculty/sokal/transgress_v2/transgress_v2_singlefile.html.

[379] Snow, Charles P. (ed.) "The Two Cultures. The Rede Lecture." The Two Cultures. Cambridge: Cambridge University Press, 1998: 1–52.

[380] Sokal, Alan, and Jean Bricmont. Impostures Intellectuelles. Paris: Odile Jacob, 1997. For a resume of the above-mentioned conflict see Herbrechter, Stefan. Posthumanismus. Eine kritische Einführung. Darmstadt: Wissenschaftliche Buchgesellschaft, 2009: 129f.

[381] cf. Belting, Hans. Bild und Kult. Eine Geschichte des Bildes vor dem Zeitalter der Kunst. München: Beck, 1990: 510–45.

[382] Lessing, Gotthold Ephraim. Laokoon: oder uber die Grenzen der Mahlerey und Poesie. Berlin: Christian Friedrich Voss, 1766. Laocoon: or, The limits of Poetry and Painting, translated by William Ross. London: Ridgeway, 1836.

[383] An argument which would follow from the ideas put forward by Belting, Hans. Bild-Anthropologie. München: Fink, 2001.

[384] Dunér, David. "Astrocognition: Prolegomena to a Future Cognitive History of Exploration." (in this volume) suggests that such a comparative analysis will shed light on the specific or typical characteristics of our species.

[385] Codignola, Luca. "Future Encounters: Learning from the past?" Humans in Outer Space – Interdisciplinary Odysseys. Eds. Luca Codignola and Kai-Uwe Schrogl. Vienna: SpringerWienNew York, 2009: 14–21.

[386] Kues, Nikolaus von. De docta ignorantia. Die belehrte Unwissenheit, vol. 2. Hamburg: Meiner, 1967. 101. translated by the author.

[387] Cf. Heuser, Marie-Luise. "Transterrestrik in der Renaissance. Nikolaus von Kues, Giordano Bruno, Johannes Kepler." Von Menschen und Ausserirdischen. Eds. Michael Schetsche, and Klaus Engelbrecht. Bielefeld: transcript, 2008: 55–80.

[388] "Spacearts – Die Space Art Datenbank." Spacearts. 21 Jan. 2010. http://www.spacearts.info.
[389] Ibid.
[390] "The First Interplanetarian Exhibition Space on Mars." Mars Patent. 21. Jan 2010. http://www.mars-patent.org.
[391] Reiche, Claudia, and Helene von Oldenburg. "The Mars Patent." The Mars Patent. 27 Jan. 2010. http://www.mars-patent.org/mes/mes.htm.
[392] Pasco, Xavier. "Controlling the Freedom of Using Space: The White House Policy Dilemma." Yearbook on Space Policy 2006/2007: New Impetus for Europe. Eds. Kai-Uwe Schrogl, Charlotte Matthieu, and Nicolas Peter. Vienna: SpringerWienNewYork, 2008: 197–210. 199.
[393] Ibid. 199f.
[394] Ibid. 199f.
[395] The same of course goes for texts, cf. Landfester, Ulrike. "Missing the Impossible: How We Talk and Write about Space. Humans in Outer Space – Interdisciplinary Odysseys." Eds. Luca Codignola and Kai-Uwe Schrogl. Vienna: SpringerWienNewYork, 2009: 94–106.

3.2 Music and the outer space – the means of universal communication or a form of art?

Anna G. Piotrowska

Studying music in outer space entails perceiving music in two prevailing perspectives: treating music as a form of artistic expression and as a means of communication (although this is not a uniquely human phenomenon as many animal species use sounds for interacting). The first understanding of music as a form of art is also deeply immersed in futurologist speculations concerning the expansion of music and is based on contemporary trends reflecting the musicological state of the art. The second outlook (connected with tendencies to rethink the process of musical signification)[396] directly refers to the historical attempts to answer the basic question: what is music and why, as well as consequently how, music – understood as a communication tool – could be employed as a universal language system in space exploration. This perspective corresponds with musicological assumptions of researching the past development of musical styles rooted in methodological practices established in the 19th century historically oriented discipline initially called *Musikwissenschaft*.

3.2.1 Music as an artistic expression in outer space

Predicting the advancement of music is a frustrating challenge like foreseeing the future of any art. Nevertheless, already in 1968 the composer, conductor, and author Ralph Alan Dale clearly postulated the "need to envisage the future not as fanciful science fiction but in a scientific manner".[397] In his vision, future music was characterised by a number of specific features including a very high level of technology, utilisation of science, dissolution of coercion, lack of egocentricity, spontaneity, universalisation of the artistic – scientific fusion, potential reconjugation of mind and emotion, elimination of social and economic class, fusion of arts themselves, no division of music into folk and art forms, etc.[398]

Although it is difficult to predict the exact paths of the music development beyond some oversimplified suggestions, the progress in music may, in fact, follow two basic trajectories: revolutionary or evolutionary. In the first case the sudden change of the paradigm will occur entailing stylistic alterations, new aesthetical presumptions and cultural evaluation, etc. The revolutionary path of expansion totally escapes any attempts to speculate on the parameters and the role of music in the future, both on Earth and in outer space. However, an evolutionary way of progressing suggests a promising possibility for trail blazing ideas of today to find their ultimate realisation in the music produced and entertained by future generations. Thus it seems justified to assume that advances in technology[399] will create more opportunities for active participation in music making (under the condition that music in outer space will continue to provide participation in recreational and therapeutic opportunities, be it in the form of composing or performing). Relying on technology can be connected with the fact that traditional instruments – often big and difficult to carry – are unlikely to be taken by the first settlers in their equipment. Their dependency on the use of computers in composing and performing music will deepen since the accelerated exploration

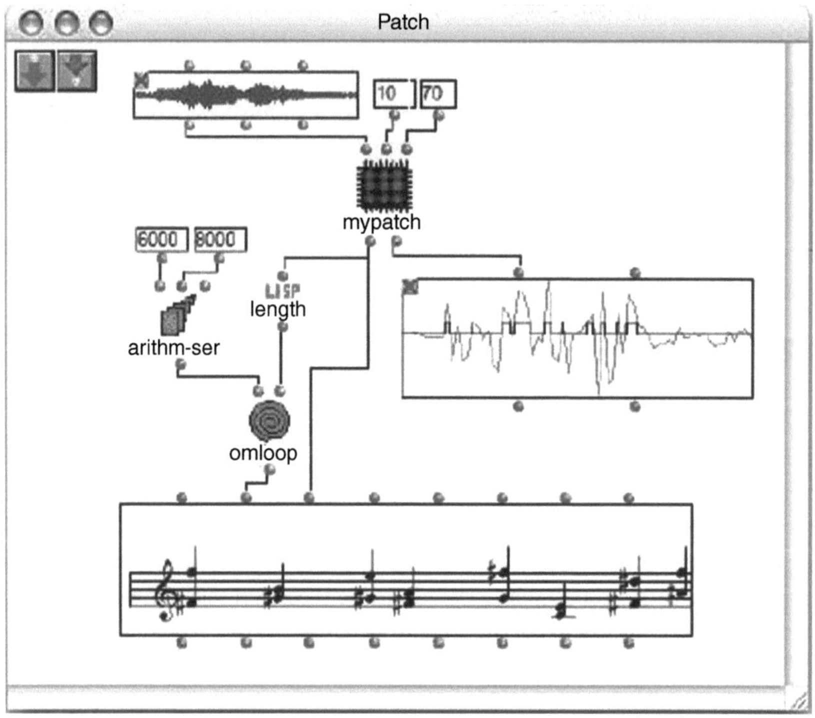

Fig. 4. *Open Music – Visual programming for music composition (source: "Open Music – A visual Programming Language." Open Music. 15 Jan. 2010. http://recherche.ircam.fr/equipes/repmus/OpenMusic).*

of robotics and machine-orientated movements is observed even now in many different backgrounds and disciplines of arts, not only music.[400]

Successful merging of two relatively distant areas – namely music and robotics – can ultimately enable creative combination of artistic and engineering practices without requiring from musicians prior knowledge of robotic technology. Some scholars insist that in future artists will use robots as sheer means of expression without having to posses the technological expertise to programme the robot completely from scratch.[401]

Technological advances will enable musicians to interact in new, unprecedented ways: for example forming the so called "Fragmented Orchestra", understood as a one distributed musical instrument combining live audio streams from geographically distant sites, with the critical role of timing.[402] Performing music will then mean not only playing (e.g. classical instruments or recorded tracks) but also generating it.

Relying on technology as a new source of musical sounds does not only solve the possible problem of lack of traditional instruments (or inability to play them in case they actually are transported) in outer space or facilitate the situation where the instruments brought from Earth represent completely incompatible sounds, but it will also provide new opportunities for composing and performing music by fusing different musical styles. As much as performing music will take new forms of music making, also composing in outer space will change its meaning, resembling more producing and mixing musical material (broadly defined) rather than conceiving new melodies, rhythms, harmonies, etc. The already widespread technique of sampling, i.e. exploitation of an already existing portion of musical material and using it in a new, innovative form (but not as a mere quotation) opens up infinite possibilities of rearranging and creative utilisation of any musical material inherited from the whole legacy of humankind – be it European or not, classical, folk, or popular tradition, etc. The musical heritage available in recorded versions will serve as a starting point for further development of musical ideas, its potential hidden in the unifying effect regardless of the initial origin of the musical input.

Alongside people using computers to create music, it is plausible to imagine specially programmed computers and biomorphic robotics able to evolve music in the realm of interactive music systems.[403] Although Artificial Life (A-Life) and Evolutionary Algorithms (EA) might provide completely new techniques for composing music (different musical applications ranging from original systems for music making such as algorithms, notably L-systems and genetic algorithms,[404] to innovative ways of performing music),[405] they, nevertheless, will always be limited by the previous music knowledge of producers. Consequently machines will either follow certain stylistic patterns or simply imitate them.

Fig. 5. *Milan Guštar, Abacus (2002) – algorithmic serial minimalistic composition for solo piano. The picture taken from the back cover of the CD is made up of 3856 pieces going towards the centre coloured according to the interval series (source: "Milan Guštar." 2002. Milan Gustar. 15 Jan. 2010. http://www.uvnitr.cz/mg/abacus/abacus.html).*

Another criticism may be that this type of music is viewed as mechanical: unlike conventionally conceived and performed music reflecting the human origin in its imperfection, the music produced with the help of computers or robots could be described as lifeless. The accent on the human dimension will be then shifted towards potential listeners interpreting final sounds according to their expectations, cultural background, etc. Hence music generated in outer space will automatically fall into the orbit of different understandings of its significance.

3.2.2 Music as a means of communication in outer space?

Given music in outer space could be employed as a form of communication, the lack of agreement on the definition of what it actually means or entails, may prevent from too farfetched assumptions. In fact anthropologists of music have spent considerable time and effort trying to grasp the communicative function of music.[406] While taking the declaration that music can be a *universal language* as an initial statement, it remains open to further speculation whether or not music could serve as the best or at least one of the best systems of communication in outer space. Not only does it raise the question of communication with the extraterrestrials[407] but it also reveals an unsolved problem of interpreting music created by different peoples between themselves. Furthermore, two important reservations should be added. The first refers to the variety of musical styles elaborated within human culture preventing agreement on one "official" style that could be representative for the human race out there. The second alludes to the fact that European philosophical thought has been – since ancient times – preoccupied with the seemingly neutral and objective nature of music embodied in its mathematical and physical parameters.[408] Yet in fact, even a very brief systematisation of historical perspectives on the definition of music (presented below) reveals the multitude of different conceptualisations of the idea of music in various epochs that, nevertheless, could be narrowed to two dominant modes of perception: the aesthetical aspect and the one more concerned with the objective characteristics of music hidden behind the numbers and mathematical schemata. Both approaches to defining music, however, aim at grasping the universality of music.

3.2.2.1 Aesthetical definitions of music as a universal art

One of the recurrent theme in writings on music has been its acclaimed universal character, the trait most probably sought after in outer space, especially in the light of multitude of cultures represented there. The process of conceptualisation of music as a thing (*res facta*) and a perfect object (*opus perfectum*) was initiated as early as the 15th century. Two centuries later Descartes' concept of affects was adopted resulting in perceiving music as a language of human emotions.[409] The assumption that music is able to express emotions stemmed from general agreement on the social character of the musical communication process and acknowledgment of the fact that musical sound structures are socially structured. Soon the concept of music as a form made up of smaller constructive units was accepted and popularised by August Wilhelm Schlegel, Adolf B. Marx, Hugo Riemann, and Hugo Leichtentritt. In the theories of music formulated at the turn of the 18th and 19th centuries, musical units were labelled as *Satz* (sentence), *Einschnitt* (unit, chunk) or as poetic ideas (themes) linked by sound-bridges. An interesting proposal was forwarded by Jean-François Sudre (1787–1862) in his 1866 book entitled *Langue musicale universelle* which resulted from his research to create an artificial language "solresol" based solely on seven different syllables represented – among others – as musical notes of different pitches.[410] Although theorists never reached definite conclusions, the very idea of music as a language hints at the possibility of applying music as a form of communication in outer space.

Human factor and the uniqueness of musical styles were underlined in the 19th century definitions of music, referring to the interpretative abilities of the humankind. The postulated common plane of similar connotations and associations attributed to music may refer to seeing music as a form of unifying experience in outer space. The minds of theoreticians of music in the second half of the 19th century were especially captured by the idea of biological evolution. Some intellectuals used terms such as *Keim* (grain), *Entwicklungsmotive* (evolutionary motif), while others discussed music in reference to the notion of *cellule – motif* (cell – motif). Aesthetic discussion about music was also influenced by physics, especially Isaac Newton's mechanics and the concept of force and energy as well as philosophical pantheism.[411] On the other hand, the development of orchestral music stimulated theoretical reflection upon instrumentation where music was described with such terms as *Klangfarbe, Farbentönung* (sound – colour), *Klangfarbenmelodie* (sound – colour-melody). Already in the 18th and 19th centuries composers began to attach programmatic titles to their instrumental compositions including additional texts inside their scores in the form of literary programmes referring to visual or emotional experiences enabling the decoding of musical connotations.[412]

The abundance of forthcoming interpretations of music leaves even more space for its new applications not only in future, but also in radically changed situations. In the 20th century reflection on music was dominated by theories inspired by set theory, structuralism, generative linguistics and semiotics. Close relations between music and language as functional systems were put forward. Striking similarities were observed between Noam Chomsky's generative grammar and his distinction between deep versus surface structures and the concept of *Forderung* (foreground structure) and *Ursatz* (fundamental structure) proposed by Heinrich Schenker for analysing music.[413] Musicologists were also inspired by Ferdinand de Saussure's distinction of language into *langue* and *parole* and consequently examined an abstract system of norms present in various composers' scores in order to determine and describe historical styles.[414] Finnish theorist Eero Tarasti[415] adapted Charles S. Pierce and Charles W. Morris's semiotics in his reflections on music, while American musicologist Leonard B. Meyer made distinctions between meaning embodied in music (*musical discourse*) and designative meaning which refers to objects or events in the non-musical domain (*discourse about music*).[416]

Postmodern approaches that preferred objectiveness and impersonality radically turned away from the European tradition of identifying the notion of "music" with the human voice expressing the mystery of human existence. *Bel canto* lost its impact, acoustic effects imitating murmur, rustle, etc. were appraised and the historic necessity of dissonance and aggressive sounds, understood as the effects of the human made "machine world" or a cry (*Urschrei*) of pantheistic Nature, was proclaimed. Additionally, the traditional model of shaping musical time was rejected. According to *Lebensphilosophie* a piece of music should be the progression of unrelated sound "moments" or the permanent, non-theological change of musical events without clearly marked beginning, end or climax. This new option for shaping musical time (and consequently form) radically denied the classical metaphor of the music work as a proportional and preferably symmetrical architectonical edifice and negated music as *l'art dramatique*.

Developing since the 1980s, cognitive musicology rejected the concept of an abstract and autonomous musical work and the idea of the objective and impersonal character of commenting on music phenomena, stressing instead the fact that music is conceived by human minds and reflection on it should concentrate on commentators' perceptions. Hence music, as a product of human creativity affected by subjective interpretation escapes the ultimate definition: remaining in the private sphere of an individual, it continues to be his or her intimate property in any conditions, including outer space experience.

3.2.2.2 Recurrent idea of mathematical nature of music

The link between mathematics and music as an explanation of the universality of music has been frequently drawn upon revealing the hidden potential of music as a form of communication in outer space. The application of more and more complicated mathematical formula and logical relations between abstract mathematical elements in reference to musical phenomena has always been perceived as enabling better understanding of the essence of the acclaimed universality of musical language. However, total reliance on mathematical modelling and narrowing musical phenomena to mathematical rigor often resulted in the neglect of other approaches.

Pythagoras and his pupils attested that some intervals deserved the name "perfect" while others did not – depending on the mathematical reduction of the string. Mathematics was applied to musical phenomena such as consonance or the magnitudes of certain intervals.[417] Pitches were defined successively by specifying the ratio between the length of the string segment producing a given pitch and the length of the segment producing the other pitch.[418] Consonances were limited to the octave (1:2), fifth (2:3) and fourth (3:4) while other intervals were classified as dissonances. Acoustically pure, perfect octaves and fifth served as means of obtaining further intervals. Hence "Phytagoreanism may be defined as the belief that all reality – including music – inheres in numbers and their relationships".[419] In this line of reasoning the speculative assumption was made that numbers are responsible for perfection in music – the idea was inherited later by a medieval philosopher named Boethius, among others, who transmitted this speculative thought to the Latin Middle Ages in his *De institutione musica* (early 6th century).[420] In 1627, *L'harmonie universelle* by Marin Mersenne (1588–1648) talked about the laws relating to the vibrating string arguing that its frequency is proportional to the square root of the tension, and inversely proportional to the length, to the diameter and to the square root of the specific weight of the string, under the condition that other conditions remain the same even when one of above mentioned quantities is altered. Other mathematical discoveries and developments of the late 16th and early 17th century (such as innovation of decimal fractions or logarithms) were also introduced into the debate over the nature of music. The 20th century theorists focused on the logical relations between groups of notes (called series or sets) generated in so-called serial music. Modern serialists strongly believed that the perfect set of numerical formulas could be applied in reference to musical pitches and forms as well as other parameters including the instruments (or precisely their timbre). Traditional notes in the score came to be treated as abstract elements of mathematical set theory.

The physical nature of music was also often discussed, attesting apparently objective character of music. It was discovered that the Earth is continuously producing a relentless hum made of an undefined number of notes, although only very tender scientific instruments such as seismometers can detect the emitted sounds. This fact corresponds with the ancient belief in the so-called harmony of the spheres that claimed that the planets and stars moving according to mathematical equations can produce music. The close relations between music and the cosmos were underlined in a 1617 book by Robert Fludd (1574–1637) entitled *De Musica Mundana* where he referred to the neo-platonic legend saying that "the universe is constructed according to mathematical harmonies which can be expressed in musical ratios".[421] He portrayed the Pythagorean monochord as God's instrument representing the cosmic harmonies encapsulated in the arithmetic ratios and structures of the heavens. Cosmos was imagined as a transcendental One divided into a series of realms and the division of the monochord reflected the musical scale. Importantly, all parts of the universe were interrelated, influencing each other, and the very notion of harmony referred to the ideal state of perfect relations. Vibrations of objects were understood in relation to the concept of sympathy, i.e. an action carried out in one part of the universe had its repercussions in another. The correspondence between macrocosms and microcosms was expanded to the human body, which was perceived also as a musical instrument producing sounds as a result of certain actions: the perfect balance between various elements such as mind, brain or nerves assured the harmony that was understood as the condition of good health.

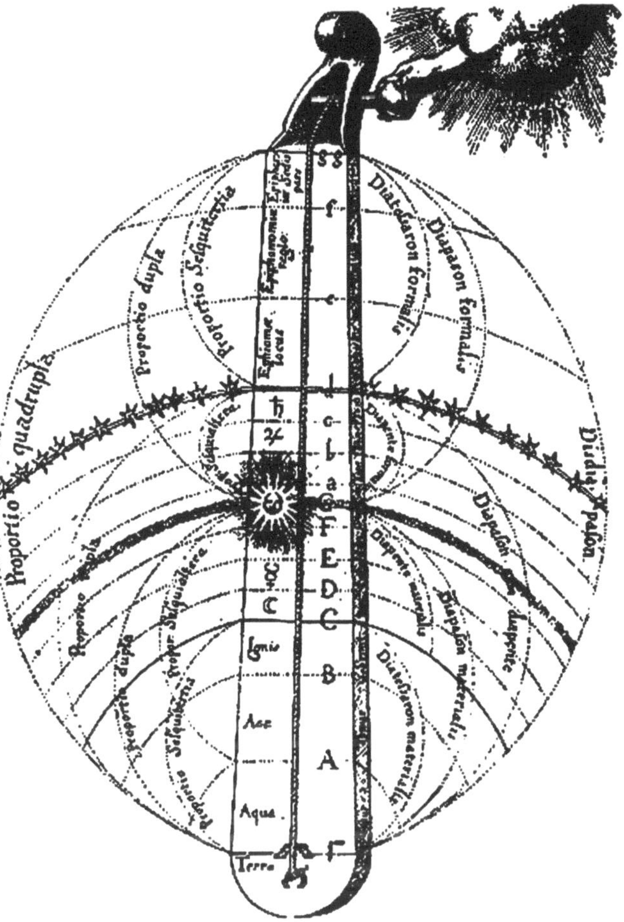

Fig. 6. *Robert Fludd's Celestial Monochord (1617) (source: "The Celestial Monocord." The Celestial Monocord. 15 Jan. 2010. http://www.celestialmonochord.org).*

Other thinkers seemed alluded to the idea of music objectivity and its immediate links with cosmic laws. Johannes Kepler (1571–1630) turned his attention to harmony and attempted to explain the proportions of the natural world, particularly the astronomical and astrological aspects, in terms of music. Kepler's interest in music began with the publication in 1596 of his *Mysterium Cosmographicum* where he considered the geometrical basis of the universe and claimed that harmonies might, consequently, be also grounded in geometry rather than in arithmetic as previously thought. In the fifth book of his 1619 *Harmonices Mundi,* he posited that harmony resulted from the tones made by the souls of heavenly bodies. Trying to shed more light on the problem of the hidden harmonies of the

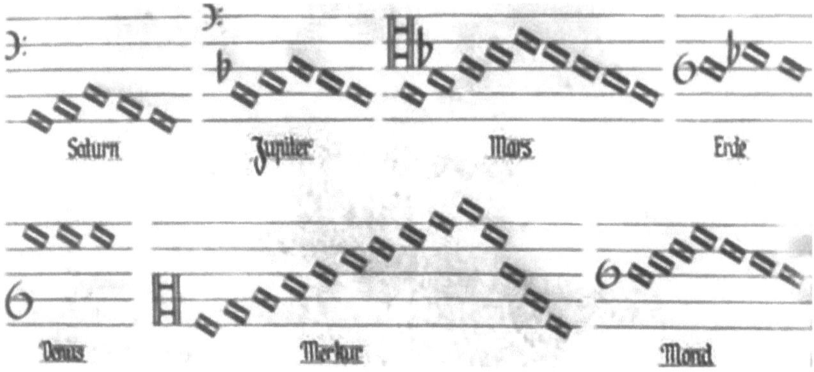

Fig. 7. *Johannes Kepler's music of the sphere (1619). (source: Astrocultura UAI. 15 Jan. 2010 http://astrocultura.uai.it/astroarte/musica/img/keplermusice.jpg).*

macrocosms and microcosms, Kepler pointed to the physical harmonies in planetary motion claiming that harmonies were "embodied in the angular motion of the planets as seen from the sun.[422] The astronomer calculated the intervals based not on the speed of the planets but by taking into account the difference between maximal and minimal orbital velocities observed from the sun.[423] Orbital distance measured from the sun accounted for such musical phenomena as musical scale or major–minor modes. While orbiting around the sun, accelerating and slowing down, the planets followed the elliptical motion expressing all the intervals of the scale by producing a range of pitches. Rooted in the Ancient concept of *musica universalis* (music of the spheres) that was interested in proportions in the movements of celestial bodies as a form of music, Kepler's theory attested to the fact that although real, the harmonies were actually soundless. Although they were not audible due to the lack of air, according to Kepler the sound that would have been produced in optimal circumstances would resemble the roar of sirens (rising and falling effect).

3.2.3 Conclusions

Despite the heated discussions as mentioned above, there has never been consensus on how music can be understood and defined. In that light, the question of the use of music as a universal language could be reformulated from searching for solutions as to how music could be used as a means of communication in outer space to the question of the role music may play in outer space: for example as a means of psychological device soothing the awkwardness of the unusual experience. That psychological function of music was well recognised in the early 20th century in cinemas where music was introduced – among other reasons – as a tool to minimise viewers' fear. Music as a familiarised factor was supposed to balance the discomfort of encountering new situations. Consequently, rather than deconstructing music as a coded form of communication, its future in outer space may be defined as a psychological means of facilitating encounters with the unknown: well chosen repertoire may help both the settlers and future space tourists. However, in that case, its primary function will be again defined as a form of artistic expression ultimately proving perceiving music as a form of art, unavoidable while discussing music as a means of communication.

Reflecting on music in outer space proves that despite the posing of future oriented problems the main questions to be answered in the realm of musicology remain the same as the ones asked today and in the past: in order to be able to provide any further speculations the discipline needs to search deep inside its own history, reformulate its methodological tools and resolve the hierarchy of inquiries.

[396] Feld, Steven. "Communication, Music, and Speech about Music." Yearbook for Traditional Music 16 (1984): 1–18.

[397] Dale, Ralph Alan. "The Future of Music: An Investigation into the Evolution of Forms." The Journal of Aesthetics and Art Criticism 26.4 (1968): 477–88. 477.

[398] Ibid.

[399] Risset, Jean-Claude. "Computing Musical Sound." Mathematics and Music. A Diderot Mathematical Forum. Eds. Gerard Assayag, Hans Georg Feichtinger, and Jose Francisco Rodrigues. Berlin-New York: Springer, 2002.

[400] Wilson, Stephen. Information Arts: Intersection of Art, Science, and Technology. Cambridge, MA: The MIT Press, 2001.

[401] Miranda, Eduardo R. and Tikhanoff, Vadim. "Musical Composition by an Autonomous Robot: An Approach to AIBO Interaction." Proceedings of TAROS 2005 – Towards Autonomous Robotic Systems, 12–14 September, 2005, London, UK. Plymouth: Faculty of Technology, University of Plymouth. 19 June 2009. http://cmr.soc.plymouth.ac.uk/publications/MirandaTikhanoff_robot_music.pdf.

[402] Jones, Daniel, Tim Hogson, Jane Grant, John Matthias, Nicholas Outram, and Nick Ryan. "The Fragmented Orchestra." Proceedings of the 9th International Conference on New Instruments for Musical Expression (NIME'09). 4–9 June 2009, Pittsburgh, PI, USA. 25 Aug. 2009. http://cmr.soc.plymouth.ac.uk/publications/NIME_09.pdf.

[403] Miranda, Eduardo R. and Drouet, E. "Evolution of Musical Lexicons by Singing Robots." Proceedings of TAROS 2006 Conference – Towards Autonomous Robotic Systems, 4–6 September 2006, Surrey University, Gilford (UK). 25 Aug. 2009. http://cmr.soc.plymouth.ac.uk/publications/Miranda_TAROS06.pdf.

[404] Miranda, Eduardo Reck. "On the evolution of music in a society of self-taught digital creatures." Digital Creativity 14.1 (2003): 29–42.

[405] Coutinho, Eduardo, Marcelo Gimenes, Joao M. Martins, and Miranda, Eduardo Reck. "Computational Musicology: An Artificial Life Approach." Plymouth: School of Computing, Communications & Electronics University of Plymouth. 26 Aug. 2009. http://cmr.soc.plymouth.ac.uk/publications/ComputationalMusicology.pdf.
[406] Seeger, Charles. "Music as a Tradition of Communication Discipline and Play." Ethnomusicology 6.3 (1962): 156–63.
[407] Schetsche, Michael T. "Encounters among the stars – Exosociological consideration." (in this volume).
[408] Music and Mathematics, from Pythagoras to Fractals. Eds. John Fauvel, Raymond Flood and Robin Wilson. Oxford, New York: Oxford University Press, 2003.
[409] Portnoy, Julius. The Philosopher and Music. A Historical Outline. New York: Da Capo Press, 1980: 148.
[410] Gajewski, Boleslas. "Grammar of Soleresol or the Universal Language of François Sudre." 27 Aug. 2009. http://mozai.com/writing/not_mine/solresol/.
[411] 19th century English psychologist Edmund Gurney (1847–1888) entitled his treatise *The Power of Sound* (1880) while German philosopher Friedrich Hausegger (1837–1899) discussed *Musik als Ausdruck* (Music as Expression, 1855) and Swiss musicologist Ernest Kurth (1886–1946) proposed such terms as *Entwincklungmotive* (developing motif) or *melodische Kraft* (melodic force) to describe music.
[412] for example *Führer durch den Konzertsaal* (Guide through the concert hall) by Herman Kretzschmar (1848–1924).
[413] Heinrich Schenker. Free composition. New York: Longman, 1979.
[414] Rom Harré. "Is there a semantics for music." The Interpretation of music: philosophical essays. Ed. Michael Krausz. Oxford: Clarendon Press, 1993: 103–21.
[415] Eero Tarasti. A theory of musical semiotics. Bloomington: Indiana University Press, 1994.
[416] Leonard B. Meyer. Style and music: theory, history, and ideology. Philadelphia: University of Pennsylvania Press, 1989.
[417] Mathieses, Thomas J. "Greek Music Theory." The Cambridge History of Western Music Theory. Ed. Thomas Christensen. Cambridge: Cambridge University Press, 2004: 109–35.
[418] Herlinger, Jan. "Medieval Canonics." The Cambridge History of Western Music Theory. Ed. Thomas Christensen. Cambridge: Cambridge University Press, 2004: 168–92.
[419] Ibid. 169.
[420] Enrico Fubini. History of music aesthetics. Translated by Michael Hatwell. London: Macmillan, 1990: 72–6.
[421] Gouk, Penelope. "Role of Harmonics in the Scientific Revolution." The Cambridge History of Western Music Theory. Ed. Thomas Christensen. Cambridge: Cambridge University Press, 2004: 223–42. 229.
[422] Ibid. 233.
[423] Ibid.

3.3 From space suits to space couture: a new aesthetic

Mark Timmins

3.3.1 Prologue

Somewhere in the near future at the launch of a European mission into space....

10...9...8...7...6...4...3...2...1... ignition... fire... ESA we have lift off...

The scene changes back to the studio where the pundits and space experts from around the world are discussing the successful launch and the new super-comfort design clothing that the astronauts are wearing on this, the first mission following the establishment of the aesthetic and design studio of ESA and the mission control management team.

To fully appreciate the significance of this mission we need to go back to the publication of the ground breaking book by the European Science Foundation entitled "Humans in Outer Space – Interdisciplinary Perspectives" where the issue of the critical need for embedding aesthetics into all aspects of the environment for humans in outer space was discussed for the first time.

3.3.2 Introduction

As we are about to enter a new age of long range space exploration, space settlements and space tourism, where the old rules of machines and the human interface are slowly but surely being broken down, we urgently need to discuss the aesthetics and design of garments. No longer can the relationship of man and machine, be won by the machine, in a very engineered, short time scale dash into near and orbital space. Humans are going to be in space for periods of time beyond weeks and into months and years. No longer will the humans' total time in the vehicle be taken up with minutely ordered instructions from mission control. It could in actual fact become very boring and one key way to offset this boredom is to introduce an element of personal expression, comfort, colour and texture by giving the astronauts the daily choice of clothing. This desire for personal expression will be magnified by the issue of space tourism, where non military-trained, ordinary (rich) citizens of Earth will wish to express their wealth and taste by wearing specifically designed garments for micro and zero gravity that will not work on the gravity rich planet we call the Earth.

To understand where we need to go in the future, in terms of the relationship between aesthetics and the critical engineering aspects of space exploration, we have to look at the past. This exploration of the past can be analysed both in practical terms of space suits, the astronaut and the portrayal of humans in outer space in the popular media of film, literature and fashion couture, as well as in some of the future-casting ideas of the last century.

3.3.3 Space clothing, fashion, couture and its portrayal in film, literature and popular culture of the 20th century

The design and aesthetics of clothing humans in outer space not only directly impinges on astronauts in the traditional sense as they undertake short and longer term missions but it also affects the mass of humanity left behind on our planet Earth.

The space design aesthetic has periodically over the last century influenced both fashion and popular culture, With films such as "Robbie the Robot", "2001 A Space Odyssey", "Buck Rogers in the 21st Century", "E.T.", "Barbarella", "Tron",

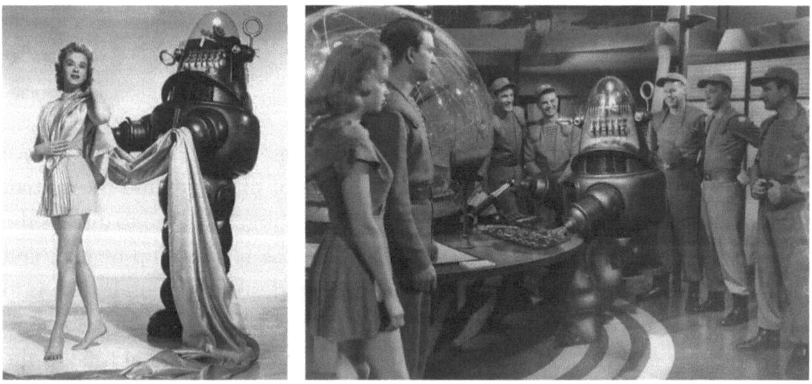

Fig. 8. *Robbie the Robot and Woman (source: "O PLANETA PROIBIDO, O FILME . . . " Outracoisa. 22 Mar. 2010. http://www.outracoisa.com.br/2008/05/20/o-planeta-proibido-forbidden-planet-1956-o-filme) and Robbie and Men of "Forbidden Planet" (source: "Drexfiles." 2009. Drexfiles. 22. Mar. 2010. http://drexfiles.files.wordpress.com/2009/03/robbie.jpg).*

Fig. 9. *1930s comic book illustration (source: unknown), "Just Imagine" (source: "She blogged by Night Gallery." She blogged by night gallery. 22 Mar. 2010. http://photos.shebloggedbynight.com/images/A_3/5/2/2/12253/JustImagine_still_e524a.jpg).*

"Men in Black", (as well as others too numerous to mention that the reader considers as classics) the Hollywood film industry constantly returns to this theme and re-injects a new vision to excite the audience about living or travelling in space and how that might affect our future. Below are some salient images to remind us of some of these classic films and images from the last 100 years.

However, it is not only Hollywood that has utilised space as a source of inspiration. The largest explosion of this phenomenon was in the febrile late '60s when the world got space fever: the excitement of the Russian Sputnik and the launch of the American Apollo series of spacecraft, which culminated with the first astronauts landing on the Moon on the 20th of July 1969.

Fig. 10. *"The Day the Earth Stood Still" (source: "Movie Posters." Movie Posters 22 Mar. 2010. http://www.movie-poster.ws/movies/scifi/images/dayearthstoodstill.jpg. and "Flash Gordon" (source: "Archives." Dial B For Blog. 22 Mar. 2010. http://www.dialbforblog.com/archives/361/flash_gordon1940.jpg).*

Fig. 11. *Captain Kirk from the 1960s TV series "Star Trek" (source: "Star Trek." Edit International. 22 Mar. 2010. http://www.editinternational.com/images/gallery/10-kirk_low.jpg).*

The famous quote "one small step for man and one large one for mankind" by Neil Armstrong became one of the iconic space related sayings, along with Star Trek Captain Kirk and his phrase "to boldly go where no man has gone before".

All of which combined to inspire fashion designers to create futuristic space inspired silhouettes. The two leaders of this fashion movement were two French fashion designers, Courrèges and Pierre Cardin. In 1968 Courrèges hit the headlines with his "SPACE AGE" collection. His designs at this time, were functional, uncluttered, and futuristic. Andy Warhol said that "Courrèges' clothes are so beautiful, everyone should look the same, dressed in silver. Silver merges into everything (. . .)".[424]

Pierre Cardin a year later launched his iconic space inspired collections and now the world could dress as if they were the crew of the Star Trek Enterprise.

This excitement with the possibilities of space couture, as opposed to clothes influenced by space, has begun again in the last three years with the support of both ESA and JAXA in the form of a space couture design competition and a series of public workshops exploring the possibilities of zero and low gravity space fashion.

The first event was sponsored by ESA and held in Norway and became known as the "Space Base Stockholm (Rymdbas Stockholm) Event" of 7–12 December 2007. This Space Base played films from the 1920s to the 1990s that explored some of the issues in science fiction and space as well as costume and fashion and

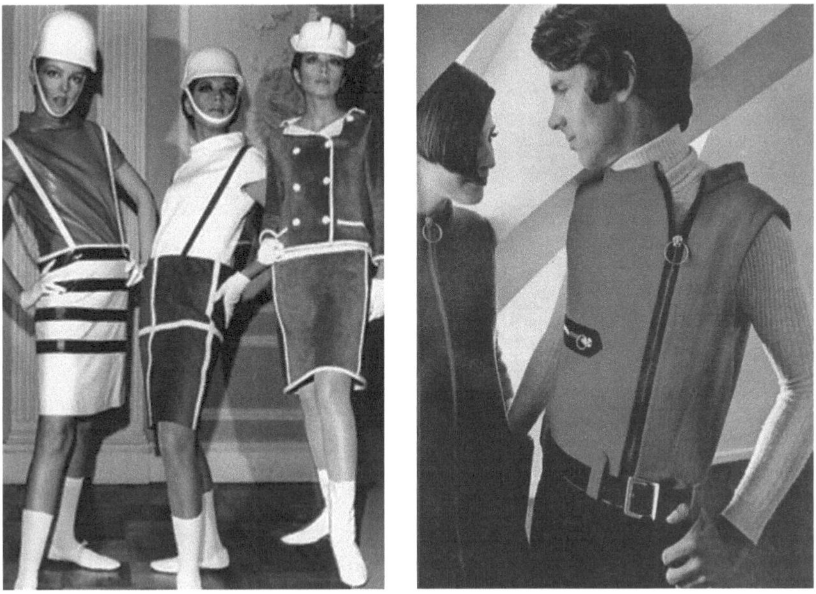

Fig. 12. *Courrèges' couture womens-wear (1968) (source: "Die Bedeutung von Courrèges." QVEST 22 Mar. 2010. http://www.qvest.de/2009/10/die-bedeutung-von-courreges/), Pierre Cardin's Men's Wear (source: "Pierre Cardin's Space Age Fashion." 12 Nov. 2009. Revel in New York. 22 Mar. 2010. http://www.revelinnewyork.com/blog/11/12/2009/pierre-cardins-space-age-fashion).*

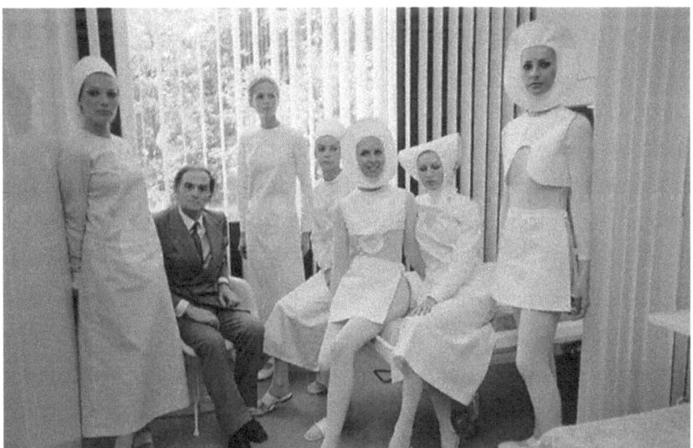

Fig. 13. *Piere Cardin's couture men's and nurses uniforms (source: "Pierre Cardin's Space Age Fashion." 12 Nov. 2009. Revel in New York. 22 Mar. 2010. http://www.revelinnewyork.com/blog/11/12/2009/pierre-cardins-space-age-fashion).*

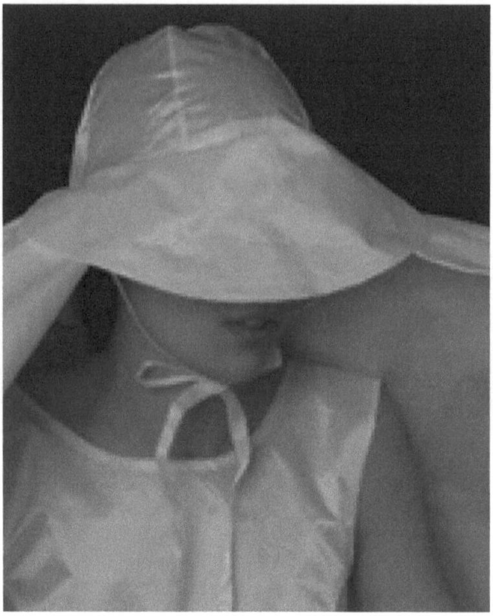

Fig. 14. *Have we moved much further on in fashion since the era of Pierre Cardin? (source: "Space inspires fashion." 26 Jan. 2007. European Space Agency. 22 Mar. 2010. http://www.esa.int/esaCP/SEMZVGSMTWE_FeatureWeek_0.html).*

then an invited team of designers and school children spent workshops designing, constructing, presenting and recording their creations. This coincided with the ISS mission of the Swedish astronaut Christer Fuglesang.[425]

The second event was sponsored by the Japanese Space Agency JAXA together with Rocketplane – a company developing spaceships for space tourism. "The Space Couture Design Contest" 2008 was organised by Ms. Onuki, a fashion designer based in Tokyo. This was an open competition and attracted over 350 entries, again showing a public awareness and desire to solve some of the issues of how, aesthetically, fashion and couture might function in space.

Finally ESA has dipped its own toe into the fashion market with a competition launched by the astronaut Frank de Winne for European primary school children to design a T-shirt to be worn by Frank de Winne in the ISS. 2000 entries were received and three winners chosen, again demonstrating the powerful mix of space, aesthetics and fashion.

So we can see that the concept of fashion in space is striking a chord with professional fashion designers, children and companies. This excitement however masks the real restrictions on all of the quoted fashion references, namely the restrictions of designing in a full gravity environment.

Fig. 15. *JAXA's brown suite (source: "Spacewear fashion show: looking fly in zero-g." 2 Nov. 2006. Pink Tentacle. 22 Mar. 2010. http://pinktentacle.com/2006/11/spacewear-fashion-show-looking-fly-in-zero-g/).*

Fig. 16. *Samples of the ESA T-Shirt Competition Winners (source: "Fly your picture on the International Space Station – your pictures!." May 2009. ESA. 22 Mar. 2010. http://www.esa.int/esaHS/SEMP9R4DHNF_index_mg_1.html).*

3.3.4 Space suits for astronauts

To begin, if we look at the clothing worn by astronauts it splits into two distinct areas: space suits and then garments/clothing worn for comfort under the Space suits. One can further distinguish between garments worn by the ISS crew members and shuttle astronauts.

The space suit is very much about protection from cold, vacuum, radiation, heat, abrasion and micro asteroid attack and is usually very restrictive and performs like a suit of armour. It looks and operates in a very mechanical manner and is visually not unlike a mini personal spaceship that is usually used for extra vehicular activities and has a strong protective element.

NASA and MIT have been exploring more modern, less restrictive space suit designs under the leadership of Dava Newman, a professor of aeronautics and astronautics and engineering systems at MIT in America. This will allow more freedom from the restrictive nature of the multilayered engineering garments to something more fluid and responsive (cf. Figure 18).

This freedom is going to become more critical as we move into Moon colonies and continue the Mars exploration, as well as in the more innovative Earth orbit architecture. The space suit is critical to allow us to develop more complex structures in space.

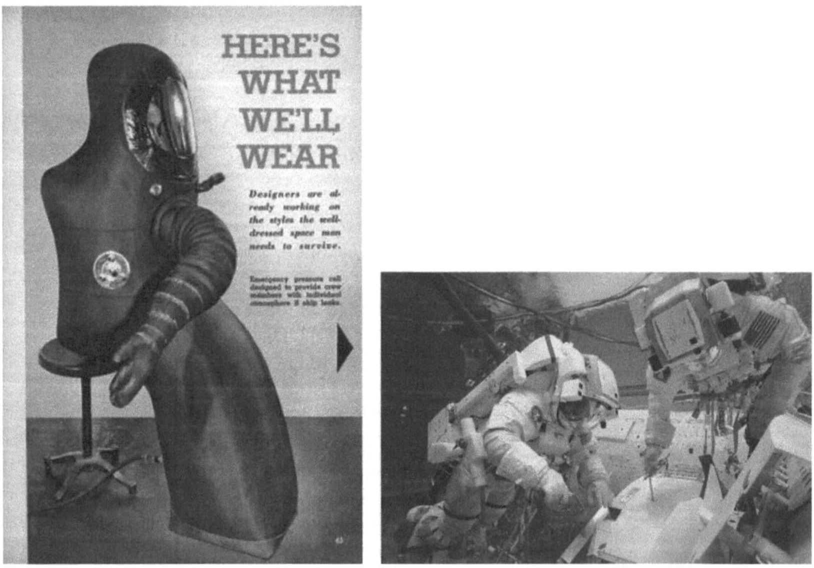

Fig. 17. *A poster from the 1940s (source: unknown) and contemporary ESA astronauts (source: "Gesucht! Europas neue Astronauten gesucht!" ESA. 22 Mar. 2010. http://www.esa.int/esaKIDSde/SEM9K51YUFF_LifeinSpace_1.html).*

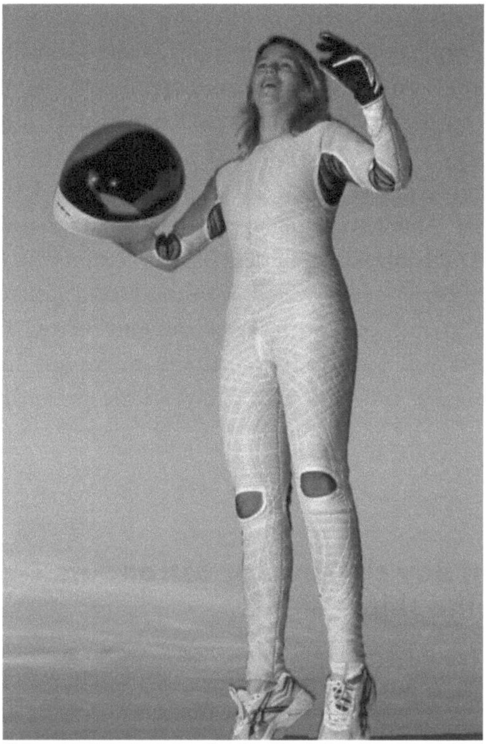

Fig. 18. *Dava Newman, MIT/NASA space suit (source: "Extra-Vehicular Activity (EVA) Research @ MVL – BIO Suite Overview." Man Vehicle Laboratory. 22 Mar. 2010. http://mvl.mit.edu/EVA/biosuit/index.html).*

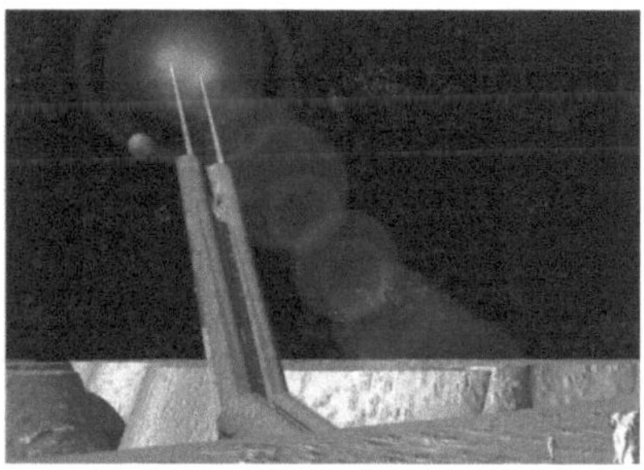

Fig. 19. *Futurist Imagery of the Bigelow space inflatable hotel (source: Bigelow Aerospace. 22 Mar. 2010. http://www.bigelowaerospace.com) and the NASA Space hotel (source: National Cheng Kun University. 22 Mar. 2010. http://www.phys.ncku.edu.tw/~astrolab/mirrors/apod_e/image/0107/maanhotel_rombaut_big.jpg).*

However being outside in space is not the only aspect of human existence out there and we need to look closely and carefully at the garments worn within a pressurised environment of the space station/hotel/moon colony/Mars extended mission craft, which will allow the astronaut to function efficiently both physically and mentally.

It is important to recognise that we need to look at the day-to-day clothing that astronauts wear. Think of getting up to put on what you wore yesterday and the day before and probably what you slept in. It is certain that a certain amount of boredom will creep in. However if we introduce a differently patterned, coloured garment, that is comforting, well designed, fitted and aesthetically pleasing, astronauts will have the opportunity to express individuality and creativity. At the same time this will allow a little personal control in the machine.

3.3.5 Present day clothing for astronauts inside the ship

The current context of clothing worn by astronauts within a pressurised environment is very interesting with the three major space agencies NASA, JAXA and Roskosmos providing a range of (what is considered appropriate) clothing for astronauts on extended missions to the International Space Station.

3.3.5.1 Russian space clothing from Roskosmos

In-flight clothing has been developed by the Kentavr-Science, Ltd. company, in consultation with the Institute of Medical and Biological Problems. There are 21 items of clothing to choose from, including underwear, socks and lingerie (for women). The clothing kit consists of the following cotton items: boxer shorts, T-shirts, polo shirts, shorts, socks, disposable paper underwear, and a warmer knitted outer operators suit. The T-shirts are available in a range of colours allowing the cosmonaut an element of personal expression.

Russian Cosmonauts are provided a set of undergarments consisting of a shirt, underpants and two pairs of socks (in a plastic bag with the size indication). Thirty pairs of socks and sixty sets of underwear are provided for a six-months flight per crewperson. Cosmonauts are thus supposed to change their underwear once every seven days and their socks every three days.

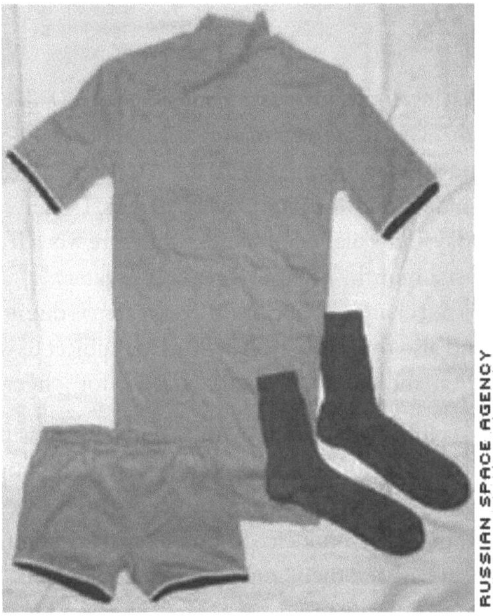

Fig. 20. *Russian Space agency typical underwear (source: Russian Space Agency).*

It is also interesting to note that the women cosmonauts are not forced to wear unisex underwear but can choose more feminine lingerie type items: bras, T-shirts and bikini-type underpants, edged with lace and made of cotton. A weekly set of underwear is provided. Disposable underpants can be changed each day. The idea is to make "Women on board feel like women, not just astronauts or cosmonauts".

Cosmonauts are also provided a so-called "casual wear set" that is intended to provide body temperature comfort for crew during their stay in the International Space Station (ISS) (with an ambient air temperature range of 20–30°C) and is used for everyday wear. As part of this set astronauts are given polo shirts with zippers made from cotton stockinet and shorts with zippered pockets. Each package also contains six cotton T-shirts of different colours from which the cosmonaut can choose according to his/her mood.

Colour is considered an important psychological factor for long and endurance missions. Thus, garments should be appealing and in harmony with the station's interior colour scheme. Cosmonauts can choose any colour combination that appeals to them.[426] This innovation of choice demonstrates that the Russians have started to realise that personal expression is an important aspect of mental wellbeing.

3.3.5.2 U.S. space clothing from NASA

According to NASA, the clothes that astronauts wear on average are "one pair of shorts and a T-shirt for every three days of exercising. Their work shirts and pants/shorts are changed, on average, once every 10 days. Crewmembers generally get a new T-shirt to wear under their work shirts every 10 days. Underwear and socks are changed every other day, but Polartec^{TM} socks, which are worn if a crewmember's feet get cold, must last a month. They also get two sweaters."[427] The reason that astronauts in space do not change their clothing very often is due in the main part to the cost of delivery and also that the astronauts are not subject to wild variations of temperature or exertion and clothing remains cleaner for longer.

> "When a piece of clothing has been worn as many times as possible, it's placed in a bag for disposal. Very little clothing is brought home by space station crewmembers. Most of it is eventually placed in the Progress resupply vehicle before it undocks from the space station. The dirty clothing and other garbage then burns up with the Progress when it re-enters the Earth's atmosphere."

In contrast to space station crewmembers, shuttle astronauts change their outfit every day. Depending on their duty they can choose to wear either trousers or shorts and have the option of wearing long- or short-sleeve polo shirts or rugby shirts, as well as sweaters and pullovers. The clothes of shuttle astronauts have plenty of pockets and velcro. The pockets and velcro help them keep everything they're working with near them.[428]

Fig. 21. *Crew in grey polo-shirts (source: NASA. 3 July 2010. http://spaceflight.nasa.gov/gallery/images/shuttle/sts-98/hires/sts098-365-0034.jpg).*

Fig. 22. *Space Shuttle Clothing (source: NASA. 22 Mar. 2010. http://spaceflight.nasa.gov/gallery/images/station/crew-2/hires/iss002e5335.jpg).*

3.3.5.3 Space clothing from JAXA

The Japanese space agency has also gone down a similar route, incorporating into its day-to-day clothing; shorts, T-shirts, boxer shorts, socks and polo shirts. Interestingly, the Japanese kit, unlike the Russian and American clothing, has started to introduce stripes giving the whole collection a more sporty feel. The

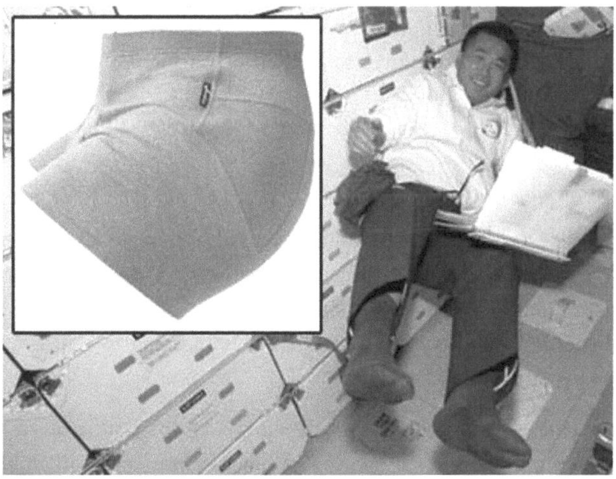

Fig. 23. *Anti-Microbial Boxers (source: "Space underpants sell, won't smell." 20 Feb. 2010. Collect Space. 22 Mar. 2010. http://www.collectspace.com/ubb/Forum14/HTML/000818.html).*

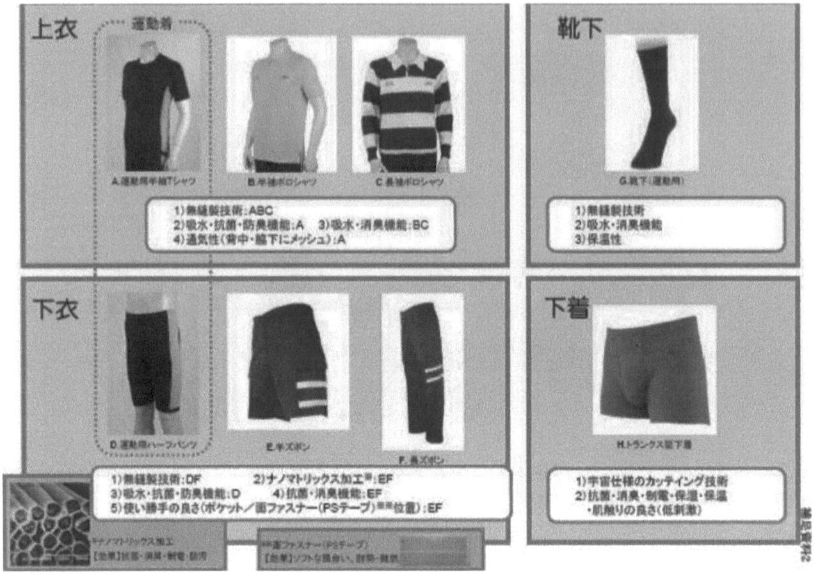

Fig. 24. *JAXA clothing kit (source: "Japanese Astronaut Tests Stinkless Space Undies." 19 Mar. 2008. Live Science. 22 Mar. 2010. http://www.livescience.com/blogs/2008/03/19/japanese-astronaut-tests-stinkless-space-undies).*

Japanese have also been experimenting with anti-microbial clothing with extended wearer trials being conducted in space in 1999. This research has had a twofold outcome; ensuring that the astronauts could remain healthy in extended wear, close fitting underwear, as well as reducing the amount of clothing that an astronaut requires to take into space.

3.3.5.4 European space clothing from ESA

The only reference that the author could find that ESA are addressing some of the issues was one in 2002 on a short project entitled VEST that researched into the development of new garments that should be comfortable, provide for bodily hygiene and thermal stability, and be aesthetically pleasing to improve the crews' psychological and physiological state.

> "The advantages of such clothing ranges are not confined to orbit. Terrestrial activities, such as sports and outdoor pursuits, can all benefit from enhancements to the antibacterial and thermo-regulatory treatment of the fabrics."[429]

So as we can see, the major space agencies are aware that astronauts need to wear clothing other than that of a space suit. However the issues of choice, aesthetic design, new and comforting materials, pleasure and individuality seem to be unimportant in achieving an almost military like uniformity. We have to be aware that living inside a machine is not natural for a human however well trained, and we do have to utilise fashion design and couture to allow for a more harmonious existence between man and machine. This leads very clearly to the importance of aesthetics and space couture.

3.3.6 The importance of aesthetics and space couture

As we can see, apart from fleeting references to the selection of colour and to the fact that women can choose lingerie with lace (which is an interesting reference to fashion couture terminology), very little design or aesthetics goes into the development of the clothing needs of long stay astronauts. To provide comfort for long stay astronauts in terms of clothing, garments need to be made of materials that offer something more than just keeping out the cold and being lightweight. As we have seen from the American astronauts clothing lists, two sweaters are provided. There is no reason why these cannot be made of cashmere, a perfect fibre for space, which is light-weight with a very good ability to wick, takes moisture away from the body and leaves the wearer warm and comfortable as well as having the added aesthetic advantage that the colours can be beautiful and the garment feels amazing. This same result of colour, comfort and personal expression can be achieved by looking at fibres such as silk, bamboo, ramie and linen, and modern garment construction technologies including fully fashioned knitwear, sonic welding, couture cut garments (that are personally tailored to the astronauts) and whole garment knitting technology which eliminates seams and associated discomfort.

Alongside this voyage of fashion discovery, we need research time and money to explore new materials (such as light emitting, charging, nano materials, shape memory, bio-mimetics, biodegradable, plasma – treated this list is not exhaustive) that could form a major part of the astronaut's wardrobe. There has been a lot of research into sportswear, which could be highly applicable to the development of new space wear. This research is starting to be undertaken as in the case of the J-clothing line designed for the Japanese space agency JAXA by the Japan Women's University in Tokyo, which explores anti bacterial and anti odour properties.

Aesthetics is critical as it covers many different functions and applies from the visual through to tactile, design, texture, engineering, architecture, interior design, smells, memories as well as other human factors.

These technical innovations are, however, only one aspect of couture/fashion design and in Europe we have a collection of fashion research universities that are unparalleled in world research terms.

A focus on aesthetics will set Europe ahead of other space programmes and will become increasingly important as we look at long term missions including flights to Mars, lunar colonies and orbiting scientific space stations like the International Space Station (ISS). One of the consequences of long distance endurance space flight is that astronauts will need some personal control of their environment, as they cannot just stop the spaceship at the end of the day and go home.

However we need to also go one step further in the life cycle of the aesthetically well designed astronaut's wardrobe and think about end of life. Currently we have a very inelegant system:

> "The Progress (re supply vessel) is then loaded up with trash including dirty laundry . . . the dirty clothes can sit around for a while on the station before they can be disposed of . . . the Progress is undocked and de-orbited on a course over the Pacific and the craft burns up in the atmosphere."[430]

Other attempts have been made to look at the recycling of garments in space with experiments on germinating seeds on worn clothing by NASA and Roskosmos. Researchers have started to look at decomposing paper and cotton undergarments with bacterium, and experiments are being undertaken with ultraviolet to kill the bacterium that cause the smells, in addition to washing and dry cleaning experiments.

The design of astronaut clothing is however only one aspect of the environment that directly affects the well being of the astronaut, and this is the other area where aesthetics and design have a major part to play.

Ever since the beginning of the space race, the engineering of the craft and safety of the crew and astronauts have rightly been the most important aspects of getting spacecraft off the ground to undertake the mission and safely return to Earth. This has resulted in brutal metal boxes and tubes painted in drab greys, which have become machines not for living but for accommodating and only just tolerating humans.

This imbalance can now can be addressed as we have started to recognise what is critical in terms of safety and we are now at the stage where there is an opportunity for aerospace companies and agencies to look towards a more human friendly environment that explores softer design issues including comfort, functionality

Fig. 25. *Inside the International Space Station: an ESA astronaut (source: ESA. 22 Mar. 2010. http://esamultimedia.esa.int/images/s100e5356.jpg).*

and colour. (In this last case we may need to look at some of the pioneering work on colour research undertaken by Rudolph Steiner following the work of Wolfgang Goethe to understand how colours, the interaction and placement of colours, can affect peoples moods, feelings of wellbeing, calmness, invigoration and happiness.)

These issues of how humans interact with space technology are starting to be recognised through some recent work of the Imperial College London and the Royal College of Art, where the lecturer Daniele Bedini (of Imperial College London) got students to explore various issues of how the design mechanics interface should sit within the new environment of space hotels.

However it is not just about design mechanic interface but should also incorporate a third leg of Aesthetics to produce objects, processes and environments that not only function but will encourage people to use them and remove some of the stress of living within a very restricted environment. This issue grows to be of more importance as space missions become longer and we look towards permanent occupation of moon bases and trips to Mars.

3.3.7 Aesthetics and space couture

The issue of aesthetics and couture also applies to space tourism, whether short-term hops undertaken by the Virgin Galactic and space plane organisations into orbital weightlessness or the longer stay tourism of Earth orbit and lunar hotels where we will see six or seven star luxury requirements. These facilities will have to include appropriately designed environments and staff uniforms, and guests will be demanding space couture that functions under and exploits weightlessness and low gravity, to produce dramatic fashion statements that would not function in the gravitational environment of Earth.

This will require new computer non-gravity environments to allow fashion designers to create in confidence rather than trying to guess what a fabric or material will do under micro or zero gravity. Designers will be able to create garments that respond to puffs of air, magnetic fields or change colour under different environmental factors such as heat or ion streams. Couture garments designed specifically for the space tourist can explore (without cost being an issue) new materials and silhouettes exploiting innovative construction technologies as well as new materials, including magnetics, ceramics, shape memory, communications technologies as well as developing a new zero gravity design aesthetic. Then we could think about the first Olympic games on the moon and the spectacular things we could do with space sport couture. This is an opportunity to tap into the creativity of Europe's design schools and lead the way out of Earth's gravity and atmosphere into a fashion aware, superbly designed, space future.

3.3.8 The way ahead

As has been touched upon in the preceding text, we are rapidly approaching a new era in not just state sponsored space exploration but the private sector is now developing new markets in space tourism and starting to think about the recreational prospects of space. The next major stage after short weightless flights in a jumpsuit on the Virgin Galactic space-craft, would seem to be space hotels either in orbit or on the Moon. Hilton hotels have invested in the region of half a million dollars in concept fees to a team of architects to come up with concepts for both orbiting hotels and Moon-based buildings. The American aerospace company Bigelow Space (which is part of a much larger entertainment group with Las Vegas casinos) is also developing inflatable technologies to build an inflatable Earth orbiting hotel.

Fig. 26. *Space hostesses from the film the "Fifth Element" (source: Gaumont Films).*

Fig. 27. *"2001 A Space Odyssey" space hostesses (source: The Invisible Agent. 22 Mar. 2010. http://theinvisibleagent.files.wordpress.com/2009/06/stewardess1938982585_76df1c0422.jpg).*

All of this is followed by the tourism infrastructure: staff, food, entertainment, money, gambling, communications, sport, visas, quarantine, health, green issues, waste, souvenirs, and of course couture garments that only work in zero gravity. This development is starting to be considered in the short to medium term but we should also be looking at the longer term including voyages to Mars and permanent settlements on the Moon, which pose greater challenges for the aesthetic design engineering interface.

These challenges for fashion, couture and clothing in the future build on the challenges and successes of space tourism and are starting to be explored by private research companies such as Phillips of Holland with garments like the "emotion garment" which, in response to a set of instructions, can squeeze, tickle, or send shivers down the wearers back.

Within garments we will need to be exploring: body enhancement, exoskeletons, protection, emotion, radiation protection, energy generation, detection, communication, location, invisibility, camouflage, thermo protection, medication,

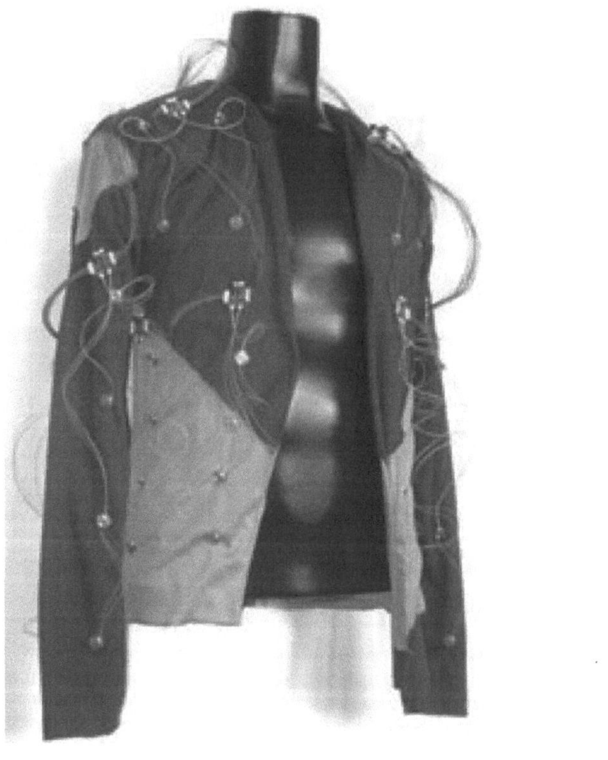

Fig. 28. *Phillips "emotions jacket" (source: "How to create emotional immersion." Philipps. 15 Mar. 2010. http://www.research.philips.com/newscenter/topics/20090415-emotionsshirt.html).*

personal expression, morphism and bio-mimetics which will allow humans to thrive in outer space working in zero and low gravity environments.

As a matter of urgency we also need to discuss the aesthetics and design of not only garments but also the machines for both transport and living that we use to break out of Earth's atmosphere.

In conclusion, we are about to enter a new age of long range space exploration, colonisation and tourism where the old rules of machines and the human interface are slowly but surely being broken down. We need as a matter of urgency to discuss the aesthetics and design of not only space couture garments and the amazing design possibilities of couture garments interacting with both micro and zero gravity, but also the day to day workers, whether they be involved with space tourism or space exploration. This is an opportunity for the European Space Agency to provide a new, more creative and visionary view about how we as Europeans can lead in the "Humans in Outer Space" journey.

[424] "André Courrèges." Fashion Model Directory. 28 Aug. 2009. http://www.fashionmodeldirectory.com/designers/courreges/.
[425] For more information on this event cf. "Space inspires fashion." 26 Jan. 2007. European Space Agency. 22 Mar. 2010. http://www.esa.int/esaCP/SEMZVGSMTWE_FeatureWeek_0.html.
[426] "Space Clothing." RuSpace. 15 Nov. 2009. http://suzymchale.com/kosmonavtka/clothing.html.
[427] "Space Wear." 4 Feb. 2003. NASA. 28 Feb. 2010. http://spaceflight.nasa.gov/living/spacewear/index.html.
[428] "Space Wear." 24 Feb. 2003. NASA. 22 Mar. 2010. http://spaceflight.nasa.gov/living/spacewear/index.html.
[429] "Mission Experiments in Focus." 23 Apr. 2002. ESA. 22 Mar. 2010. http://www.esa.int/esaHS/ESANK2G18ZC_astronauts_0.html.
[430] "Astronauts' Dirty Laundry." 10 Apr. 2003. NASA. 15 Mar. 2010.

3.4 Looking back, looking forward and aiming higher: next generation visions on humans in outer space

Agnieszka Lukaszczyk, Bejal Thakore and Juergen Schlutz

3.4.1 Introduction

The launch of the first artificial satellite, Sputnik, by the Soviet Union on 4 October 1957 marked the start of the space age. This single event spurred a change in the way nations viewed national achievement, technological prowess and the strategic value of space technology. Just weeks later, it was followed by the U.S.' launch of the Explorer-I satellite and triggered both orbital missions and robotic landers to near solar system planets and beyond, as well as extensive utilisation of satellites for Earth observation, communication and other services until today. In 1961, the flight of Yuri Gagarin pioneered human space flight and took the space age to yet another level, extending the frontier of human exploration and the search for life in to space. The early Russian cosmonaut and American astronaut missions, the historic lunar landings during the Apollo programme, and the establishment of operational platforms for space sciences and research in Earth orbit leading to the International Space Station (ISS) are but a few of the many achievements of the space world in the last 52 years.

At the time of the 50th anniversary of the Sputnik launch, in 2007, the Space Generation Advisory Council took a first step into preparing and shaping the next 50 years through contributions from the young generation. Partnering with the Frederick S. Pardee Center for the Study of the Longer-Range Future, the Boston University Center for Space Physics, The Planetary Society and the Secure World Foundation, a series of surveys and online discussions collected some 750 key visions from approximately 276 delegates worldwide.[431] The study, largely directed by the delegates themselves, was the first of its kind and uniquely aimed at:

- Introducing new space enthusiasts from under-represented areas and allowing the flow of fresh ideas to be featured within input to several different space leaders from industry and agencies;
- Engaging the next generation of space enthusiasts into a discussion that was futuristic yet with tangible deliverables;

- Encouraging new space advocates to embark on research and further investigating how to bring the new visions to fruition in the near future;
- Utilising a unique approach of working cooperatively with modern communication methods while driving contributions by the competition for a unique prize;
- Providing the opportunity to present the results of the surveys and online discussions to the Boston University Conference and
- Creating a first attempt at drafting a cohesive worldwide policy statement that reflects the new mandate for upcoming international space activities from the youth of today.

In order to incorporate up to date considerations and to keep the recommendations relevant in the constantly changing political, social, and technical environments of the international space actors, educational institutions and the broad public, SGAC conducted a second shorter survey on the potential of space solutions to address global challenges on Earth in early 2008. The results of the second survey were briefly presented to the audience of the 12th Symposium of the International Space University (ISU).[432]

The relevance of the SGAC recommendations to the space community has already been confirmed by the international feedback and invitations to contribute to agency and industry discussions, while it encourages SGAC to continue the activities in capturing the dynamics of a transforming space sector relevant to the feasibility and desirability of the future visions.

This paper summarises the methodologies and outcomes of the surveys, highlighting the unique character of the SGAC 50 years vision project, which aimed to illustrate various ideas young people have around the world regarding the human space activities, their presence in the outerspace, and how it will affect the life ion Earth.

3.4.2 About The Space Generation Advisory Council (SGAC)

The Space Generation Advisory Council in Support of the United Nations Programme on Space Applications aims to represent young people (students and young professionals) between the ages of 19 and 35 to the United Nations (UN), other international organisations, governments, space agencies as well as research institutions and industry. SGAC was created in 1999 at the Third United Nations Conference on the Exploration and Peaceful Uses of Outer

Space (UNISPACE III) where space leaders from all over the world came to the conclusion that youth should have a voice at the UN, providing the unprecedented views of men and women who grew up in times of natural progress of space exploration. Therefore parallel to UNISPACE III, the Space Generation Forum took place wherein 160 young people from 60 countries worked together and produced and adopted the Declaration of the Space Generation. In order to fulfill its mandate, SGAC has been given permanent observer status to the UN Committee on the Peaceful Uses of Outer Space (COPUOS) and consultative status with the UN Economic and Social Council (ECOSOC) so that it can regularly report its activities to the UN.

SGAC has a true global network with over 4000 members worldwide who communicate and work together via e-mail lists, online forums and projects, local, regional and international workshops and conferences as well as various grassroots activities.[433] The organisation has two Regional Coordinators (RCs) each from the six UN regions, and currently 66 so-called National Points of Contact (NPoC). The focus of the organisation is to impact on policymaking and thus advise political decision makers based on the opinions and ideas of the youth of the world. The policy input includes regular reports to the Member States of UN COPUOS, including the UNISPACE III Action Teams. In addition, SGAC gives input to national and international space agencies as well as governmental entities dealing with space. For instance SGAC was the only Non-Governmental Organisation (NGO) invited to the Space Summit in 2002 in Houston, which gathered major space companies, high-level space experts and heads of space agencies to discuss the future of space. More than that, SGAC was invited by the European Commission to provide the perspective for the future European space workforce during the consultation process for the Green Paper on the European Space Policy.[434]

Having such an impact on policy making motivates the members of SGAC to actively participate in its activities on national, regional, and international levels. Members can address the issues they care about and have their voices heard because they know that through SGAC they have access to the UN and various high-level space persons. For instance, in 2007 the members of the SGAC South American region wrote reports on the space situation in their respective countries, which were then delivered to the COPUOS delegates.

In addition to policy focus, SGAC carries out a number of projects around the world including projects on education, space security and outreach programmes.[435] SGAC sends its members to local schools to teach astrophysics, astrobiology, astronomy, science and technology, sustainable development etc. Also the organisation organises space awareness days, movie nights, workshops, conferences, space parties (Yuri's Night), and technical activities. Hands-on activities such as the Under the African Skies project, help to build knowledge and confidence but, most importantly, they allow SGAC to teach people in developing countries about the impact of space technologies on their day to day lives and how they could benefit even further from a number of space applications.

Because of its good relationships with the UN Office for Outer Space Affairs (OOSA) and the United Nations Educational, Scientific and Cultural Organization (UNESCO), SGAC is able to send some of its members from developing countries, fully funded, to a number of international conferences and workshops. This gives those young people the opportunity to interact and learn from experts in the field. But most of all, it allows them to present their ideas and thoughts in international forums where they can gather advice and input from people they would normally have no access to.

In summary, due to its international mandate, SGAC is able to carry out many useful activities around the globe. Its network has grown significantly in the past 10 years, thus it likes to call itself a "network of networks" as it not only represents its members on the international level but it also works as an umbrella for other student and youth groups around the world.

Fig. 29. *A selection of active space advocates and supporters at the 10th anniversary of SGAC in Vienna, 2009 (source: SGAC).*

3.4.3 Aiming ahead: The 50 years survey

Held in two consecutive surveys in 2007 and 2008 respectively, the 50 years vision surveys and online discussions provided the means for collecting data and interacting with the SGAC community. The contributing young students and professionals incorporated their personal area of interest or expertise and to some extent also their socio-demographic backgrounds into the responses, making the compilation of these visions and the collective recommendations extremely rich in terms of their international appeal and multidisciplinarity.

3.4.3.1 The first surveys: methodology and impact

In early 2007, the initial round of surveys asked for a brainstorming of visions for the next 50 years from a large target group. The study was conducted in three phases, starting with the first online survey. A project website was hosted at the SGAC website,[436] where delegates could register to participate and become active in online forums with others who made inputs to the first survey. As the survey collected responses, a team of delegates devised a strategy towards refining and analysing the diverse results that would lead to understanding how the second survey could best build upon the inputs received. This team of delegates, working remotely via the online forums, carried out the analysis of the results and insights into what key aspects need to be reflected in the final output, specific to the Boston Conference input.

One could even reason that the series of events that marked the history of the last 50 years of space did so due to the expectations and aspirations of young motivated scientists, engineers, business and legal professionals. This study already highlights the interests and aspirations of the younger space community and sets a dialogue amongst the interested individuals to understand how they can work together better to bring their visions into reality. To derive specifics for each vision, the survey asked for its contributors to keep the following questions in mind:

- What will space look like in the next 50 years?
- Will humans be able to function in the outer space on more permanent basis?
- How can we learn from our history on Earth as we move into this next frontier?
- How can we create opportunities for sustainable, beneficial and effective use of space?
- Who will decide these questions?

- How will conflicts of interests be settled?
- What new systems, structures and paradigms do we need as we begin this new adventure?

These open ended questions helped drive the online discussions that allowed for the contributors to expand on their visions and find a path to feasibility and to understand if the timeline they had initially attached to their vision was realistic and what externalities it depended on.

The online discussions were pertinent to understanding what these aspirations were. But as Nobel Laureate Neil's Bohr once said, "Prediction is always hard, especially about the future".[437] This study is not to be taken as a mere predictions tool. The younger space professionals via their online deliberations have set important milestones that space agencies should be including in future space programmes.

Some 750 visions of the next 50 years from 276 individual contributors have been generated and compiled, impacting not only the direct technologies for implementation of space programmes, but also addressing many aspects of our social, cultural and political lives on Earth in the relation to space exploration. With the successful execution of the surveys, the Boston conference itself has provided valuable experience for the delegates who attended and several online participants still retain the motivation to build upon the results and present these key visions of the new generation to various space agencies as well as organisations.

3.4.3.2 The second survey: Rationale and methodology

Recommendations from the 50 years survey, Part 1, included suggestions to match the visionary space activities and technology [predicaments] [predictions?] to roadmaps and inputs of other international societies in the same field. This led to an analysis and integration of the United Nations Millennium Development Goals (MDG)[438] set by the UN Millenium Declaration (UN, 2000).[439] The UN MDGs were put in place to address the most pressing challenges and thus formulate an agenda on which the countries of the world can focus to improve the quality of human life by the year 2020. When doing so, the focus of the second survey automatically became more concise on aspects of space solutions to global challenges and potential near-future solutions or early steps.

The context of the MDGs, stemming from the work of several international committees via multiple expert panels over an extended period of few years, is somewhat larger than the SGAC vision project mandate. Thus, the original list of biggest challenges used in the survey was reduced to make it more relevant to our audience and to the application of space technology or space based data. Engineering solutions within the MDGs like "[e]nsure environmental sustainability" and "[d]evelop a global partnership" are examples of applicable elements in the list, whilst the listing of biggest challenges used in our survey did not list gender inequality, child and maternal health. Despite the deviation from the MDGs, giving the youth space community an awareness of the UN MDG and setting the challenge to work towards solutions for some of these by aiding with space technology and its applications may inspire several young students and professionals to further work and research. Of the respondents only 33% were aware of the UN MDG Goals.

The particular questions to guide the input to the second survey included the following:

- What are the biggest challenges facing us globally?
- What are the challenges you think we are closest to solving?
- What is the role of space technologies in helping solve the global challenges?
- What are specific recommendations we can take to policy makers?

The aim of this particular study as the extension of the 50 years visions was not to suggest that space applications and technologies would have a unique selling proposition in solving every Earthly challenge, but if applied to some challenges, they may offer means of additional information, new approaches and accelerated progress. An overwhelming majority (82%) of the respondents felt that space technologies could play a part in solving most of the global challenges.

The preliminary results of this survey were presented at the ISU's 12th annual symposium in 2008 on "Space Solutions to Earth's Global Challenges". This survey, which ran with a shorter time line than the previous one, attracted responses from 159 individuals and resulted in specific recommendations for how space solutions can help address global challenges.

3.4.4 Visions abound: Results from the 50 years survey

The first surveys' results and key visions show a variety of [aspects of] [approaches to] space exploration and applications for a continued human reach beyond this planet. Physical means such as human missions play a role as much as ideological impacts on international relations and education, to name just a few. Classification

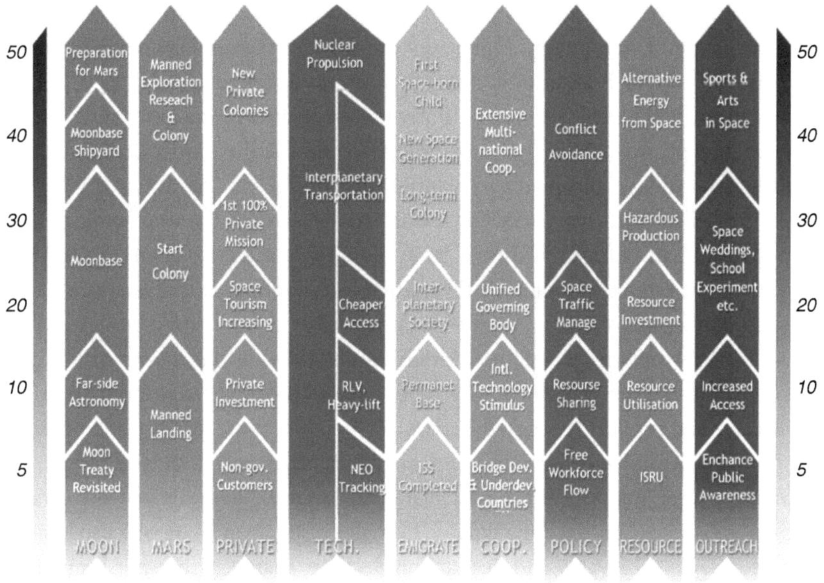

Fig. 30. *The SGAC 50 years visions roadmap (source: SGAC).*

of these visions into categories addressing implementation and potential impact together with a timeline in the form of a target achievement date enabled the drawing of [allowed to obtain] a roadmap as shown in Figure 30 Here, vertical tracks show non-exclusive strategies from near to long term development in the timescales of 5–50 years from the time of the survey. The visions have been categorised according to those relating to (from left to right) Moon, Mars, development of private space activities, technology development, humanity's "emigration" into space, international cooperation, international policy formulations, space resource utilisation, and public outreach and education.

Apart from the roadmap representation, the key visions were themed into three main categories, which will be treated separately in the following sections.

3.4.4.1 Improvement of human survival probability

Obviously, strong support can be found for the improvement of human survival probability. The specific contributions within the 50 years visions highlight and emphasise the sustained exploration of the Moon and the long-term utilisation and extension of near Earth orbiting facilities (along with continued exploitation of the International Space Station) leading up to the establishment of a permanent lunar outpost and settlement. This paves the way for further expansion towards Mars and allows for humanity to become a multi-planet species.

This support for exploration and potential emigration into space in its constructive and positive drive is closely linked to the fragility and vulnerability of Earth and its unique biosphere to external threats such as Near Earth Objects (NEO) as well as the threats posed by the destructive nature of war technology and weapons that can be used against a large mass of humanity. There was strong support for the establishment of a fully ratified international space treaty prohibiting space weapons by 2010. A "self-sustained lunar colony" (by 2045 A.D.) as shown in the roadmap as well as Mars missions provide the means of directly improving human survival probability. The required precursor exploration needs a global drive and support through the space perspective on the fragility of humanity and through the advancement of political systems that enable greater peace.

Whilst a large majority supported sustained lunar exploration and the establishment of a self sustained lunar colony, human presence on Mars was also largely supported as an achievable and desirable milestone, including but not limited to the search for life and the investigation of living conditions on different planetary bodies.

3.4.4.2 Improvement of human quality of life and the environment

Several contributors suggested that space activities have a significant impact on the quality of life globally. Technology development and its utilisation in several fields related to improving our lives, security and access to resources are direct benefits but there is also the indirect benefit of improvement of political systems.

"Improvement of technology" as used here, is a broad term that relates to technology cited in the survey responses such as that employed in the fields of disaster management, e-health and telemedicine, tele-education and products derived from an application of technology developed for space activities (spin-offs). Strong support for the continuation and full utilisation of the International Space Station (ISS) can also be related to this context of technology development as several responses stated its "utilisation" for the betterment of life sciences and medicine back on Earth.

In relation to the depletion of natural and metallic resources as populations grow and energy demands escalate, there was very strong support for the role of space in monitoring resource management on Earth as well as for access to space resources to improve quality of life on Earth. Potential resources cited include the free and constant solar energy, metallic elements from Near Earth Asteroids, hydrogen and helium-3 from lunar regolith and others to ease the burden on current Earth assets to support human activities on Earth as well as in space.

3.4.4.3 Improvement of human knowledge and understanding

There was a strong mandate for space agencies to create more awareness that space technology and activities are for the benefit of humanity. Specifically, contributions stated that there needs to be more awareness about scientific platforms in space, the experiments that have been conducted and their applicability to research and advancement on Earth. This can be achieved through relating technological developments taking place via research in space to daily life issues and the concerns of the general public in order to stimulate increased public awareness and motivation. One key vision that stands out called for "creating a strong perception that space is for the benefit for all by providing global environmental effects of human activities by 2015". This would require the integration of several climate change models with earth process models along with extensive use of remote sensing data from space in conjunction with *in-situ* measurements from a variety of platforms leading to a global carbon emissions picture.

In addition, some key visions suggested that space programmes and roadmaps should be developed in conjunction with decadal surveys of learned societies from all fields as space has the potential to affect, enhance or even accelerate their goal achievements.

3.4.5 Observations from the 50 years visions survey Part 1

The contributors to the surveys and discussions showed a superb understanding and support of the context of space exploration. Whilst the majority of our contributors were young students and professionals within the 18–35 age range, they supported space exploration and space activities and envisioned long-term space programmes, yet information to the wider public as to why the particular path was chosen is crucial to continued support and sustainability. This is exemplified by, for instance, the association of human survival probability with destruction of Earth's biosphere by external and natural threats (NEO impact) as well as threats carried by human activities (wars). This young audience showed good awareness of the challenges facing us and sought some solutions that could be found in space for these challenges, in particular the global threat of war, the digital divide and the need to make education available worldwide, finding a solution to our energy demands, controlling climate change and finding solutions to life-threatening diseases.

3.4.6 Back to Earth: Results from the second 50 years survey

Inspired by some of the recommendations described above and with specific reference to the questions in Sect. 3.4.2, responses to the above questions were sought in a continued though smaller survey process. The participants were asked to choose what they considered to be the three most pressing global challenges. As for the biggest challenges, the choices were set in line with the MDGs as shown in Figure 31. A majority of the respondents considered climate change and the global energy crisis to be the most pressing challenges facing our world. When asked to rank the 2nd and 3rd biggest challenges, respondents chose lack of clean drinking water and making education accessible to all, respectively.

Reducing the scope down from 50 years to the next 20 years, the challenges of climate change and solving the energy crisis get even stronger support as the most pressing issues of our world by the respondents (Figure 32).

The contributors were also asked what challenge they considered to be closest to solving. In this context, 26% of the respondents considered a solution to the energy

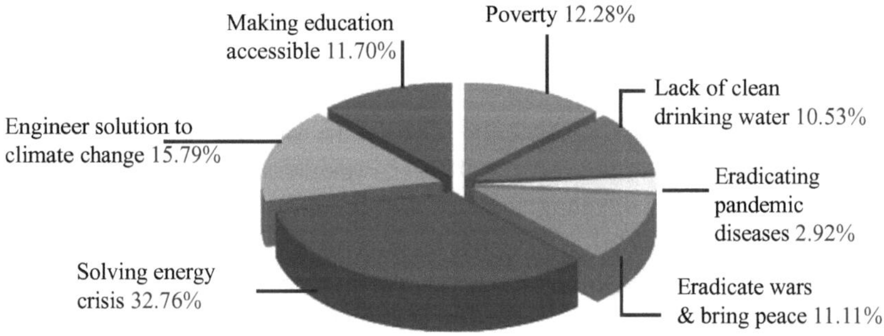

Fig. 31. *The distribution of biggest global challenges as derived from the contributions to the SGAC 50 years visions survey, Part 2.*

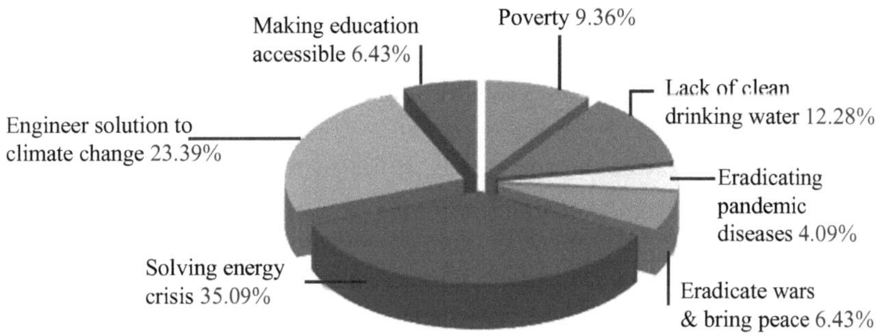

Fig. 32. *The most pressing global challenge in the next 20 years as derived by the contributions to the SGAC 50 years visions survey, Part 2.*

crisis most feasible, 22% considered us closest to a solution for providing clean drinking water (including sanitation and hygiene) and 19% felt that global access to education is achievable in the near term. This analysis clearly indicates a lack of important international decisions and activities towards the pressing challenge of climate change.

The important role of space technologies in the solution of these global challenges is confirmed by 82% of the respondents. Interestingly, support for space solutions for generation and provision of energy (45%) and for monitoring and controlling climate change (18%) was very strong, followed by helping make education accessible globally (15%) as shown in Figure 33.

Respondents indicated strong support for space solar power and access to space resources to solve the energy crisis. One key vision that emerged from responses was that the demonstration of space solar power as a means for providing electricity to remote and underdeveloped parts of the world could be

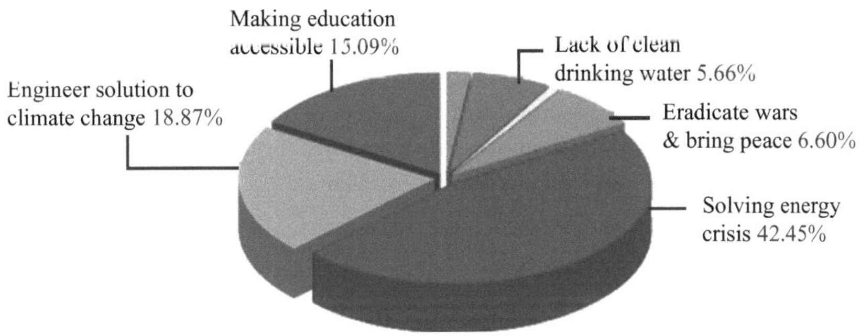

Fig. 33. *Responses onthe use and applicability of space technologies to solutions to the biggest global challenges as derived from the contributions to the SGAC 50 years visions survey, Part 2.*

demonstrated as early as 2015. Also, in response to what recommendations can be given to be worked on in the next five years, one area suggested was "Testing and designing technologies for mining Helium 3 on the Moon and for Orbital Microwave Solar Energy plants".

Significant portion of the survey was dedicated to a futuristic outlook on humans living in the outer space. The results of this part of the survey clearly illustrated that a large number of the respondents believes that eventually humans will be able to establish settlements on Moon, Mars and perhaps other planets or asteroids. These settlements were foreseen not only to be used for scientific research but allowing humans eventually emigrate to destinations outside of the planet Earth.

3.4.7 Biggest challenges: Further recommendations for the next 50 years

When asked to elaborate on how we can solve these global challenges, several contributors recognised how space technologies can play a major part in providing solutions and support. The key recommendations were grouped into three areas and are summarised below:

1. Human settlements on the Moon:

   - Establishment of a permanent Moonbase where humans could live and run experiments on continuous basis. Such Moonbase would eventually grow to a place where humans would not only work (scientists) but could live; thus, be able to sustain themselves and their families.
   - Creation of a shipyard on the moon, which would make moon not only a destination to travel to from the Earth but also a stop to refuel/recharge to go further in the outer space. This would be a significant milestone for humans not only in space exploration but in colonizing parts of the solar system.
   - Preparations for Mars: if humans reach the state of sustaining themselves on the Moon they could eventually aim for Mars.

2. Mars:

   - Once reaching Mars and sustaining themselves on the Moon humans would eventually colonize Mars. Mars would become home for number of humans. First children would be born, social and political infra-

structures would be created. Mars would no longer be a distant planet but a destination to visit.
- Significant scientific research would be conducted on Mars to sustain the humans living on the planet but also to lead to knew discoveries.

3. Private entrepreneurship:

- Development of private space companies and corporations, which would conduct business in the outerspace without government interference. Space tourism would be accessible to an average citizen of the Earth. Visiting Moon, Mars, or an asteroid would be available to masses.
- First hotel in space would be build. It would function similarly to the international Space Station – orbiting around the Earth. It would eventually be accessible to an average human.

4. Emigration to the outer space:

- Once permanent settlements are created on Moon and Mars humans would be free to emigrate there and become residents. The notion of citizenship would change forever. Humans would no longer be citizens of just countries but also planets.
- Human society would be revolutionized, as it would no longer be bound to the planet Earth. True interplanetary society would exist, which would have an impact on everything we know today: culture, religion, social standards, etc.
- First children would be born in space and true space generation would eventually exist. Those humans would grow up in an environment where interplanetary travel is possible and humans are living on various planets.

5. Space based solutions for global challenges:

- Development of technology to use ore and energy sources available beyond Earth to solve the energy crisis. An active research base into the different options available to us for spacecraft propulsion can be achieved by 2012, space solar power platforms by 2015. Research into this could have offshoots into more efficient energy production and management.
- Improved monitoring of Earth to capture climate change effects along with better monitoring of the Sun and understanding of planetary processes on Mars and Venus.

- Improvement of rapid response to disasters: Earth based/space based, especially for "developing" or "third world countries" which do not have indigenous tools.
- Improvement of communication (and IT) infrastructure for worldwide education, dissemination of information and creation of free distant-learning for everyone in remote, under privileged and under resourced regions.
- Enhancement of space research in order to find solutions applicable on earth, e.g. "space based vaccines in an isolated environment (a drug to cure AIDS)?"
- Development of a new generation of space propulsion and launch means that would give cheaper and easier access to space in order to harvest and to exploit resources available in space.

6. Increasing human security and quality of life:

- Increased use of space assets to determine and locate resources such as arable land in terms of agriculture in order to solve the problems of hunger.
- Bridging gender inequality by greater involvement and participation of women in all domains.
- An increased effort develop collaborative NEO tracking and mitigation initiatives.
- Definition of and commitment to aggressive goals for space exploration (by individual nations or the international community). Any spin-offs or benefits generated by pursuing these goals should be widely publicised.

7. Policy and international cooperation:

- Formalisation and acceptance of a "space code of conduct" and the drafting of an international declaration of human intent to establish a peaceful space faring civilisation.
- Establishment of a space code of conduct and a global system for space situational awareness with a jurisdiction comprising the Earth and Moon.
- Policy formulation to ensure safety and security in space, with a secondary objective to reduce and, if possible, make military industrial complexes less desirable.
- Establishment of an international Moon/Mars base with operations similar to the ISS.

- Continuation of the search for outer planets with life as this can have a profound effect to unify all of humanity.

It is noteworthy that amongst the contributed recommendations for the visions there was still strong support for space exploration and especially for lunar exploration. Several contributors commented that using space and planetary exploration as understanding survival strategies with limited resources was crucial to understanding how to manage the burden of human activities on planetary resources. Some outstanding visions also noted that "Space allows each of us to reexamine the Earth from a new vantage point that reveals its vulnerability and thus our tenuous grip with survival in the Universe". The respondents recognised that the (UN MDG) goals are not particularly space-related, but a worldwide space venture could help to unify public and political support for them. Furthermore, the "overview effect" of space activities helps realisation of the diminutive size and fragility of our home planet and encourages striving for actions and developments towards achievement of the goals.

One key vision chosen here summarises how suggestions made in the surveys considered important externalities, in this case growing population which was not listed as one of the challenges:

> "We must introduce a renewable energy source in order for world economies to grow, for the growing middle class in emerging markets to have increased wealth, and for people who have dreamed all their lives of owning a vehicle to finally realise that dream. This is why clean technology represents the biggest opportunity of an era and why long-term thinking will be a critical tool for those participating in this massive industrial transformation."

3.4.8 Conclusions and early efforts

The development and formulation of a vision that coherently expresses the suggestions of a global audience without national or institutional boundaries is an achievement in itself. In deliberations on the future for space exploration, the ideas of young people should be heard and considered in the consultation process, since these are the future leaders and major actors in space industry and government agencies. They will live through the long-term results of policy decisions made today and the recommendations combine the idealism and vision of young people with the realism gained from analysis of the historic achievements and the first professional steps within the space sector.

Whilst our study here presents contributions from the next generation of space professionals and leaders, they should not be mistaken as a prescription only for government agencies, but we truly believe that they can be treated as suggestions to any space actor. They give directions, but industry and agencies need to agree on solutions and analyse tradeoffs for their implementation in order to create sustainable space programmes with wholehearted public support. The framework of recommendations provided in the 50 years vision project can facilitate a space programme that will first and foremost enable the advancement of the human conditions of living and the associated actions towards long-term progress.

So far, in addition to the Boston University conference, the visions and recommendations have been successfully presented to the AIAA,[440] UN COPUOS as a general statement, to the then President of the Republic of India, Dr. A. P. J. Abdul Kalam,[441] to the Royal Aeronautical Society, to the international audience at the ISU Symposium 2008 and at the 59th International Astronautical Congress, within the framework of an industry future visions workshop[442] and to selected individuals within space agencies and industry [by December 2008].

Many of the issues addressed within the visions may seem to relate to a far future, however, considering these as contributions of a new generation and asserting key timescales to solving some of these challenges, the work on realisation and implementation of the 50 years vision starts now. Changing social and political conditions as well as cost and risk assessment will induce variations and changes to the target dates, also altering the perception and importance of single goals or recommendations. Hence, it is highly essential that international and intergovernmental organisations, in particular the UN, space agencies, industry and educational institutions, all help to spread true facts about the real impact of each and every global threat.

One observation found in comparing the results in the SGAC 50 years survey Part 1 as opposed to Part 2, is that knowledge regarding how space technology is applied for earth applications and how it affects day to day lives is not sufficiently communicated, particularly in contrast to exploration activities. A potential explanation could be the increased media efforts to ensure public and political support for these exploration missions, while Earth-related missions are much less visible in their implementation and the utilisation of their results often linked to the initial space programme in a limited way.

We recommend that the international space community takes these recommendations seriously and, through endorsement of these recommendations, encourages young students and professionals to work on the biggest global challenges while, in doing so, being reminded of the role of space technology and space derived information.

Acknowledgements

We would like to thank Kat Koderre, Tiffany Fierson, Alex Karl as well as all other SGAC members who have contributed to the 50 Year Visions project. We would also like to express our gratitude to Dr. Mark Lupisella, Dr. Ray Williamson, Mrs. Cynda Collins Arsenault of the Secure World Foundation and Lou Friedman of the Planetary Society for their continual support, guidance and encouragement of our youth contribution.

[431] Thakore Bejal, Chris Boshuizen, Tiffany Firestone, William Marshall, Robbie Schingler, and Heng Shi. "Vision for the Next 50 Years of Space Exploration on the Occasion of the 50th Anniversary of Space Exploration Recommendations from Students and Young Space Professionals." Report to UN COPUOS. Vienna, 2007.

[432] Thakore, Bejal. "Looking Back, Looking Forward and Aiming Higher: The Next Generation's Visions of the Next 50 years in Space." Presentation. Space Solutions to Earth's Global Challenges. ISU Symposium 2008, Strassbourg, France, Feb. 2008.

[433] Cf. the Space Generation Advisory Council. Website: http://www.spacegeneration.org/About.

[434] "About the Space Generation Advisory Council." Space Generation Advisory Council. 15 June 2009. http://www.spacegeneration.org/About.

[435] Current projects: Space Policy, Near Earth Objects, Disaster Management, YGNSS.

[436] SGAC. Website: www.spacegeneration.org.

[437] Ackerman, G. A. "It is hard to predict the future: the evolving nature of threats and vulnerabilities." Rev. sci. tech. Off. int. Epiz., 25.1 (2006): 353–360.

[438] For the progress on the MDGs. see: http://www.un.org/millenniumgoals/.

[439] UN 2000, United Nations General Assembly Resolution 2 session 55 United Nations Millenium Declaration on 8 September 2000, available online. http://www.un.org/millennium/declaration/ares552e.htm.

[440] (Kathleen Coderre, personal communication).

[441] (personal conversation).

[442] (Juergen Schlutz, personal representation).

3.5 Humans in outer space: Existential fulfilment or frustration? Existential, psychological, social and ethical issues for crew on a long-term space mission beyond Earth orbit

Berna van Baarsen

"Great minds have purposes,
little minds have wishes".
Washington Irving (1783–1859)

"*The end is close. On 14 July, at 14.00 Moscow time, we will land again on 'Earth'*".[443] After 105 days of confinement inside the isolation facility at the Russian Institute for Biomedical Problems (IBMP) in Moscow, the crew, 4 Russians and 2 Europeans, will see the Sun again.[444] This closed the pilot stage of the Mars520 mission simulation that was launched by the Russian IBMP institute, in cooperation with the European Space Agency (ESA), to increase understanding of the psychological and medical aspects of long-duration spaceflight.

In Europe, no less than five and a half thousand women and men volunteered to be part of this Mars500 study. What inspired them to apply? Why do scientists aspire to study the opportunities for going to Mars? What makes astronauts decide to leave the security of their houses, leave behind their beloved ones, and travel to the Moon? For as long as humans have existed they have been searching for challenges beyond known borders. Is it about knowledge and learning? Is it curiosity? Or do these challenges have their roots in more fundamental human structures such as purpose and meaning of life? In the present paper we discuss the importance of purpose and meaning of life in the context of Space exploration, in addition to what is already known from Space research literature and casuistry.

3.5.1 Purpose and meaning of life

Purpose and meaning of life are important concepts in theories on existential coherence.[445] Purpose in life aims at finding and evaluating the gaps in life, such as

Fig. 34. *The Challenge of a new point of view (source: Ning. 27 Jan. 2010. http://api.ning.com/files/Bjn*ipUFS1faYi9XIn72taUSWx5zRbVYzf2XXajOtf5*IeI6QcqpH5FhQYkqr1UCVzYs3Rk-Bi7iSs6wzkoG83YeTLkAk-4f/anewpointofview.jpg).*

finding a life partner or a challenging job or vocation, learning a new language, fighting a life-threatening illness, or adopting new values and motives.[446] Closing these gaps in order to achieve a coherent and meaningful life is at the core of our existence. It is about making our dreams come true; to meet our expectations of what life is about. It has to do with success and failure, with fulfilment and frustration, and with experiences of joy, enlightenment, love and hope, but also displeasure, confusion, regret, and despair. The meaning of life, or fulfilment, can be derived from experiencing "nature and culture" or "by experiencing another human being in his very uniqueness – by loving him".[447] Like the satisfaction and happiness that people can experience when they have shared something important with others, such as the experience of crew members in the Simulation of Flight on International Crew on Space Station (SFINCSS) isolation experiment who, during the "Closure Session", expressed how surprised and pleased they were to learn and share novelties about each other.[448]

3.5.1.1 Understanding or urge

The framework of existential coherence helps to explain how health and the ability to function, influences quality or meaning of life from an existential perspective. There are roughly two "pathways" to reach existential coherence in life.

The first pathway is *mental* in focus; we perceive the missing pieces in our lives, we try to understand them and, if possibly, try to act upon them. We perceive, we process, and we find meaning.[449] Individuals who try to mentally understand the

Fig. 35. *Apollo 15 Lunar Module Pilot James Irwin salutes the U.S. flag (source: Apollo Image Gallery. 27 Jan. 2010. http://www.apolloarchive.com/apollo_gallery.html).*

missing parts in their life may be called *knowledge seekers*. They come into action, or intend to do so, depending on their talents and other resources. The knowledge seeker may be the scientist who wants to improve his knowledge and expertise, who wants to meet his quota in high standing scientific, peer-reviewed publishing, or who wants to become a prominent leader in the scientific community.

In the second pathway, our perception of the gaps in our lives generates a *physical or emotional* urge to close them. The feeling of urge and allowing oneself to react on it brings about an impact on self, others and the world. Eventually it may even create new urges that lead to new actions and impacts, and so on. Individuals who act on the basis of such an urge may be called *thrill seekers*. They can be the bungee jumpers who want to go for the experience beyond the predictable and the safe. To move on an urge is said to be beautiful because it takes us right into the flow of life, not straining in the action, but simply enjoying the sensation of the moment and letting the resulting impact be part of the next moment's urge.[450]

3.5.1.2 Front-runners

Although we are all part of the grand web of life by participating in the exchange of information, knowledge, inspiration, emotions, love and support, and we all have our dreams and try to act out of Understanding or Urge, some of us seem to be particularly brave and outgoing and can have much more impact than others. These exceptional knowledge (understanding) and thrill seekers (urge) could be seen as *Front-runners* who are in search of great challenges, pushing personal and public frontiers, and establishing new private and worldly goals. Examples of exceptional knowledge seekers are Alexander von Humboldt (1769–1859), Albert Einstein (1879–1955) and Sergei Korolev (1907–1966) who devoted their lives to science in the pursuit of new knowledge. An example of an outstanding thrill seeker is Yuri Gagarin (1934–1968), the first human to travel in space. On April 12, 1961, some minutes before the launch, Gagarin described that for him to take part in a never encountered duel of nature was a challenge that went beyond the ordinary:

> "Dear friends, known and unknown to me, my dear compatriots and all people of the world! Within minutes from now, a mighty Soviet rocket will boost my ship into the vastness of outer space. What I want to tell you is this. My whole life is now before me as a single breathtaking moment. (. . .) I don't have to tell you what I felt when it was suggested that I should make this flight, the first in history. Was it joy? No, it was something more than that. (. . .) To be the first to enter the cosmos, to engage single-handed in an unprecedented duel with nature – could anyone dream of anything greater than that? (. . .) After all, in all times

> and epochs the greatest happiness for man has been to take part in new discoveries."[451]

After the flight, Gagarin kept being exposed to high occupational risks; among other things he re-qualified as a fighter pilot and while on a routine training flight he died at 35 years of age in a MiG-15UTI jet crash near Moscow.

3.5.2 Talents and constraints

3.5.2.1 Earth-related research

Our ability to function, i.e. to perceive, to think, to imagine, to experience, to reflect and to adapt, enables us to express our purpose in life and to use our talents. However, Understanding and Urge do not automatically lead to the achievement of our objectives. In our endeavours to reach our purpose, and to ultimately give our lives meaning, we have to deal with opportunities and restrictions. Studies on professional training, on academic and health care careers, on coping with illness and dying, have revealed various person specific and social-environmental factors. *Successful (coping) behaviour* has been related to (perceived) health, personality (e.g. high self-esteem, talent), emotions (e.g. happiness), social attitudes and skills necessary for the acquisition and development of social contacts and support (e.g. faith, low social anxiety), and personal standards (e.g. low expectations), as well as effective support systems, adequate information and communication, technology, and favourable job characteristics.[452]

Purpose and meaning of life has been connected to group membership, affect balance, self-confidence, happiness, sociability, self-esteem, mental health, responsibility and self-control, dedication to a cause, and clear purposes.[453]

An interesting example is the specialist work in a paediatric intensive care unit in a medical hospital.[454] For paediatric intensive care unit fellows their work gives meaning to their life because it is frequently life saving; their work makes a difference between life and death. However, intensive care fellows also face external stresses from their work environment, because of its complexity, the long working hours, the limited resources, the limits of science and technology, and the (relative) isolation from family and friends. They confront deeper inner feelings regarding their own high expectations, fallibility, anger, sense of loss, loneliness, limited control and the need to work under time pressure and closely with tense, grieving families. They may struggle in finding ready answers to existential questions such as: "Why do some children die while other children live?".

3.5.2.2 Space related research

In analogy to the example above, Antarctic and space research have shown that similar factors seem to operate in the extreme, isolated conditions inherent in long term space spaceflight. For instance, personal traits seem to become more explicit and accentuated during confinement.[455] Also, the importance of management of emotions and building of supportive and constructive relations between crewmembers[456] has been established. The presence of differences in cultural background and communication styles impacts evaluations of social support[457] and the functioning of the personal network.[458] This can be clearly seen in the different needs for and expectations of personal and confidential contact, expressed by crewmembers from different nationalities.[459] Palinkas, Johnson and Boster[460] reported a significant decrease in the extent to which crewmembers were satisfied with the support obtained both from family and friends as well as from other crewmembers. Other researchers found however that support from family and friends on Earth has a positive influence on mood and mission performance,[461] and is even more significant than potential support from crewmembers.[462] Further, tensions in interpersonal relations affect the ability to interact.[463] Blaming or scapegoating a less popular crewmember, which has occurred in past expeditions (1991, 1998), may lead to insomnia, depression and agitation (i.e. the "long-eye" phenomenon").[464] Mood problems (e.g. anxiety, depression, suspicion) and sleep disturbances have been reported during long-term missions.[465] Researchers have mentioned the importance of sensory deprivation and inputs such as the variety of food choices.[466] Also, the impact of bad news from home on personal and professional performance behaviour has been discussed.[467]

The following case is of relevance in this respect. During his journey to the ISS in October 2007, the Malaysian astronaut Sheikh Muszaphar Shukor (Amus) was not informed of the tragic news that his younger brother Ajil, with whom he was extremely close, had had an accident and slipped into a coma. Upon his return to Earth, he learned that his brother had died and had written a book about his brother astronaut. Amus completed the book with stories that were written by himself and his family, and dedicated it to his brother.[468]

The ethical dilemma in this and similar cases is of course the weighing of the pro's (e.g. right to be informed) and con's (e.g. concern, anxiety, mission performance and safety) in deciding whether or not crewmembers should be informed about bad news from home. Although delayed information may be good for performance and mission success, disclosure gives the opportunity to share emotions with family and/or crewmembers and hence to give bad news a place or meaning in life. In agreement with Kanas and Manzey,[469] it would be advisable to address these issues before the mission in the informed consent briefing, in order to

	Topics/Concepts			
	ADAPTATION TO EXTREME ENVIRONMENTS			
	WELL-BEING/QUALITY OF LIFE – Health (physical/mental) – Personality traits/Coping style/Talent – Emotions/Feelings – Perceptions/Expectations – Purpose in life/Meaning of life		TEAM-WORK AND SOCIAL INTERACTION – Compatibility – Multi-nationality – Language – Gender	
Tools (Selection/Preparation/Training)	To prevent	To sustain/increase	To prevent	To sustain/increase
Communication systems – Verbal – Non-verbal	Aggressiveness Defensiveness	Friendliness Inviting	Misunderstanding/Conflict Antipathy	Understanding/Cooperation Sympathy
Information systems	Ignorance	Awareness	Uncertainty	Confidence
Values/belief systems – Culture/Religion – Gender – Team	Prejudice Stereotyping Insincerity Irresponsibility	Beauty Equality Integrity Responsibility	Disrespect Shame Suspicion Blame	Respect Pride Trust Praise
Person (psychological-existential) systems – Attitude – Disposition – State of mind – Purpose in life – Meaning of life	Prejudice Indifference/Spiritlessness Instability Frustration Regret	Open-mindedness Understanding/Urge Steadiness Fulfilment Joyfulness	Intolerance Lack of interest Irritability Demotivation/Ordeal Boredom	Tolerance Curiosity Patience Dedication/Challenge Inspiration

Social support systems				
– Individual	Emotional loneliness	Belonging	Loneliness	Solitude/Privacy
– Family	Social loneliness	Support	Concern/Despair	Hope
– Team	Discomfort	Comfort	Isolation (scapegoating)	Integration
Technology and Life support systems	Errors/Distraction	Accuracy/Alertness	Inefficiency	Efficiency
Professional expertise				
– Skills	Incompetence	Competence	Loss of control/Doubt	Control/Confidence
– Decision making	Indecisiveness	Deliberation	Bias (groupthink)	Group reflection
– Emotional intelligence	Unresponsiveness/ Immaturity	Empathy/Wisdom	Group stagnation	Group development
– Leadership	Rigidity/Ambiguity	Flexibility/Determination	Discouragement	Encouragement
– Fellowship	Selfishness/Suspicion	Humility/Confidence building	Opposition	Involvement
Habitat/Spacecraft	Lack of privacy	Privacy	Discomfort	Comfort
Job characteristics (monotony/workload)	Out of control	Collected	Heteronomy	Autonomy
Sensory inputs				
– Food	Tastelessness/Monotony	Appeal/Variety	Separateness	Collective
– Light/Temperature/Noise	Intrusiveness/ Overwhelming	Non-disturbance	Lack of balance	Balance
– Games	Boredom	Relaxation	Withdrawal/Passivity	Engagement/Activity
– Earth signals-vision/Time delay	Existential loneliness	Embeddedness	Abandonment	Interaction/Communication

Fig. 36. *Overview of psychological, existential, social and environmental factors, their effects on individual and team well-being, performance and interaction, and tools for selection, preparation and training crew members for long-term spaceflight.*

give ground-personnel the necessary background data to act in agreement with the preferences of the crewmember and/or her/his family.

3.5.3 Adaptation to extreme environments: fulfilment or frustration

The extent to which personal and environmental resources are available determines success and failure. Success or *fulfilment* (i.e. meaning of life, challenge) and failure or *frustration* (i.e. anger, sense of loss, fallibility, blame, limited control and boredom) are two sides of the same coin. People who have found meaning or purpose in their lives experience existential fulfilment. They see the demands of life as challenges instead of barriers. On the other hand, those who have high expectations, who feel they have not been able to give their lives any meaning or who feel empty, lonely or isolated from others, may develop a sense of existential vacuum or frustration,[470] that ultimately leads to depression, self-derogation, and risky behaviour.[471] Excessive knowledge and thrill seekers may be particularly prone to existential frustration or to boredom if they are not talented or resilient enough, or if their environment cannot or does not support them in their search for new understanding, a new urge or impact. When job characteristics are extremely stressful, such as in situations of life and death, or when small teams have to work in highly autonomous conditions where team members are fully dependent on each other in managing technology, tasks, communication (delays), group interaction, support, and daily (inter)personal and occupational hazards, successful adaptation and goal achievement can be very difficult. For instance, in the Mars520 simulation mission, the long-lasting isolation and the time delay between the "spacecraft" and "Earth" may increase feelings of loneliness and abandonment that may subsequently have a detrimental effect on cognitive performance.[472] Especially when questions of purpose and meaning of life arise, matters of challenge, dedication, and inspiration on the one hand, and torment, ordeal, de-motivation, and boredom on the other, may become crucial.

3.5.4 Interdisciplinary framework

Asking the question "What is it exactly that inspires people to act?" is also asking the question "What motivates them to continue?" Although it is tempting to explain this in terms of Understanding and Urge, these are not the only instruments determining what encourages people to keep going. As has been noted, in reaching fulfilment and in preventing frustration, various individual, team, job, and environmental/habitat factors come into play. Figure 36 gives an overview of individual and team characteristics that, together with suitable preparation and training protocols, may be effective in preventing frustration and sustaining or increasing fulfilment.[473] Figure 36 has been specifically developed for small and highly autonomously working teams in confined, isolated conditions typical of long-term spaceflights. Figure 36 is not inclusive and gives no empirical data on the interrelationships between the proposed factors, tools and outcomes that may over time serve as causal factors themselves. Figure 36 highlights the need that success or fulfilment of a mission/purpose should be interpreted within an interdisciplinary psychological-existential-social-environmental framework whereby outcomes should be evaluated on both individual and team levels.

3.5.4.1 Selection

Space agencies will greatly benefit from the insights from psychological and existential studies on the effects of extreme and isolated working and living conditions on Earth. The acknowledgement of meaning of life as a vital human aspect is significant for the selection, training, and coaching of future space travellers.

First, selection procedures might gain in efficiency when the two U's: Understanding and Urge, are added as personality variables. From the perspective of long-term space missions beyond low Earth orbit, front-runners are of special importance because through their talents, their imagination, their perseverance and their intellectual and courageous contributions, they are able to impact not only themselves, but also society as a whole. However, if front-runners are our future space travellers, we need to decide whether the thrill seeker or the knowledge seeker has more of the required qualities and strengths. Which one would be the most responsible, the most up to his or her task? The ultimate goal for the thrill seeker is experiencing and controlling the risks and dangerousness of the chosen adventure and becoming more skilled at it. For the knowledge seeker, on the other hand, the challenge is the creation of new understanding. From examples from the past, we may infer that Understanding and Urge both matter. Christoffel Columbus (1451–1506) was eager to find a shorter course of navigation but he was also a scientist in that he systematically recorded his every day experiences in his logbook. And while Gagarin seemed to enjoy the thrill of being sent on risky but challenging flights, he also contributed to new knowledge and insights by reporting what he saw during his flight:

Fig. 37. *Columbus taking possession of the new country (source: "Christopher Columbus." Pictorial Americana – Selected Images from the Collection of the Library of Congress. 27 Jan. 2010. http://www.loc.gov/rr/print/list/picamer/paColumbus.html).*

> "The first thing that struck me [. . .] was that you can see everything in detail from space. I saw rivers, lakes, coastal lines, islands, forest tracts – all nice and clear. And striking was the sky, black with stars, with a very bright sun. [. . .] The Earth's halo was startling too – a blue, very thin hazy band by the horizon as if covered in veil. The gigantic ball of the Earth is all over, if you look at it through the porthole. Your eyes are just not enough to embrace the whole of it."[474]

Particularly in extremely confined, isolated and risky space enterprises where the patience, perseverance, and wisdom of astronauts and cosmonauts are frequently put to the test, it is interesting to study which of Urge or Understanding best describe the characteristics of successful front-runners. Moreover, considering the increased weight of personality traits and crew interaction during confinement, crew members should be selected as a team.[475] Individual dispositions should be attuned to one another in a way that reasoning, deliberation, and an open, critical attitude are encouraged. The selection of teams that are open-minded to discussion and reflection is significant for long-term space flights because they may be less prone to groupthink.[476]

3.5.4.2 Training and coaching

Second, training procedures will gain in effectiveness when, next to technical information and job-related factors, relevant psychological, existential and social (family-team) aspects are taken into account. It is important to acknowledge the vulnerable side of people and to be attentive to the possible harmful effects of existential distress that may express itself through statements or questions such as: "Why am I in space and not on Earth?", "What is the meaning of being so far from home?". Coping with unknown, perhaps even dangerous, situations is a unique personal affair and should be adequately prepared for. Individual crewmember support should be provided early on in the training period.[477] A person-existential-social approach in training and coaching crewmembers may create open-mindedness and lower the risk of intolerance and irritability among the crew. Also, personal dedication and team motivation (e.g. joint goals) must be safeguarded because they influence intentions and behaviour, and are basic to Understanding and Urge.

Fig. 38. *David Scott Station 6 Oanframe (Apollo 15) (source: Apollo Image Gallery. 27 Jan. 2010. http://www.apolloarchive.com/apollo_gallery.html).*

3.5.4.3 Counter measures

Finally, meaning and purpose in life are helpful concepts in developing and testing adequate counter measures against frustration and boredom. A study among college students[478] showed for instance that students who spend more time exercising, studying, spending time with friends and talking with professors outside of class, experience a higher purpose in life than students who spend more time watching TV and playing video games. Thus, in this study, social and campus activities seem to be more related to personal fulfilment than the more isolating activities. Findings like these elicit new questions about the value of computer games or other passive entertainment products for astronauts. Moreover, they stress the necessity of comparing different countermeasures and determining which ones are most efficient in preventing boredom, distraction, and loneliness.

3.5.5 Discussion and conclusions

Humans in outer space are no dream but reality. Humans, as long as they exist, will walk on the Earth; they will explore the Moon and will travel beyond the Moon's orbit. It is not a question if, but when, the first humans will be sent on an interplanetary flight.[479] In exploring space humans "recognize the need for research on and development of efficient sustainable food, water, and air-production, recycling and safety systems".[480] But for successful space exploration and survival, the importance of the human psychological and existential factor should be acknowledged as a priority. Existential fulfilment or frustration as outcomes of meaning and purpose of life are key elements of human space flight not only for the astronauts involved but also in a more philosophical sense for humankind as a whole. Examples from life on Earth and in space show that an individual's Understanding or Urge in giving meaning to life is of vital importance in our preservation of life. It is true: without food and water people die. But without meaning or purpose in life there is no hope and ambition, and without hope and ambition there is no life either.[481]

Although space activities are often described in terms such as "challenge", "excitement", "adventure", "destiny", "curiosity", "expansion", "exploration", and "spirit",[482] these concepts have not been captured in a theoretical psychological-existential framework. The absence of such a framework implies that today it is difficult to make quantitative and qualitative predictions of human factors on motivation, performance and adaptive behaviour, in spite of the fact that they could be the determining factors for mission failure or success. Although much effort has been put in predicting and explaining successful human performance under extreme conditions, we need more psychological research in space. An overview of both person and environment related resources and their effects on individual and team well-being and performance, would be a helpful tool in finding answers to fundamental questions that deal for instance with existing ("What do we know?") and future knowledge ("How can we imagine the unknown?"). Moreover, space agencies would greatly benefit from the insights from psychological and existential studies on the effects of extreme and isolated working and living conditions on quality and meaning of life.

A multidisciplinary review of available research on meaning and purpose of life and the relevance of these studies for humans in outer space would be very beneficial. Historical research on written log and personal stories of isolated and lonely voyages of discovery may give us valuable information that, interpreted within our current theoretical knowledge, will be of use in future exploratory expeditions. With the outlook to guide future psychological research on the existential and interpersonal aspects of space flight missions, we may achieve new insights into the relationship between humans, between humans and machines, and also between humans and maybe even other forms of life in outer space.

Thanks to individual front-runners who seek the borders of their own knowledge, capacity and courage in order to find meaning in their lives, we are able "to follow the thrust of *joint* scientific and cultural curiosity" and "define European *psychological, existential and ethical* values and priorities" [emphasis added].[483] In understanding the human factor in space travelling and technology, psychology is indispensable. We need new paradigms in which psychological and existential values are integrated with technology, health and safety regulations, habitat requirements and designs, physiology, medical-ethical issues, and interpersonal and communication factors.

Human space endeavours are deeply rooted expressions of human existence and meaning of life. The two U's, Understanding and Urge, conceptualise the characteristics of the women and men, who will represent humankind on this journey.

---

[443] "Mars500 Diary: The End is Close." 14 July 2009. European Space Agency. 15 Jan. 2010. http://www.esa.int/esaMI/Mars500/SEMGW57CTWF_0.html.

[444] "Media Opportunity: Crew Completes 105-day Simulated Mars Mission 14 July." 19 June 2009. European Space Agency. 15 Jan. 2010. http://www.esa.int/esaHS/SEMJTC0P0WF_index_0.html.

[445] Ventegodt, Soren, Trine Flensborg-Madsen, Niels-Jorgen Andersen, and Joav Merrick. "The Life Mission Theory VII. Theory of Existential (Antonovsky) Coherence: A Theory of Quality of Life, Health, and Ability for Use in Holistic Medicine." The Scientific World Journal 5 (2005): 377–389.

[446] see for example Ventegodt, Soren, Trine Flensborg-Madsen, Niels-Jorgen Andersen, and Joav Merrick. op. cit.; Frankl, Viktor E. Man's Search for Meaning. New York: Washington Square Press/ Pocket Books, 1984.
[447] Frankl, Viktor E. op. cit. 115.
[448] Kass, Rachel and James Kass. "Team-Work during Long-Term Isolation: Sfincss Experiment GP-006." Simulation of Extended Isolation: Advances and Problems. Ed. Victor M. Baranov. Moscow: Firm SLOVO, 2001: 124–47. 142.
[449] Frankl, Viktor E. op. cit.
[450] Ventegodt, Soren, Trine Flensborg-Madsen, Niels-Jorgen Andersen, and Joav Merrick. op. cit 379.
[451] Available from http://johnkemeny.com/blog/?p=431, 130809.
[452] see for example Tazelaar, Fritz, and Reinhard Wippler. "Problemspezifische Anwendungen der allgemeinen Theorie mentaler Inkongruenzen in der empirischen Sozialforschung." Soziale Bedingungen – individuelles Handeln – soziale Konsequenzen. Eds. Günther Büschges and Werner Raub. Frankfurt am Main: Lang, 1985: 117–79; Weiss, Robert S. "Loss and Recovery." Handbook of Bereavement: Theory, Research and Intervention. Eds. Margaret S. Stroebe, Wolfgang Stroebe, and Robert O. Hansson. Cambridge: University Press, 1993: 271–84; Dykstra, Pearly A. and Jenny De Jong Gierveld. "The Theory of Mental Incongruity, with a specific Application to Loneliness among Widowed Men and Women." Theoretical Frameworks for Personal Relationships. Eds. Ralph Erber, and Robing Gilmour. NJ: Lawrence-Erlbaum, 1994: 235–59; Dykstra, Pearl A. "Loneliness Among the Never and Formerly Married: The Importance of Supportive Friendships and a Desire for Independence." Journal of Gerontology: Social Sciences 50B (1995): 321–9; Thoits, P. A. "Stress, Coping and Social Support Processes: Where are we? What next?" Journal of Health and Social Behavior (Special Issue) (1995): 53–79; Lazarus, Richard S. "The Role of Coping in the Emotions and how Coping Changes over the Life Course." Handbook of Emotion, Adult Development, and Aging. Eds. Carol Magai, and Susan H. McFadden. San Diego: Academic Press, 1996: 289–306; De Jong Gierveld, Jenny. "A Review of Loneliness: Concept and Definitions, Determinants and Consequences." Reviews in Clinical Gerontology 8 (1998): 73–80; van Baarsen, Berna. "Theories on Coping with Loss: The Impact of Social Support and Self-Esteem on Adjustment to Emotional and Social Loneliness following a Partner's Death in Later Life." Journal of Gerontology: Social Sciences 57B (2002): 33–42; van Baarsen, Berna. "Suffering, Loneliness and the Euthanasia Choice. An Explorative Study." Journal of Social Work in End-of-Life & Palliative Care 4 (2008): 189–213.
[453] see for example Yarnel, T.D. "Purpose-in-life test: Further correlates." Journal of Individual Psychology 27.1 (1971): 76–9; Pearson, Paul R. and Brian F. Sheffield. "Purpose in Life and the Eysenck Personality Inventory." Journal of Clinical Psychology 30.4 (1974): 562–4; Simmons, Dale D. "Purpose in Life and the Three Aspects of Valuing." Journal of Clinical Psychology 36.4 (1980): 921–2; Lazarus, Richard S. and Anita Delongis. "Psychological Stress and Coping in Aging." American Psychologist 38 (1983): 245–54; Harlow, Lisa L., Michael D. Newcomb, and P. M. Bentler. "Depression, Self-Degradation, Substance Use and Suicide Ideation: Lack of Purpose in Life as a Mediational Factor." Journal of Clinical Psychology 42.1 (1986): 5–22. 21; Zika, Sheryl and Kerry Chamberlain. "On the Relation between Meaning in Life and Psychological Well-Being." British Journal of Psychology 83 (1992): 133–45.
[454] Jellinek, M. S., Todres, I. D., Catlin, E. A., Cassem, E. H., and A. Salzman. "Pediatric Intensive Care Training: Confronting the Dark Side." Critical Care Medicine 22.1 (1994): 179–80.
[455] Kraft, Norbert, Natsuhiko Inoue, Koh Mizuno, Hiroshi Oshima, and Chiharu Sekiguchi "Psychological testing/trait and group dynamics during 110-day confinement in an isolated environment." Simulation of Extended Isolation: Advances and Problems. Ed. Victor M. Baranov. Moscow: Firm SLOVO, 2001. 191–202.
[456] Kass, Rachel and James Kass. op. cit.
[457] Palinkas, Lawrence A., Jeffrey C. Johnson, James S. Boster, Rakusa-Suszczewski, Valeri P. Klopov, V. P., Xue Quan Fu, and Usha Sachdeva. "Cross-Cultural Differences in Psychosocial Adaptation to Isolated and Confined Environments." Aviation, Space, and Environmental Medicine 75 (2004): 973–80.

[458] Kanas, Nick, Vyacheslav Salnitskiy, Ellen M. Grund, Vadim Gushin, Daniel S. Weiss, Olga Kozerenko, Alexander Sled, and Charles R. Marmar. "Social and Cultural Issues during Shuttle/MIR Space Missions." Acta Astronautica 47 (2000): 647–55.
[459] Gushin, Vadim I., N. S. Zaprisa, J. M. Pustinnikova, T. M. Smirnova, and I. I. Popova. "Characteristics of Russian and Non-Russian Crewmembers' Communication with External Parties under Prolonged Isolation." Simulation of Extended Isolation: Advances and Problems. Ed. Victor M. Baranov. Moscow: Firm SLOVO, 2001: 85–100.
[460] Palinkas, Lawrence A., Jeffrey C. Johnson, and James S. Boster. "Social Support and Depressed Mood in Isolated and Confined Environments." Acta Astronautica 54 (2004): 639–47.
[461] Kanas, Nick, and Dietrich Manzey. Space Psychology and Psychiatry. California: Microcosm Press; London: Kluwer Academic Publishers, 2003.
[462] Palinkas, Lawrence A., Eric K. E. Gunderson, Albert W. Holland, Christopher Miller, and Jeffrey C. Johnson, J. C. "Predictors of Behavior and Performance in Extreme Environments: The Antarctic Space Analogue program." Aviation, Space, and Environmental Medicine 71 (2000): 619–25.
[463] Cazes, Geneviève, Elisabeth Rosnet, Claude Bachelard, Christine Le Scanff, and Jean Rivolier, "Group Dynamics during the EXEMSI Isolation Study." Advances in Space Biology and Medicine 5. Ed. Sjoerd L. Bonting. Greenwich: JAI Press, 1996: 245–62.
[464] Kanas, Nick and Dietrich Manzey. op. cit. 83.
[465] Palinkas, Lawrence A., Jeffrey C. Johnson, and James S. Boster. op. cit.; Kanas, N. "Psychiatric Issues Affecting Long Duration Space Missions." Aviation, Space, and Environmental Medicine 69.12 (1998): 1211–6; Mizuno, K., Inoue, N., Kraft, N., Oshima, H., and C. Sekiguchi. "Relationship between Aerobic Work Capacity and Mood during Sfincss-99 Project." Simulation of Extended Isolation: Advances and Problems. Ed. V. M. Baranov. Moscow: Firm SLOVO, 2001: 370–87.
[466] see for example Kanas, Nick, and Dietrich Manzey. op. cit.; Jorgensen, Jesper. "Sensory Deprivation – a Challenge for Space Architects." SAE Technical Paper Series 2005–01–2912. SAE International, 2005.
[467] see for example Kanas, Nick and Dietrich Manzey. op. cit.; Oberg, James E. Red Star in Orbit. New York: Random House, 1981; Kelly, A. D., and Kanas, N. "Communication between Space Crews and Ground Personnel: A Survey of Astronauts and Cosmonauts." Aviation, Space, and Environmental Medicine 64 (1993): 795–800.
[468] Shukor Al-Masrie and Sheikh Mustapha. Reaching for the Stars. Malaysia: MPH Publishing, 2008.
[469] Kanas, Nick and Dietrich Manzey. op. cit. 159.
[470] Frankl, Viktor E. The Doctor and the Soul. New York: Vintage, 1959.
[471] see for example 38, 39, Molasso, William R. "Exploring Frankl's Purpose in Life with College Students." Journal of College & Character VII.1 (2006): 1–10. 7.
[472] van Baarsen Berna and F. Fabio Ferlazzo. "The Effects of Group Dynamics and Loneliness on Cognitive and Emotional Adaptation to Extreme Confined Environments." Presentation. Investigator Working Group meeting in the context of the IBMP/ESA Mars500 Programme. Cologne, Germany. 28–29 Apr. 2008; van Baarsen, Berna, Fabio Ferlazzo, Denise Ferravante, Francesco Di Nocera, Jesper Jörgensen, Jan H. Smit, Marijtje A. J. van Duijn, Anna-Maria Giannini, André Kuipers, and Joop van der Pligt. "Digging into space psychology and isolation: The Mars520 LODGEAD study. Preliminary results of the Mars105 pilot study." The 60th International Astronautical Congress, Daejeon, Republic of Korea, 12–16 Oct. 2009.
[473] van Baarsen, Berna. "Further Elaboration of the Scheme developed by the ESA Research Topical Team – Weiss K., Kanas, N., Schevchenko, O., Ferlazzo F., van Baarsen B., Greenberg, N., Whiteley I., and G. G. De la Torre – Psychosocial and Neurobehavioural Aspects of Human Space Flight." Topical Team Meeting. London, United Kingdom, 9 Apr. 2009.
[474] "Yuri Gagarin in the Voice of Russia Radio Archives." The Voice of Russia. 15 Jan. 2010. http://www.vor.ru/Space_now/First_in_space/Space_next107_eng.html. 130809.
[475] Kraft, N., Inoue, N., Mizuno, K., Oshima, H., and C. Sekiguchi. op. cit.

[476] Sandal, Gro M. "Psychosocial Issues in Space: Future Challenges." Gravitational and Space Biology Bulletin 14.2 (2001): 47–54.
[477] Baranov, Victor, Evgeny Demin, Vadim Gushin, Mark Belakovsky, Chiharu Sekigushi, Natsuhiko Inoue, Koh Mizuno, Michel Vachon, and Leena Tomi. "SFINCSS-99 (Simulation of Flight on International Crew on Space Station): Experience and lessons learned." Simulation of Extended Isolation: Advances and Problems. Ed. Victor M. Baranov. Moscow: Firm SLOVO, 2001: 524–30.
[478] Molasso, William R. op. cit.
[479] Manzey, Dietrich. "Human Missions to Mars: New Psychological Challenges and Research Issues." Acta Astronautica 55 (2004): 781–90.
[480] Lukaszczyk, Agnieszka. "Aiming Ahead: Next Generation Visions for the Next 50 Years in Space. Humans in Outer Space – Interdisciplinary Odysseys. Eds. Luca Codignola, and Kai-Uwe Schrogl. Vienna: SpringerWienNewYork, 2009: 44–53. 46.
[481] Boss, Pauline. A Review of Loss, Trauma, and Resilience: Therapeutic Work with Ambiguous Loss. New York: Norton, 2006.
[482] see for example "Mars500 diary: The end is close." 14 July 2009. European Space Agency. 15 Jan. 2010. http://www.esa.int/esaMI/Mars500/SEMGW57CTWF_0.html; Codignola, Luca and Kai-Uwe Schrogl. Eds. "The Vienna Vision on Humans in Outer Space." Humans in Outer Space – Interdisciplinary Odysseys. Vienna: SpringerWienNewYork, 2009: 227–33; Landfester, Ulrike. "Missing the Impossible: How We Talk and Write About Space." Humans in Outer Space – Interdisciplinary Odysseys. Eds. Luca Codignola, and Kai-Uwe Schrogl. Vienna: SpringerWien NewYork, 2009: 94–106; Bohlmann, Ulrike M. "The Need of a Legal Framework for Space Exploration." Humans in Outer Space – Interdisciplinary Odysseys. Eds. Luca Codignola, and Kai-Uwe Schrogl. Vienna: SpringerWienNewYork, 2009: 182–95.
[483] Codignola, Luca and Kai-Uwe Schrogl (eds.). "The Vienna Vision on Humans in Outer Space." Humans in Outer Space – Interdisciplinary Odysseys. Vienna: SpringerWienNewYork, 2009: 227–33. 231.

CHAPTER 4

ANNEX

4.1 Useful web-addresses related to human exploration

European Centre for Space Law (ECSL)	http://www.esa.int/SPECIALS/ECSL/index.html
ESA's research as part of human spaceflight and exploration	http://www.esa.int/esaHS/research.html
The Erasmus Centre of ESA's Directorate of Human Spaceflight	http://www.esa.int/SPECIALS/HSF_Research/index.html
ESA's ISS Tracker	http://wsn.spaceflight.esa.int/?pg=iss
The Webstreaming Network, the Multimedia Streaming Portal of ESA's Directorate of Human Spaceflight	http://wsn.spaceflight.esa.int/
Exploring Mars: ESA's Mars Express	http://www.esa.int/SPECIALS/Mars_Express/index.html
ISS Legal Framework Overview	http://www.esa.int/esaHS/ESAH7O0VMOC_iss_2.html
Mars 500 Information Kit	http://esamultimedia.esa.int/docs/Mars500/ESA_Mars500InfoKit.pdf
The SETI Institute	www.seti.org
Music Space	http://www.hobbyspace.com/Music/ Music Space celebrates music that is written for space exploration. Or is about space exploration. Or is played in space. Or makes you feel like you are in space. Or is actually from space.

4.2 The Vienna Vision on Humans in Outer Space

Vienna Vision on Humans in Outer Space

Leonardo da Vinci's man on Mars. Courtesy J.-C. Worms. Image credits: 1. Human figure in a circle, L. da Vinci, ca. 1485-90, www.davincisketches.com 2. Image of Mars, ESA © 2007 MPS for the OSIRIS team

www.esf.org

www.espi.or.at

Humans in Outer Space: Interdisciplinary Odysseys

The European Science Foundation (ESF) has organised the first comprehensive trans-disciplinary dialogue on humans in outer space. This dialogue goes further than regarding humans as better-than-robot tools for exploration. It investigates the human quest for odysseys beyond Earth's atmosphere and reflects on the implications of the findings of extraterrestrial life.

The inherent human curiosity for exploring the unknown is at the heart of this dialogue, and is addressed through collaboration between the ESF Standing Committee for the Humanities (SCH) and the ESF European Space Sciences Committee (ESSC), in cooperation with the European Space Agency (ESA) and the European Space Policy Institute (ESPI) in Vienna. Recently the "Athens declaration" enabled by the ESSC established a scientific framework for defining Europe's exploration programme.

The Vienna Vision on Humans in Outer Space was developed at the "Humans in Outer Space" conference, held in Vienna on **11 – 12** October 2007 locally organised by ESPI with the support of the Austrian Ministry for Transport, Innovation and Technology (BMVIT). This vision provides a European perspective in identifying the relevant needs and interests linked with space exploration. It is presented to several European and international fora, in order to make it a useful element for the position-finding and decision-making process.

European Science Foundation — ESPI — esa — bmvit

Humans in Outer Space: The way forward for the next 50 years

Space age has reached its 50th anniversary. Development of robotic exploration to distant planets and bodies across the solar system, as well as pioneering human space exploration in Earth orbit and the Moon, paved the way for ambitious long-term space exploration. Europe has always played a significant role in the endeavours of humankind to explore other worlds and to understand the Universe in which we live.

Today, space exploration goes far beyond a merely technological endeavour, as its further development will have a tremendous social, cultural and economic impact. Space activities are now entering an era where the contribution of the humanities - history, philosophy, anthropology, the arts as well as the social sciences - political science, economics and law - will become crucial for the future of space exploration. Now that the awareness for the societal complexity of activities in space is growing internationally, it is vital that Europe, with a stronghold in natural sciences as well as its identity firmly rooted in the humanities and the social sciences, grasps the opportunity to involve their specific knowledge(s) in the long-term planning of exploration undertakings.

The next generation may be given the opportunity to explore new places and discover new worlds. Those adventures will be driven by the human desire of quest for knowledge and human curiosity. They will provide a main opportunity for equitable international cooperation. Humans divided on Earth will hopefully unite in space as citizens of one planet.

First Odyssey

Humans in Earth orbit:
What effect does it have?

Home – Earth is a fragile oasis in the vastness of the solar system and it needs to be protected from natural and man-made threats. Once in space, humans are no longer just citizens of individual countries, but also of the planet Earth caring for its overall global sustainability.

Progress – Human space flight is a major source of innovation. It can benefit societies around the world with a variety of technological spin-offs and scientific research possibilities; it has even been argued that it could help to overcome the limits for growth on several levels. Through endeavours such as the International Space Station new partnerships are built, which can cultivate international cooperation in a spirit of friendship and mutual understanding.

Technology – Humans increasingly rely on technological advancements in their everyday lives. Relationship between humans and machines will reach new dimensions, and in the process may make it necessary to readjust our notion of 'humanity'. Space applications can have a positive impact on the quality of life on Earth and eventually beyond. Through television and internet everyone can virtually experience space flight or the vistas of planetary surfaces. In the near future space tourism may no longer be a dream but become a possibility to those interested.

Law – The legal framework for space activities needs to be further developed in a way that cultivates peaceful uses of outer space and equal rights for all humankind. Human rights will also have to be considered, as new moral challenges will face humanity.

Second Odyssey

Humans in space exploration:
What effects will it have?

Humanity – In the new era of technological advancements, the human factor is essential. Without human presence in space, spaceflight and exploration will lack an important dimension. Global cooperative endeavours will allow fostering the further development of collaboration among peoples, societies and cultures.

Discovery – Space exploration allows for discovery in two ways: It makes it possible to search for specific things, i.e. new energy resources, but it also opens up the opportunity to follow the thrust of scientific and cultural curiosity. The latter is one of the most inspiring traits of humankind since the beginning of its history and it should lead again to incredible discoveries.

Culture – Space exploration is a challenging, cooperative endeavour that offers opportunities to further strengthen European ties and define European values and priorities. The identity of Europe is constituted by its specific cultural approach towards both scientific and moral issues, and it will be this angle which will influence societal development as well as serve as inspiration for the younger generations.

Rights – Through space exploration, new partnerships will form. This will call for a proper legal framework serving to peacefully regulate issues such as space traffic management. Furthermore planetary protection needs to be elaborated with international partners concerning forward and backward contamination, and Europe must play an influential role in that context.

Third Odyssey

Humans migrating the Earth:
How will it affect human thought?

Habitat – Driven by curiosity and in order to extend opportunities, humans may eventually search for settlements outside our planet. What is unimaginable today may become necessary in the future. The first child to be born in space will mark the dawn of a true space generation.

Encounters – Humans should be open to the idea of possible encounters with other forms of life in outer space, either through the discovery of life in the solar system (extinct or extant), or through the reception of extraterrestrial radio signals. A new era will begin should humans realise that they are not alone in the universe. Such discovery may likely cause the development of a new collective identity for humanity.

Belief system – What people believe in, and how such beliefs are structured, has a strong binding force on societies, on Earth and eventually beyond. Human belief systems, whether religious or secular, change in the context of new living environments, and in contact with other forms of life and societies. As the merely technological or political approach will no longer be sufficient in dealing with such contacts, the humanities and the social sciences will gain in importance.

Adapting – Past encounters that took place on Earth show that human beings did eventually adapt to unforeseeable realities, although often at very great costs. While the first effects of an encounter between humans and extraterrestrial life are unpredictable, humans need to be aware that they will be held morally, economically and politically accountable for their choices.

4.3 Summary Report of the Review of U.S. Human Space Flight Plans Committee

The U.S. human spaceflight programme appears to be on an unsustainable trajectory. It is perpetuating the perilous practice of pursuing goals that do not match allocated resources. Space operations are among the most complex and unforgiving pursuits ever undertaken by humans. It really *is* rocket science. Space operations become all the more difficult when means do not match aspirations. Such is the case today.

The nation is facing important decisions on the future of human spaceflight. Will we leave the close proximity of low-Earth orbit, where astronauts have circled since 1972, and explore the solar system, charting a path for the eventual expansion of human civilization into space? If so, how will we ensure that our exploration delivers the greatest benefit to the nation? Can we explore with reasonable assurances of human safety? And, can the nation marshal the resources to embark on the mission? Whatever space programme is ultimately selected, it must be matched with the resources needed for its execution. How can we marshal the necessary resources? There are actually more options available today than in 1961 when President Kennedy challenged NASA and the nation to "land a man on the Moon by the end of the decade."

First, space exploration has become a global enterprise. Many nations have aspirations in space, and the combined annual budgets of their space programmes are comparable to NASA's. If the United States is willing to lead a global programme of exploration, sharing both the burden and benefit of space exploration in a meaningful way, significant benefits could follow. Actively engaging international partners in a manner adapted to today's multi-polar world could strengthen geopolitical relationships, leverage global resources, and enhance the exploration enterprise. Second, there is now a burgeoning commercial space industry. If we craft the space architecture to provide opportunities to this industry, there is the potential – not without risk – that the costs to the government would be reduced. Finally, we are also more experienced than in 1961, and able to build on that experience as we design an exploration programme. If, after designing cleverly, building alliances with partners, and engaging commercial providers, the nation cannot afford to fund the effort to pursue the goals it would like to embrace, it should accept the disappointment of setting lesser goals.

Can we explore with reasonable assurances of human safety? Human space travel has many benefits, but it is an inherently dangerous endeavor. Human safety can never be absolutely assured, but throughout this report, it is treated as a *sine qua non.* It is not discussed in extensive detail because any concepts falling short in human safety have simply been eliminated from consideration.

How will we explore to deliver the greatest benefit to the nation? Planning for a human spaceflight programme should begin with a choice about its goals – rather than a choice of possible destinations. Destinations should derive from goals, and alternative architectures may be weighed against those goals. There is now a strong consensus in the United States that the next step in human spaceflight is to travel beyond low-Earth orbit. This should carry important benefits to society, including: driving technological innovation; developing commercial industries and important national capabilities; and contributing to our expertise in further exploration. Human exploration *can* contribute appropriately to the expansion of scientific knowledge, particularly in areas such as field geology, and it is in the interest of both science and human spaceflight that a credible and well-rationalized strategy of coordination between them be developed. Crucially, human spaceflight objectives should broadly align with key national objectives.

These more tangible benefits exist within a larger context. Exploration provides an opportunity to demonstrate space leadership while deeply engaging international partners; to inspire the next generation of scientists and engineers; and to shape human perceptions of our place in the universe. The Committee concluded that the ultimate goal of human exploration is to chart a path for human expansion into the solar system. This is an ambitious goal, but one worthy of U.S. leadership in concert with a broad range of international partners.

The Committee's task was to review the U.S. plans for human spaceflight. In doing so, it assessed the programmes within the current human spaceflight portfolio; considered capabilities and technologies a future programme might require; and considered the roles of commercial industry and our international partners in this enterprise. From these deliberations, the Committee developed five integrated alternatives for the U.S. human spaceflight programme. The considerations and the five alternatives are summarized in the pages that follow.

4.3.1 Current programmes

Before addressing options for the future human exploration programme, it is appropriate to discuss the current programmes: the Space Shuttle, ISS and Constellation, as well as the looming problem of "the Gap".

4.3.1.1 Key questions to guide the plan for human spaceflight

The Committee identified the following questions that, if answered, would form the basis of a plan for U.S. human spaceflight:

1. What should be the future of the Space Shuttle? 2. What should be the future of the International Space Station (ISS)? 3. On what should the next heavy-lift launch vehicle be based? 4. How should crews be carried to low-Earth orbit? 5. What is the most practicable strategy for exploration *beyond* low-Earth orbit?

The Committee considers the framing and answering of these questions individually, and in a consistent way, to be at least as important as their combinations in the integrated options for a human spaceflight programme.

4.3.1.2 Space Shuttle

What should be the future of the Space Shuttle? The present plan is to retire it at the end of FY 2010, with its final flight scheduled for the last month of that fiscal year. Although the current Administration has relaxed the requirement to complete the last mission before the end of FY 2010, there are no funds in the FY 2011 budget for continuing Shuttle operations.

In considering the future of the Shuttle, the Committee assessed the realism of the current schedule; examined issues related to Shuttle workforce, reliability and cost; and weighed the risks and possible benefits of a Shuttle extension. The Committee noted that the projected flight rate is nearly twice that of the actual flight rate since return to flight after the *Columbia* accident. Recognizing that undue schedule and budget pressure can subtly impose a negative influence on safety, the Committee finds that a more realistic schedule is prudent. With the remaining flights likely to stretch into the second quarter of 2011, the Committee considers it important to budget for Shuttle operations through that time.

Although a thorough analysis of Shuttle safety was not part of its charter, the Committee did examine the Shuttle's safety record and reliability. New human-rated launch vehicles will likely be more reliable once they reach maturity, but in the meantime, the Shuttle is in the enviable position of being through its infant mortality phase. Its flight experience and demonstrated reliability should not be discounted.

Once the Shuttle is retired, there will be a gap in America's capability to launch humans into space. That gap will extend until the next U.S. human-rated launch system becomes available. The Committee estimates that, under the current plan, this gap will be at least seven years long. There has not been this long a gap in U.S. human launch capability since the U.S. human space programme began.

Most of the integrated options presented below would retire the Shuttle after a prudent fly-out of the current manifest, indicating that the Committee found the interim reliance on international crew services acceptable. However, one option does provide for an extension of Shuttle at a minimum safe flight rate to preserve U.S. capability to launch astronauts into space. If that option is selected, there should be a thorough review of Shuttle recertification conducted to date and overall Shuttle reliability to ensure that the risk associated with that extension would be acceptable. This review should be performed by an independent committee, with the purpose to ensure that NASA has met the intent behind the relevant recommendation of the Columbia Accident Investigation Board.[484]

4.3.1.3 International Space Station

In considering the future of the International Space Station (ISS), the Committee asked two basic questions: What is the outlook between now and 2015? Should ISS be extended beyond 2015?

The Committee is concerned that the ISS, and particularly its utilization, may be vulnerable after Shuttle retirement. ISS was designed, assembled and operated with the capabilities of the Space Shuttle in mind. The present approach to its utilization is based on Shuttle-era experience. After Shuttle retirement, ISS will rely on a combination of new, and as yet unproven, international and commercial vehicles for cargo transport. Because the planned commercial resupply capability will be crucial to both ISS operations and utilization, it may be prudent to strengthen the incentives to the commercial providers to meet the schedule milestones.

Now that the ISS is nearly completed and is staffed by a full crew of six, its future success will depend on how well it is used. Up to now, the focus has been on assembling ISS, and this has come at the expense of using the Station. Utilization should have first priority in the years ahead.

The Committee finds that the return on investment of ISS to both the United States and the international partners would be significantly enhanced by an extension of ISS life to 2020. It seems unwise to de-orbit the Station after 25 years of assembly and only five years of operational life. *Not* to extend its operation would significantly impair U.S. ability to develop and lead future international spaceflight partnerships. Further, the ISS should be funded to enable it to achieve its full potential: as the nation's newest national laboratory, as an enhanced test bed for technologies and operational techniques that support exploration, and as a framework that can support expanded international collaboration.

The strong and tested working relationship among international partners is perhaps the most important outcome of the ISS programme. The partnership expresses a "first among equals" U.S. leadership style adapted to today's multi-polar world. That leadership could extend to exploration, as the ISS partners could engage at an early stage if aspects of exploration beyond low Earth orbit were included in the goals of the partnership agreement.

4.3.1.4 The Constellation Programme

The Constellation Programme includes: the Ares I launch vehicle, capable of launching astronauts to low-Earth orbit; the Ares V heavy-lift launch vehicle, to send astronauts and equipment to the Moon; the Orion capsule, intended to carry astronauts to low-Earth orbit and beyond; and the Altair lunar lander and lunar surface systems astronauts will need to explore the lunar surface. As the Committee assessed the current status and possible future of the Constellation Programme, it reviewed the technical, budgetary and schedule challenges that the programme faces today.

Given the funding originally expected, the Constellation Programme was a reasonable architecture for human exploration. However, even when it was announced, its budget depended on funds becoming available from the retirement of the Space Shuttle in 2010 and the decommissioning of ISS in early 2016. Since then, as a result of technical and budgetary issues, the development schedules of Ares I and Orion have slipped, and work on Ares V and Altair has been delayed.

Most major vehicle-development programmes face technical challenges as a normal part of the process, and Constellation is no exception. While significant, these are engineering problems, and the Committee expects that they can be solved. But these solutions may add to the programme's cost and/or delay its schedule.

The original 2005 schedule showed Ares I and Orion available to support ISS in 2012, only two years after scheduled Shuttle retirement. The current schedule now shows that date as 2015. An independent assessment of the technical, budgetary and schedule risk to the Constellation Programme performed for the Committee indicates that an additional delay of at least two years is likely.[485] This means that Ares I and Orion will not reach ISS before the Station's currently planned termination, and the length of the gap in U.S. ability to launch astronauts into space will be no less than seven years.

The Committee also examined the design and development of Orion. Many concepts are possible for crew-exploration vehicles, and NASA clearly needs a new spacecraft for travel beyond low-Earth orbit. The Committee found no compelling evidence that the current design will not be acceptable for its wide variety of tasks in the exploration programme. However, the Committee is concerned about

Orion's recurring costs. The capsule is considerably larger and more massive than previous capsules (e.g. the Apollo capsule), and there is some indication that a smaller and lighter four-person Orion could reduce operational costs. However, a redesign of this magnitude would likely result in over a year of additional development time and a significant increase in cost, so such a redesign should be considered carefully before being implemented.

4.3.2 Cabability for launch to low-Earth orbit and exploration beyond

4.3.2.1 Heavy-lift launch to low-Earth orbit and beyond

No one knows the mass or dimensions of the largest piece that will be required for future exploration missions, but it will likely be significantly larger than 25 metric tons (mt) in launch mass to low-Earth orbit, the capability of current launchers. As the size of the launcher increases, fewer launches and less operational complexity to assemble and/or refuel them results, and the net availability of launch capability increases. Combined with considerations of launch availability and on-orbit operations, the Committee finds that exploration will benefit from the availability of a heavy-lift vehicle. In addition, heavy lift would enable the launching of large scientific observatories and more capable deep-space missions. It may also provide benefit in national security applications. The question is: On what system should the next heavy-lift launch vehicle be based?

Potential approaches to developing heavy-lift vehicles (Figure 1) are based on NASA heritage (Shuttle and Apollo) and EELV (evolved expendable launch vehicle) heritage. Each has its distinct advantages and disadvantages.

Family			Launch mass to LEO
NASA heritage	Ares family	Ares V + Ares I	160 mt + 25 mt
		Ares V Lite	140 mt
	Shuttle derived family		100–110 mt
EELV heritage family			75 mt

Fig. 1. *Characteristics of heavy-lift launch vehicles, indicating the EELV and NASA heritage families.*

In the Ares-V-plus-Ares-I system planned by the Constellation programme, the Ares I launches the Orion and docks in low-Earth orbit with the Altair lander launched on the Ares V. It has the advantage of projected very high ascent crew safety, but it delays the development of the Ares V heavy lift vehicle until after the independently operated Ares I is developed.

In a different, related architecture, the Orion and Altair are launched on two separate "Lite" versions of the Ares V, providing for more robust mass margins. Building a single NASA vehicle could reduce carrying and operations costs, and accelerate heavy-lift development. Of these two Ares system alternatives, the Committee finds the Ares V Lite in the dual mode the preferred reference option. The more directly Shuttle-derived family consists of in-line and side-mount vehicles substantially derived from the Shuttle, providing more continuity in workforce. The development cost of the more Shuttle-derived system would be lower, but it would be less capable than the Ares V family and have higher recurring costs. The lower launch capability could eventually be offset by developing on-orbit refueling.

The EELV-heritage systems have the least lift capability, so that to provide equal performance, almost twice as many launches would be required, when compared to the Ares family. If on-orbit refueling were developed and used, the number of launches could be reduced, but operational complexity would be added. However, the EELV approach would also represent a new way of doing business for NASA, which would have the benefit of potentially lowering development and operational costs. This would come at the cost of ending a substantial portion of the internal NASA capability to develop and operate launchers. It would also require that NASA and the Department of Defense jointly develop the new system.

All of the options would benefit from the development of in-space refueling, and the smaller rockets would benefit most of all. The potential government-guaranteed market for fuel in low-Earth orbit would create a stimulus to the commercial launch industry. In the design of the new launcher, in-space stages and in-space refueling, the Committee cautions against the tradition of designing for ultimate performance, at the cost of reliability, operational efficiency and life-cycle cost.

4.3.2.2 Crew access to low-Earth orbit

How should U.S. astronauts be transported to low-Earth orbit? There are two basic approaches: a government-operated system and a commercial crew-delivery service. The current Constellation Programme plan is to use the government-operated Ares I launch vehicle and the Orion crew capsule. However, the Committee found that, because of technical and budget issues, the Ares I schedule no longer supports the ISS.

Ares I was designed to a high standard in order to provide astronauts with access to low-Earth orbit at lower risk and a considerably higher level of reliability than is available today. To achieve this, it uses a high-reliability rocket and a crew capsule with a launch-escape system. But other potential combinations of high-reliability rockets and capsules with escape systems could also provide that reliability. The Committee was unconvinced that enough is known about any of the potential high-reliability launcher-plus-capsule systems to distinguish their levels of safety in a meaningful way.

The United States needs a way to launch astronauts to low-Earth orbit, but it does not necessarily have to be provided by the government. As we move from the complex, reusable Shuttle back to a simpler, smaller capsule, it is an appropriate time to consider turning this transport service over to the commercial sector. This approach is not without technical and programmatic risks, but it creates the possibility of lower operating costs for the system and potentially accelerates the availability of U.S. access to low-Earth orbit by about a year, to 2016. The Committee suggests establishing a new competition for this service, in which both large and small companies could participate.

4.3.2.3 Lowering the cost of space exploration

The cost of exploration is dominated by the costs of launch to low-Earth orbit and of the in-space systems. It seems improbable that significant reductions in launch costs will be realized in the short term until launch rates increase substantially – perhaps through expanded commercial activity in space. How can the nation stimulate such activity? In the 1920s, the federal government awarded a series of guaranteed contracts for carrying airmail, stimulating the growth of the airline industry. The Committee concludes that an architecture for exploration employing a similar policy of guaranteed contracts has the potential to stimulate a vigorous and competitive commercial space industry. Such commercial ventures could include supply of cargo to the ISS (already underway), transport of crew to orbit and transport of fuel to orbit. Establishing these commercial opportunities could increase launch volume and potentially lower costs to NASA and all other launch-services customers.

This would have the additional benefit of focusing NASA on a more challenging role, permitting it to concentrate its efforts where its inherent capability resides: for example, developing cutting-edge technologies and concepts, and defining programmes and overseeing the development and operation of exploration systems, particularly those beyond low-Earth orbit.

The Committee strongly believes it is time for NASA to reassume its crucial role of developing new technologies for space. Today, the alternatives available for exploration systems are severely limited because of the lack of a strategic investment in technology development in past decades. NASA now has an opportunity to develop a technology roadmap that is aligned with an exploration mission that will last for decades. If appropriately funded, a technology development programme would re-engage the minds at American universities, in industry and within NASA. The investments should be designed to increase the capabilities and reduce the costs of future exploration. This will benefit human and robotic exploration, the commercial space community, and other U.S. government users.

4.3.3 Future destinations for exploration

What is the strategy for exploration beyond low-Earth orbit? Humans could embark on the following paths to explore the inner solar system:

- Mars first, with a Mars landing, perhaps after a brief test of equipment and procedures on the Moon.
- Moon first, with lunar surface exploration focused on developing the capability to explore Mars.
- Flexible path to inner solar system locations, such as lunar orbit, Lagrange points, near-Earth objects and the moons of Mars, followed by exploration of the lunar surface and/or Martian surface.

A human landing followed by an extended human presence on Mars stands prominently above all other opportunities for exploration. Mars is unquestionably the most scientifically interesting destination in the inner solar system, with a history much like Earth's. It possesses resources, which can be used for life support and propellants. If humans are ever to live for long periods on another planetary surface, it is likely to be on Mars. But Mars is not an easy place to visit with existing technology and without a substantial investment of resources. The Committee finds that Mars is the ultimate destination for human exploration; but it is not the best first destination.

What about the Moon first, then Mars? By first exploring the Moon, we could develop the operational skills and technology for landing on, launching from and working on a planetary surface. In the process, we could acquire an understanding of human adaptation to another world that would one day allow us to go to Mars.

There are two main strategies for exploring the Moon. Both begin with a few short sorties to various sites to scout the region and validate the lunar landing and ascent systems. In one strategy, the next step would be to build a base. Over many missions, a small colony of habitats would be assembled, and explorers would begin to live there for many months, conducting scientific studies and prospecting for resources that could be used as fuel. In the other strategy, sorties would continue to different sites, spending weeks and then months at each one. More equipment would have to be brought on each trip, but more diverse sites would be explored and in greater detail.

There is a third possible path for human exploration beyond low-Earth orbit, which the Committee calls the Flexible Path. On this path, humans would visit sites never visited before and extend our knowledge of how to operate in space – while traveling greater and greater distances from Earth. Successive missions would visit: lunar orbit; the Lagrange points (special points in space that are important sites for scientific observations and the future space transportation infrastructure); near-Earth objects (asteroids that cross the Earth's path); and orbit around Mars. Most interestingly, humans could rendezvous with a moon of Mars, then coordinate with or control robots on the Martian surface.

The Flexible Path represents a different type of exploration strategy. We would learn how to live and work in space, to visit small bodies, and to work with robotic probes on the planetary surface. It would provide the public and other stakeholders with a series of interesting "firsts" to keep them engaged and supportive. Most important, because the path is flexible, it would allow many different options as exploration progresses, including a return to the Moon's surface, or a continuation to the surface of Mars.

The Committee finds that both Moon First and Flexible Path are viable exploration strategies. It also finds that they are not necessarily mutually exclusive; before traveling to Mars, we might be well served to both extend our presence in free space and gain experience working on the lunar surface.

4.3.4 Integrated programme options

The Committee has identified five principal alternatives for the human spaceflight programme. They include one baseline case, which the Committee believes to be an executable version of the current programme of record, funded to achieve its

	Budget	Shuttle life	ISS life	Heavy launch	Crew to LEO
Constrained options					
Option 1: Programme of Record (constrained)	FY10 Budget	2011	2015	Ares V	Ares I + Orion
Option 2: ISS + Lunar (constrained)	FY10 Budget	2011	2020	Ares V Lite	Commercial
Moon first options					
Option 3: Baseline – Programme of Record	Less constrained	2011	2015	Ares V	Ares I + Orion
Option 4A: Moon First – Ares Lite	Less constrained	2011	2020	Ares V Lite	Commercial
Option 4B: Moon first – Extend Shuttle	Less constrained	2015	2020	Directly Shuttle Derived + refueling	Commercial
Flexible path options					
Option 5A: Flexible path – Ares Lite	Less constrained	2011	2020	Ares V Lite	Commercial
Option 5B: Flexible path – EELV Heritage	Less constrained	2011	2020	75 mt EELV + refueling	Commercial
Option 5C: Flexible path – Shuttle Derived	Less constrained	2011	2020	Directly Shuttle Derived + refueling	Commercial

Note: Programme-of-Record derived options (Options 1 and 3) do not contain a technology program; all others do.

Fig. 2: *A summary of the integrated programme options.*

stated exploration goals, as well as four alternatives. These options are summarized in Figure 2 above.

The committee was asked to provide two options that fit within the FY 2010 budget profile. This funding is essentially flat or decreasing through 2014, then increases at 1.4% per year thereafter, which is less than the 2.4% per year used to estimate cost inflation. The first two options are constrained to that budget.

Option 1. Programme of Record as assessed by the Committee, constrained to the FY 2010 budget. This option is the Programme of Record, with only two changes the Committee deems necessary: providing funds for the Shuttle into FY 2011 and including sufficient funds to de-orbit the ISS in 2016. When constrained to this budget profile, Ares I and Orion are not available until after the ISS has been de-orbited. The heavy-lift vehicle, Ares V, is not available until the late 2020s, and worse, there are insufficient funds to develop the lunar lander and lunar surface systems until well into the 2030s, if ever.

Option 2. ISS and Lunar Exploration, constrained to FY 2010 budget. This option extends the ISS to 2020, and it begins a programme of lunar exploration using Ares V (Lite). The option assumes Shuttle fly-out in FY 2011, and it includes a technology development programme, a programme to develop commercial crew services to low-Earth orbit, and funds for enhanced utilization of ISS. This option does not deliver heavy-lift capability until the late 2020s and does not have funds to develop the systems needed to land on or explore the Moon.

The remaining three alternatives are fit to a different budget profile – one that the Committee judged more appropriate for an exploration programme designed to carry humans beyond low-Earth orbit. This budget increases to $3 billion above the FY 2010 guidance by FY 2014, then grows with inflation at a more reasonable 2.4% per year.

Option 3. Baseline Case – Implementable Programme of Record. This is an executable version of the programme of record. It consists of the content and sequence of that programme – de-orbiting the ISS in 2016, developing Orion, Ares I and Ares V, and beginning exploration of the Moon. The Committee made only two additions it felt essential: budgeting for the fly-out of the Shuttle in 2011 and including additional funds for ISS de-orbit. The Committee's assessment is that, under this funding profile, the option delivers Ares1/Orion in FY 2017, with human lunar return in the mid-2020s.

Option 4. Moon First. This option preserves the Moon as the first destination for human exploration beyond low-Earth orbit. It also extends the ISS to 2020, funds technology advancement, and uses commercial vehicles to carry crew to low-Earth orbit. There are two significantly different variants to this option.

Variant 4A is the Ares Lite variant. This retires the Shuttle in FY 2011 and develops the Ares V (Lite) heavy-lift launcher for lunar exploration. *Variant 4B* is the Shuttle extension variant. This variant includes the only foreseeable way to eliminate the gap in U.S. human-launch capability: it extends the Shuttle to 2015 at a minimum safe-flight rate. It also takes advantage of synergy with the Shuttle by developing a heavy-lift vehicle that is more directly Shuttle-derived. Both variants of Option 4 permit human lunar return by the mid-2020s.

Option 5. Flexible Path. This option follows the Flexible Path as an exploration strategy. It operates the Shuttle into FY 2011, extends the ISS until 2020, funds technology development and develops commercial crew services to low-Earth orbit. There are three variants within this option; they differ only in the heavy-lift vehicle.

Variant 5A is the Ares Lite variant. It develops the Ares Lite, the most capable of the heavy-lift vehicles in this option. *Variant 5B* employs an EELV-heritage commercial heavy-lift launcher and assumes a different (and significantly reduced) role for NASA. It has an advantage of potentially lower operational costs, but requires significant restructuring of NASA. *Variant 5C* uses a directly Shuttle-derived, heavy-lift vehicle, taking maximum advantage of existing infrastructure, facilities and production capabilities.

All variants of Option 5 begin exploration along the flexible path in the early 2020s, with lunar fly-bys, visits to Lagrange points and near-Earth objects and Mars fly-bys occurring at a rate of about one major event per year, and possible rendezvous with Mars's moons or human lunar return by the mid to late 2020s.

The Committee has found two executable options that comply with the FY 2010 budget. However, neither allows for a viable exploration programme. In fact, the Committee finds that no plan compatible with the FY 2010 budget profile permits human exploration to continue in any meaningful way.

The Committee further finds that it is possible to conduct a viable exploration programme with a budget rising to about $3 billion annually above the FY 2010 budget profile. At this budget level, both the Moon First strategy and the Flexible Path strategies begin human exploration on a reasonable, though hardly aggressive, timetable. The Committee believes an exploration programme that will be a source of pride for the nation requires resources at such a level.

4.3.5 Organizational and programmatic issues

How might NASA organize to explore? The NASA Administrator needs to be given the authority to manage NASA's resources, including its workforce and facilities. Even the best-managed human spaceflight programmes will encounter developmental problems. Such activities must be adequately funded, including reserves to account for the unforeseen and unforeseeable. Good management is especially difficult when funds cannot be moved from one human spaceflight budget line to another – and where new funds can ordinarily be obtained only after a two-year delay (if at all). NASA should be given the maximum flexibility possible under the law to establish and manage its systems.

Finally, significant space achievements require continuity of support over many years. One way to ensure that no successes are achieved is to continually pull up the flowers to see if the roots are healthy. (This Committee might be accused of being part of this pattern!) NASA and its human spaceflight programme are in need of stability in both resources and direction.

4.3.6 Summary of key findings

The Committee summarizes its key findings below. Additional findings are included in the body of the report.

The right mission and the right size: NASA's budget should match its mission and goals. Further, NASA should be given the ability to shape its organization and infrastructure accordingly, while maintaining facilities deemed to be of national importance.

International partnerships: The U.S. can lead a bold new international effort in the human exploration of space. If international partners are actively engaged, including on the "critical path" to success, there could be substantial benefits to foreign relations, and more resources overall could become available.

Short-term Space Shuttle planning: The current Shuttle manifest should be flown in a safe and prudent manner. The current manifest will likely extend to the second quarter of FY 2011. It is important to budget for this likelihood.

The human-spaceflight gap: Under current conditions, the gap in U.S. ability to launch astronauts into space will stretch to at least seven years. The Committee did not identify any credible approach employing new capabilities that could shorten the gap to less than six years. The only way to significantly close the gap is to extend the life of the Shuttle Programme.

Extending the International Space Station: The return on investment to both the United States and our international partners would be significantly enhanced by an extension of ISS life. *Not* to extend its operation would significantly impair U.S. ability to develop and lead future international spaceflight partnerships.

Heavy-lift: A heavy-lift launch capability to low-Earth orbit, combined with the ability to inject heavy payloads away from the Earth, is beneficial to exploration, and it also will be useful to the national security space and scientific communities. The Committee reviewed: the Ares family of launchers; more directly Shuttle-derived vehicles; and launchers derived from the EELV family. Each approach has advantages and disadvantages, trading capability, lifecycle costs, operational complexity and the "way of doing business" within the programme and NASA.

Commercial crew launch to low-Earth orbit: Commercial services to deliver crew to low-Earth orbit are within reach. While this presents some risk, it could provide an earlier capability at lower initial and lifecycle costs than government could achieve. A new competition with adequate incentives should be open to all U.S. aerospace companies. This would allow NASA to focus on more challenging roles, including human exploration *beyond* low-Earth orbit, based on the continued development of the current or modified Orion spacecraft.

Technology development for exploration and commercial space: Investment in a well-designed and adequately funded space technology programme is critical to enable progress in exploration. Exploration strategies can proceed more readily and economically if the requisite technology has been developed in advance. This investment will also benefit robotic exploration, the U.S. commercial space industry and other U.S. government users.

Pathways to Mars: Mars is the ultimate destination for human exploration; but it is not the best first destination. Both visiting the Moon First and following the Flexible Path are viable exploration strategies. The two are not necessarily mutually exclusive; before traveling to Mars, we might be well served to both extend our presence in free space and gain experience working on the lunar surface.

Options for the Human Spaceflight Programme: The Committee developed five alternatives for the Human Spaceflight Programme. It found:

- Human exploration beyond low-Earth orbit is not viable under the FY 2010 budget guideline.
- Meaningful human exploration is possible under a less constrained budget, ramping to approximately $3 billion per year above the FY 2010 guidance in total resources.
- Funding at the increased level would allow either an exploration programme to explore Moon First or one that follows a Flexible Path of exploration. Either could produce results in a reasonable timeframe.

[484] "Prior to operating the Shuttle beyond 2010, develop and conduct a vehicle recertification at the material, component, subsystem, and system levels. Recertification requirements should be included in the Service Life Extension Program." [Columbia Accident Investigation Board, R9.2–1].

[485] The independent assessment was conducted for the Committee by the Aerospace Corporation.

4.4 The Global Exploration Strategy Framework: The Framework for Coordination (Executive Summary, May 2007)[486]

Space exploration enriches and strengthens humanity's future. Searching for answers to fundamental questions such as: "Where did we come from?" "What is our place in the universe?" and "What is our destiny?" can bring nations together in a common cause, reveal new knowledge, inspire young people and stimulate technical and commercial innovation on Earth.
The Global Exploration Strategy is key to delivering these benefits.

One of the most fundamental human characteristics is a relentless curiosity that drives us to investigate the unknown. Throughout our history, we have looked beyond our apparent boundaries to the mysteries that lie beyond.

Compelled to explore, to understand and to use the world in which we find ourselves, we have spread across continents and oceans. We have probed the farthest reaches of the planet – the frozen poles, the deep oceans, the high atmosphere.

With increasing intent and determination, we are resolved to explore our nearest companions – the Moon, Mars and some nearby asteroids. Our goal is not a few quick visits, but rather a sustained and ultimately self-sufficient human presence beyond Earth supported by robotic pathfinders.

Sustainable space exploration is a challenge that no one nation can do on its own. This is why 14 space agencies[487] have developed The Global Exploration Strategy: The Framework for Coordination, which presents a vision for robotic and human space exploration, focussing on destinations within the solar system where we may one day live and work. It elaborates an action plan to share the strategies and efforts of individual nations so that all can achieve their exploration goals more effectively and safely.

This Framework does not propose a single global programme. Rather, it recommends a voluntary, non-binding forum, the international Coordination Mechanism, through which nations can collaborate to strengthen both individual projects and the collective effort.

Robust science and technology efforts, such as the pursuit of space exploration, help to define nations and their place in the world. The number of countries involved in space exploration is growing steadily and we are entering a new era of historic significance, in which we will extend human presence beyond Earth's orbit, physically and culturally.

The Moon is our nearest and first goal. As a repository of four billion years of solar system history, it has enormous scientific significance. It is also a base from which to study Earth and the universe, and to prepare humans and machines for venturing farther into space.

Mars is also a prime target. With an atmosphere and water, it may hold key secrets to the evolution of life in our solar system. Eventually, we hope to reach other, even more challenging destinations, such as asteroids and the Moons of the giant planets.

A partnership between humans and robots is essential to the success of such ventures. Robotic spacecraft are our scouts and proxies, venturing first into hostile environments to gather critical intelligence that makes human exploration feasible. Humans will then bring their ingenuity, creativity and problem-solving skills to these destinations.

This Global Exploration Strategy will bring significant social, intellectual and economic benefits to people on Earth. We will learn about the evolution of the solar system and how to protect against harsh environments. By understanding how planets work, we learn more about our Earth. The technologies created will help build a more sustainable society.

Space exploration also offers significant entrepreneurial opportunities by creating a demand for new technologies and services. These advances will encourage economic expansion and the creation of new businesses.

Finally, this new era of space exploration will strengthen international partnerships through the sharing of challenging and peaceful goals. It will inspire people everywhere, particularly youth. It will steer many students toward careers in science and technology and provide them with challenging jobs that encourage innovation and creativity.

Opportunities such as this come rarely. The human migration into space is still in its infancy. For the most part, we have remained just a few kilometres above the Earth's surface – not much more than camping out in the backyard.

It is time to take the next step.

Chapter 1: The Global Exploration Strategy: The Framework for Coordination

> *Space exploration is essential to humanity's future. It can help answer fundamental questions such as: "Where did we come from?" "What is our place in the universe?" and "What is our destiny?" It can bring nations together in a common cause, reveal new knowledge, inspire young people and stimulate technical and commercial innovation on Earth. The Global Exploration Strategy is key to unlocking this door to the future.*

Human curiosity compels us to *explore*, to *understand* and to *use* the world in which we find ourselves. Voyages of exploration and discovery are a sign of cultural vigour; every vibrant society has looked beyond its horizons to somewhere new.

Scientific evidence suggests that modern humans emerged in ancient Africa and spread across Eurasia, beginning about a million years ago. Primitive peoples may have built rafts to sail the oceans.

During the ensuing millennia, humans have migrated throughout all the continents. Explorers and adventurers go first, embracing the risks inherent in pitting themselves against unknown and often hostile environments. Scientists and traders follow and, eventually, ordinary people move in to create permanent settlements.

The human migration into space is the next chapter in this story of exploration and expansion. It is still in its infancy; less than 50 years have passed since the first humans ventured beyond the shores of Earth into the new ocean of space. Nor have we gone far; since Yuri Gagarin's flight in 1961, almost all of the 450 human explorers of space have remained just a few hundred kilometres above the Earth's surface. Only the two dozen Apollo astronauts who visited the Moon between 1968 and 1972 have ventured farther.

Though we have barely begun this new journey of exploration, we have already learnt many of the new skills needed to live and work in space, whether we are physically present ourselves or are sending robotic spacecraft in our stead.[488]

So far, these robotic proxies have been the only ones to explore the most distant and challenging destinations far beyond low Earth orbit. They have given us a tenuous "virtual presence" throughout the solar system. The most travelled of these probes, Voyager 1, launched in 1977, is only now leaving the solar system forever.

Meanwhile, we have harnessed Earth orbit to serve society, using satellites to provide global telecommunications, navigation and environmental monitoring, deliver reliable weather forecasts, and aid emergency workers responding to natural disasters.

This document is not concerned with these proven and well-managed uses of space. Instead, the *Global Exploration Strategy* (GES) addresses a new opportunity. It elaborates a vision for globally coordinated space exploration focussed on solar system destinations where humans will someday live and work.

It also sets the stage for the discussions and hard work that will turn the vision into reality. It includes an action plan for coordinating strategies to help national space agencies[489] reach their space exploration goals more effectively and safely.

The number of countries involved in space exploration is growing steadily. Building on what has already been learnt, our overall ability to accomplish scientific, technological and human goals has never been greater.

We are now entering a new wave of space exploration, one of historic significance. The United States has developed its Vision for Space Exploration; the European Space Agency has its Aurora space exploration programme. China, India, Japan and Russia have ambitious national projects to explore the Moon or Mars, while future national missions are being discussed in Canada, Germany, Italy, Republic of Korea and the United Kingdom.

The public has marvelled as astronauts from several countries build the International Space Station, perhaps the most ambitious science and technology project ever undertaken. They have witnessed China become the third country to launch their own astronauts.

On the robotic front, the Huygens probe revealed a new world of river valleys and mountains beneath the dense haze of Saturn's moon, Titan. The Hayabusa spacecraft landed on the asteroid Itokawa to pick up samples for return to Earth, heralding a new era of round-trip exploration in interplanetary space.

Bilateral and multilateral cooperation among space-faring nations has enabled much of what has been achieved so far, and this will continue in the future. But there's never been a single, comprehensive strategy for space exploration that allows existing plans to be coordinated and new ones to be developed.

This GES *Framework for Coordination*, developed by 14 space agencies, therefore represents a new beginning. International discussions during 2006 produced a common set of space exploration themes, as elaborated in this document. The Framework makes the case for a voluntary, non-binding forum (the *Coordination Mechanism*) where nations can share plans for space exploration and collaborate to strengthen both individual projects and the collective effort.

As a voluntary mechanism, the international coordination process is open to new participants. Each will bring their own perspectives and skills and, in return, will gain access to the common knowledge and experience.

This Framework is *not* a proposal for a single programme, but recognizes that individual space exploration activities can achieve more through coordination and cooperation. Nations have varying scientific, technological and societal objectives for their space activities, and – inevitably – some can afford to do more than others.

For the foreseeable future, the Moon, Mars and near-Earth asteroids are the primary targets for human space exploration. We do not yet have the practical knowledge or skills to send humans to other exciting but more distant destinations such as Jupiter's moon Europa, or Titan and Enceladus, which orbit Saturn.

But exploring even the first group of feasible destinations will require both robotic and human missions of all sizes and complexity. A coordinated strategy will help individual nations with shared objectives to engage in joint projects that will maximise their return on investment. The scientific and technical successes – and even the failures – of each project can be used to improve the ones that follow.

The Framework calls for the development of an international exploration coordination tool to enhance mutual understanding among partners and to identify areas for potential cooperation. By jointly creating a common language of exploration building blocks, planners and engineers will be able to agree how practical features such as communications, control, life support and docking systems could be made to work together. Such "interoperability" between space vehicles will lower the risks of space exploration and could assure crew safety in case of life-threatening emergencies.

Although government agencies have led the creation of the GES, the Framework also recognises that industry will have an increasingly important role in turning the new frontiers of space to economic opportunity. It is hoped that entrepreneurs will create businesses to exploit resources or provide commercial services such as cargo transport and telecommunications. Thus, space exploration will become more sustainable and government resources may be released to push further the bounds of human knowledge.

The successful exploitation of Earth orbit over the past 30 years suggests that this is likely. In time, issues of property rights and protection of sites of interest may arise and the GES Coordination Mechanism will help address such new challenges.

The following chapters present the rationale for space exploration, organized under specific themes that highlight its benefits to society. This Framework outlines the steps humanity must take to embark on this new journey, focussing on the complementary roles of human and robotic missions.

It argues for a return to the Moon, a target of intrinsic value and an essential stepping stone to the exploration of Mars and beyond. It concludes that improved coordination of national and international efforts can make space exploration more robust and more affordable for all.

Chapter 2: Space exploration in service of society

Global-scale space exploration represents the sum of many projects undertaken nationally and internationally. But it also signifies a collective will to find answers to profound scientific questions, to create new economic opportunity and to expand the boundaries of human life beyond Earth. These goals of space exploration in the service of society are embodied in the recurring themes of the Global Exploration Strategy.

Globally, humankind is facing many pressing social, political and environmental challenges. In this context, the relevance of space exploration to society is sometimes not well understood. So, why does it matter? How can space exploration contribute to our common future?

Space exploration is today's expression of a fundamental human characteristic: our deep curiosity and a resulting imperative to explore the unknown.

This is how we gain new knowledge and skills that become part of our collective ability to solve human problems and support commercial activities in useful and unpredictable ways. The very difficulty of space exploration is what triggers human inspiration and innovation.

The first 50 years of spaceflight provide many notable examples. Satellites have revolutionized global communications and navigation, and have provided critical data on climate change. Robotic and imaging technologies and other tools developed for the demanding space environment have found important applications on Earth, such as airport security scanners and medical diagnostic instruments.

In the future, a sustained but affordable agenda of globally coordinated space exploration can serve society through:

- securing new knowledge and solving global challenges in space and on Earth through innovative technology;
- permanently extending human presence into space, physically and culturally;
- enabling economic expansion and new business opportunities;
- creating global partnerships by sharing challenging and peaceful goals and
- inspiring society through collective effort and personal endeavour.

These benefits are encompassed in five exploration themes (no prioritization is implied):

Theme 1: New knowledge in science and technology

At its core, exploration is about taking manageable risks to discover what is unknown. Significantly, much of what it reveals is unknowable in advance. This presents challenges for those wanting to weigh the risks against the returns from new investments.

This problem is as old as innovation itself; when Heinrich Hertz developed the first apparatus to transmit and receive electromagnetic waves in 1887, he hardly envisaged the vast global telecommunications networks of today, or the economic activity they sustain.

Space exploration generates new knowledge that helps us understand the solar system in relation to both the biosphere of Earth and to the vast universe beyond. Though many mysteries remain, we have made a good start with robotic spacecraft, brief human missions to the Moon and human activity in low Earth orbit.

The scientific exploration of the solar system began in the earliest days of the space age. First we sent robotic probes to the nearest planetary bodies, the Moon, Venus and Mars. We also investigated the local space environment and learnt how the Earth's magnetic field protects us from the Sun's lethal radiation and continuous bombardment by material it casts off.

These early missions significantly enriched our knowledge about each celestial body. Today, more sophisticated missions are starting to unravel the inner workings of the solar system since it formed some 4.5 billion years ago.

Systematic, science-driven space exploration reveals fundamental truths about the history of the solar system and the origin and nature of life. Both robotic and human exploration are necessary to answer the key questions.

Much of the current research focusses on two big questions: how did the solar system evolve and is there life beyond Earth? These questions are profoundly important, scientifically and philosophically. The quest to answer these questions took us first to our nearest neighbours in the solar system.

Though often called Earth's twin because of its similar size, Venus could hardly be more hostile to life as we know it. Planetary probes found hellish conditions; runaway global warming has produced a dense atmosphere of choking sulphuric acid and temperatures hot enough to melt lead. We will continue to study Venus, but it is unlikely to be a destination for human exploration.

The story on Mars is different. Although its surface today is cold and barren, evidence suggests it was once warm and wet; it also seems to have many of the raw materials for life, including the key one – water. Robotic missions have found signs of recent hydrological activity and some scientists believe there is evidence of frozen oceans.

Today, we simply do not know whether life could, did or even still does exist on Mars. But, despite the undeniable hazards of its inhospitable surface, it is, unlike Venus, a place we can contemplate visiting to look for answers.

Of necessity, the most detailed investigations will occur in locations accessible to humans, where they can undertake long-term research. However, robotic proxies will allow us to reach farther afield. Already, they have found simple carbon molecules that could be precursors to the complex chemistry of life in such diverse places as comets and the thick atmosphere of Titan, a huge Moon orbiting Saturn.

Why is our chemical make-up more akin to both the Sun and the giant planets, such as Saturn and Jupiter, than it is to the rocky planet on which we live? The answers are unclear and will only be found by mapping the distribution of these "pre-biotic" chemicals throughout the solar system, as well as in the planets forming around distant stars.

Another enduring question that space exploration seeks to answer is: how did our solar system come to be? The Moon has been described as a potentially unique museum of the history of the solar system and it may play a key role in unravelling this story.

Furthermore, if – as is currently thought – the Moon was formed by a cataclysmic collision between a Mars-sized object and the early Earth, some clues about the earliest conditions on Earth may be uniquely preserved in the Moon. The Earth's surface records little of the solar system's origins or history; its face has been remade many times by geological processes such as earthquakes, volcanic eruptions and erosion.

Global scale space exploration is an engine of scientific and technical progress. Problem-solving drives innovation and the bigger the problem, the greater the innovation.

Exploring this "planetary museum" will likely require the ability to roam over the lunar surface and to dig several hundred meters down. Both tasks will require humans and robots working in partnership, and similar techniques will be needed on Mars. Taking ice core samples, just as we do in Greenland and the Antarctic, will provide an historical record of the planet's climate. Subsurface bodies of water, if they exist, may yield life forms protected from radiation and the cold of the surface.

We have also started to investigate the leftover materials from which the solar system was built: asteroids and comets. Our interest in these bodies is scientific, economic and practical. We are interested in exploring their role in distributing pre-biotic chemicals and water through the early solar system. In addition, there are some who believe there is a realistic prospect of mapping and even exploiting these objects for their mineral resources.

More significantly, we know these objects can and do strike Earth and likely have caused several mass extinctions in the past. It could happen again. So far, humanity has been lucky, but one day our luck may run out.

Studies are underway to develop the first spacecraft capable of deflecting an Earth-crossing asteroid. A healthy space exploration programme will generate the knowledge needed to give us this ultimate insurance policy. As space scientist Carl Sagan once observed, the dinosaurs died out because they did not have a space programme.

Theme 2: A sustained presence – extending human frontiers

Since humans first emerged in ancient Africa, we have populated the most hot, cold, humid and dry regions of Earth. Our first tentative steps in space exploration have already expanded humanity's reach to the hostile environment of Earth's orbit and the Moon's surface, while robotic probes have reached out even farther.

The brief sorties by the Apollo astronauts required the ability to sustain humans for only a few days on the lunar surface; there was no attempt to establish a long-term presence or exploit local resources. Space stations like Mir and the International Space Station have extended our staying power to months and years, albeit in a manner that requires constant support from Earth.

Going farther afield and establishing long-term, self-sufficient outposts will require a significant commitment of human, scientific, technical and economic resources.

But why send humans into space at all? Why not let robots do it all? Humans have unique decision-making capabilities that allow them to respond to new situations based on previous experience and knowledge. Sending humans to live and work in space takes full advantage of the intellectual capital and real-time reasoning that only they can provide. A human can quickly find and tighten a loose bolt on a core-sample drilling rig, whereas it might take hours to programme a robot to do so, even if it had the means to sense the problem. We are a long way from having robots that can match humans, even in the lab.

We know it is possible to establish a more sustained human presence on the Moon and that it may well be possible to extract resources that will reduce dependence on Earth. Many more resources certainly exist on Mars, but the much longer travelling time makes the technical requirements more demanding and increases the risks, especially from radiation.

In the long run, having a sustained and self-sufficient presence in space will also allow humanity to maintain off-world repositories of knowledge and history.

It will almost certainly redefine our relationship with Earth in profound ways, and increase our appreciation of the rare bounty we have here. One of the great legacies of the space programme is the psychological impact of the image taken by Apollo astronauts of a vibrant Earth floating above the lifeless plains of the Moon, fragile and isolated in the emptiness of space.

Theme 3: Economic expansion

The first stages of space activity were driven by national space agencies, but business has gradually come to play a larger role. Today, a multi-billion dollar industry uses privately-owned satellites to provide voice telephone service, mobile Internet access, and high quality television broadcasting to subscribers around the world.

More recently, commercial Earth observation satellites have been launched. At first, governments were their key customers, but their client base expanded rapidly. Countless users now have satellite-based navigation equipment in their private cars and anyone can access geographical data through software tools such as Internet-based Google Earth.

> *In the past century, governments have nurtured major industries, either through investment in infrastructure, such as railways, highways and the Internet, or by becoming the first customer for services like air delivery of the mail. These investments now yield large tax returns to national treasuries. Space-based services are following the same model.*

Already, far-sighted entrepreneurs are thinking about further commercial expansion into space. As space exploration extends to the Moon and Mars, there will be potential opportunities for companies to provide crew and cargo transportation services, telecommunications and navigation systems, and space-based resource extraction and processing capabilities.

For example, Moon rocks are rich in oxygen that might be exploited to provide life support systems for lunar operations. Liquid oxygen can also be used as a rocket propellant – and it might be more economical to manufacture it in space than to lift it off the Earth.

Mining the Moon might also yield titanium – a strong but light metal favoured for high-end aerospace applications. Finally, the Moon's known abundance of Helium-3 could prove valuable if fusion reactors ever become feasible in the future.

There are also potential opportunities in commercial space tourism, both real and virtual. New telecommunications and robotic innovations create the prospect of offering customers on Earth a "virtual presence" on the Moon or Mars. For those who yearn to experience the real thing, sub-orbital spaceflight is on the verge of becoming reality. The future may also hold Earth-orbiting space hotels and excursions to the Moon.

Much of the technology for space exploration will be created by business, and business will find unexpected ways of exploiting this know-how in the wider economy. Governments can assist by stimulating links between the public and private sectors in innovative ways – prize funds are one example.

For business to be confident about investing, it needs the certainty of a long-term commitment to space exploration, the opportunity to introduce its ideas into government thinking, and the rule of law. This means common understanding on such difficult issues as property rights and technology transfer. The Coordination Mechanism foreseen as part of the Global Exploration Strategy will provide a forum to discuss these important issues.

The challenges and constraints of doing things in space stimulate creative minds. Many of the capabilities and technologies developed for the space programme probably would not have appeared in its absence, even with the same level of investment.

Space exploration brings together diverse expertise, creating opportunities for innovative ways of working. Skills required for space exploration, such as the ability to engineer very complex systems and design highly reliable mechanisms and software, are now used in the wider economy.

Some of the challenging technologies for the new era of space exploration include:

- efficient power generation and energy storage;
- space and surface transportation;
- communications and navigation;
- health care for human explorers,
- including tele-medicine;
- autonomous operation and smart decision-making for robotic explorers;
- planetary resource extraction and utilisation;
- on-orbit spacecraft servicing;
- human-robot cooperation and
- safe habitats with efficient life support and environmental control.

Development of these technologies will be driven by the constraints of space exploration, such as minimising mass and designing for reliable operation in a high radiation environment. Such attributes often lend themselves to terrestrial products and services. For example, robotic instrumentation developed to search for life on Mars is now being turned into a portable tuberculosis diagnostic machine for use in the developing world.

We will need to support human and biological life far from Earth by conserving resources and recycling as much as possible. Meeting these challenges will foster spin-off opportunities in fields such as medicine, agriculture and environmental management, and help achieve sustainable development on Earth.

In the past, human explorers have overcome complexities and uncertainties that demanded the utmost intelligence, ingenuity and innovation. Space exploration in the future will be no different. New technologies and an effective partnership between humans and machines will be key requirements in exploring and exploiting planetary surfaces to support human operations in remote locations.

Theme 4: A global partnership

Space is an unforgiving environment and no nation has the resources to take on all of its challenges at once. So space-faring nations have worked together from the earliest days in bilateral or multilateral partnerships.

The Apollo-Soyuz project in the 1970s was a striking example not just of technical cooperation, but also of political détente at the height of the Cold War. The 17-nation European Space Agency has its origin in the wish to build scientific links across the whole continent. Today, ESA's programme – which includes exploring Mars and building launchers and meteorological satellites – is far beyond the capabilities of any one European country.

The shared challenges of space exploration and the common motivation to answer fundamental scientific questions encourage nations of all sizes to work together in a spirit of friendship and cooperation.

The International Space Station programme, arguably the largest project of its type ever undertaken, has clearly demonstrated the value of a partnership approach. The U.S., Canada, Europe, Japan, and Russia have achieved together what no one nation could have accomplished alone – and, in the process, have forged strong ties, including cultural and political understanding.

Other examples of partnership abound:

- The Japan Aerospace Exploration Agency (JAXA) and NASA worked together to land the Hayabusa probe on the asteroid Itokawa. It is expected to return the first samples from the asteroid to Earth in 2010.
- Novel U.S. and European scientific instruments will soon orbit the Moon aboard an Indian spacecraft.
- The Chinese Double Star spacecraft are probing the relationship between the Earth's magnetic field and the solar wind with the help of instruments built in Europe.
- China and Russia are planning a joint mission to one of Mars' moons.
- Japan and Europe are cooperating on a mission to the innermost planet, Mercury.

These successes suggest that much more can be achieved with a global strategy for space exploration. Partnerships will enable nations to develop a common understanding of their respective interests, to share lessons learnt and thus avoid costly mistakes, and to discuss scientific results that will help in planning for the future.

Most importantly, we need a forum to discuss the essential building blocks of space exploration and practical issues such as interoperability – ensuring that different systems can work together. Internationally-agreed standards that allow a mobile phone bought in China to work in Canada or a car made in Germany to meet U.S. safety laws are critical to the global economy; they will be just as important when human activities extend beyond Earth. Complex issues such as the protection of areas of scientific importance may arise and can be discussed before they block progress.

By developing a common language of space exploration, nations can more readily share their specific objectives and enhance opportunities for joint projects. Leveraging national funds and coordinating mission objectives will enable them to build upon, strengthen, and expand existing global partnerships through space exploration.

This spirit of partnership will indirectly enhance global security by providing a challenging and peaceful activity that unites nations in the pursuit of common objectives. It is inclusive; the goal is to expand the opportunity for participation in space exploration to all nations and their citizens.

Theme 5: Inspiration and education

Space exploration catches our attention in a special way. It excites and inspires us to think about the wonders of the universe in which we live. People all over the world experience a sense of pride in unique achievements and the pain of failure when missions go awry.

In the future, new virtual reality technologies will enable them to share the excitement and wonder of exploration and discovery more viscerally than ever before. In a very real sense, they can "be there" when humans land on Mars or robots land on Europa.

One of the greatest legacies of space exploration is the role it plays in inspiring young people to think about what they want to achieve in their lives and to reach beyond the obvious. An interest in space steers many of them toward careers in science and technology and prompts them to make educational choices that will get them there. Space exploration programmes also provide a wealth of new information for educators in all disciplines, making lessons more exciting and intriguing for their students.

Space exploration creates jobs with limitless possibilities for creativity, challenge and motivation. This is a powerful magnet to attract and sustain new generations of scientists and engineers, most of whom will find careers in the wider economy.

Many countries are concerned about a decline in their scientific and technical talent. A vibrant space exploration programme can help turn this trend around.

Chapter 3: Mapping the space exploration journey

Space exploration follows a logical set of steps, starting with basic knowledge and culminating, it is hoped, in a sustained human presence in space. This journey requires a variety of both robotic and human missions. The Global Exploration Strategy provides a framework to coordinate the efforts and contributions of all nations so that all may participate in the expansion into space and benefit from it.

Since the first satellite was launched in 1957, space exploration has evolved in a characteristic way, progressing steadily from short term, very focussed missions to longer and more comprehensive ones.

During the Apollo programme in the 1960s and 1970s, humans visited the Moon for fewer than three days on each mission. With space stations such as

Salyut, Skylab, Mir and the International Space Station, we learned how to live for months in space. Russian cosmonaut Valeri Polyakov holds the record with 14 months onboard Mir.

By building upon these experiences, we are now preparing to establish a sustained human presence on the Moon and, eventually, in other parts of the solar system. The long term space exploration envisioned in this document is very different from the International Space Station. It is not a single space project but instead will comprise multiple missions and projects, large and small, to several destinations. Nations not involved in the ISS can and are making valuable contributions to space exploration.

Individual projects may emphasise certain goals more than others – for example, focussing on robotic science on Mars or testing technology needed for resource utilisation on the Moon. Each project will support the overall goal of extending the human frontier, step by step.

This diagram[490] shows how far we have progressed toward continuous living and working at key destinations. The central vertical bar in the diagram shows the threshold that must be crossed to achieve sustainable space exploration. This means not just simply tackling a new environment for a brief time, but

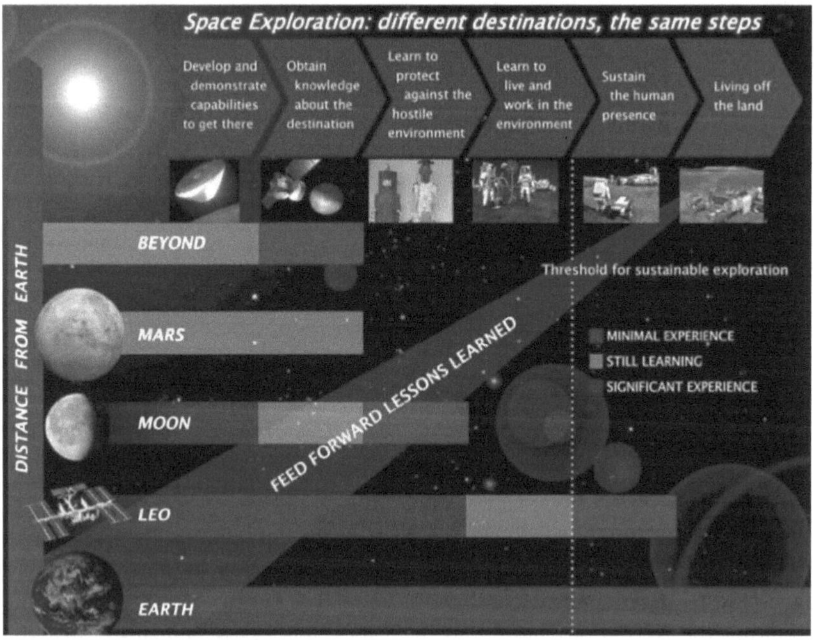

Fig. 3. *Space exploration: different destinations, the same steps.*

actually living there and using local resources, with little or no support from Earth.

We have not yet reached this level of autonomy for either robotic or human missions. Our activities in low Earth orbit approach the threshold of sustainability but crossing it remains an enormous challenge, even for robotic missions. For example, in principle, we have the technology to refuel communications satellites but we lack the infrastructure to make this a reality.

Robotic exploration is a key first step in expanding human presence into the solar system. Several generations of robotic exploration may be required to gain basic knowledge about a target destination before human exploration is useful or justified.

First, we send orbiters to remotely sense the surface and identify safe locations for landing. They are followed by landers that investigate the surface directly and then by robotic sample-return missions that carry material back to be examined in terrestrial labs.

Today, we have a limited amount of material on Earth that has been returned from outer space. Within the next decade, this knowledge will be increased by robotic missions returning material from certain asteroids and one of the moons of Mars.

The first robotic sample-return mission from the surface of Mars will likely occur around the same time that humans return to the Moon – an indication of just how large a technical challenge it represents.

Robotic probes that have explored the major bodies in the solar system have generated much valuable data but, for the more distant destinations ("Beyond", in the terms of the diagram), the knowledge accumulated so far has been limited by the constraints of our space technology.

We have glimpsed a riverbed and rocks of ice on Titan, but we do not know the composition of the ice, or if rivers still flow there. We believe the ice of Europa covers a liquid ocean, but we do not know whether it might contain life.

We have accomplished only the first tentative steps towards understanding these destinations; we cannot speculate if and when humans will reach them or what technology we will use.

This schematic picture of exploration shows that experience gained at each step of the journey enables the next one. Equally important, parallel progress toward several destinations may well generate valuable experience that is useful for all. Progress along the pathway to each destination will be assisted by increased coordination between projects.

Chapter 4: The Moon: A second home in the solar system

> *The Moon will be the first place where humans learn to live on another celestial body. Just three days from Earth, the Moon has low gravity and natural resources that make it an ideal location to prepare people and machines for venturing farther into space. As a repository of four billion years of solar system history and as a place to observe the Earth and the universe, it has great scientific potential. Exploration of the Moon will also reveal whether the resources available in space will allow humans to live off the land.*

In the 1960s, robotic spacecraft from the United States and the Soviet Union began exploring the Moon. The first soft landing was made in 1966 by the Soviet spacecraft Luna-9. It was followed by several more Soviet and U.S. lunar missions, including orbiters, sample-return missions, and rovers.

During this period, six Apollo crews also landed on the Moon and returned samples to Earth.

Thanks to these dramatic successes, lunar material could be examined in laboratories on Earth. The oldest material proved to be nearly a billion years older than the oldest known terrestrial rocks. Samples from the Moon still provide the best measurement of the age of any planetary surface.

Sustained human exploration will start on the Moon. That is where we will learn to live and work without immediate support from Earth and where we can test technologies needed for human missions to Mars and beyond.

Lunar scientific exploration will involve three types of investigations: science "of the Moon", science "from the Moon", and science "on the Moon". Science "of the Moon", which involves lunar geology, geochemistry and geophysics, will help us understand the history of the Moon. Current theories suggest the Moon was created when a body the size of Mars struck the young Earth, throwing vaporized rock into Earth's orbit. This material later coalesced into the Moon.

The Moon is an invaluable witness to much of solar system history. It has recorded this history more completely and more clearly than any other planetary body. For example, did the comets and meteorites that bombarded the Earth and the Moon in their early history contain the building blocks of life? The answer may be preserved on the pristine surface of the Moon. To make sense of the data encoded in the Moon, we may need both extensive robotic exploration and sophisticated surveying by humans at sites of high scientific interest.

Science "from the Moon" will take advantage of the Moon's lack of atmosphere and its "radio quiet" environment to provide a stable platform for observing the universe. For example, astronomers are interested in constructing a lunar-based low-frequency radio telescope to "see" signals emanating from the formation of the first stars, billions of years ago.

Science "on the Moon" will investigate the effects of the lunar environment on robotic instruments, equipment and humans. Exposure on the lunar surface to low gravity, radiation, dust, micrometeorites and wide variations in temperature will pose numerous challenges. Understanding these effects will enable engineers to develop materials and design systems for long-term use by humans in this hostile environment.

To sustain human presence beyond Earth, we must learn from science "on the Moon" how to live and work on other celestial bodies. A critical step will be to determine whether we can use the Moon's resources. For example, the ability to extract oxygen from the lunar soil might provide not only breathable air for the crew's life support system but also perhaps fuel for spacecraft.

Another priority will be to build on our experience with the International Space Station to develop efficient recycling techniques that will reduce the use of consumables such as air, power and water. This work may also teach us how to manage precious resources on Earth.

Finally, it is incumbent upon us to consider that the lunar environment is both fragile and special; we must protect and preserve it even as we explore it.

The Moon, as our closest "natural space station," is the ideal place for humanity to develop the capability to journey to Mars and beyond. The Moon is only three days travelling time from Earth, compared with a minimum six months for Mars, and the communications delay is only one and a half seconds instead of tens of minutes.

The development of transportation, life support and habitation systems, and advanced robots, can all be attempted in a challenging environment on the Moon, before they are used farther away. Human explorers will also use the Moon to develop their skills and learn how to prepare their bodies and minds for the long journey ahead.

The Moon has a strong place in the culture of many peoples and it instinctively appeals to the human imagination. It is the only celestial body that is familiar to all humanity as a "place" and not just a point of light. It is a place, moreover, that many more humans can aspire to visit in the future.

Just as the first lunar landings nearly 40 years ago enthralled an earlier generation, lunar exploration in the years to come will continue to inspire enthusiasm and creativity among future generations around the world.

Compared with the early days of lunar exploration, the more sophisticated media of today will create novel means to relate the space exploration journey to all people. Anyone may be able to participate personally in lunar robotic and human missions through virtual-presence technologies. In particular, children can be involved and will be inspired to become the explorers of the future – as scientists, engineers, teachers and entrepreneurs.

Chapter 5: To Mars and beyond

Mars is a key focus for space exploration because it has both an atmosphere and water. Increasingly complex robotic missions are already being mounted to study its geology and to search for the presence of ancient and maybe even existing life forms. As robotic capabilities reach their limits, humans will step in to unlock further secrets. Other destinations such as asteroids, comets and the moons of the giant planets are also important targets of human curiosity.

Mars engages the public's imagination as much as the Moon – perhaps even more. Millions avidly follow the adventures of little rovers that explore the Martian surface. Human exploration, when it happens, will be even more exciting.

The possibility of humans visiting, exploring and living on Mars may be the most challenging but also the most rewarding objective of space exploration in this century. Although many approaches to such a mission have been studied, its technical and financial feasibility is not yet certain and much more preparation is needed.

At present, Mars is being explored by robotic orbiters, landers and rovers. In the longer term, there are plans for ambitious robotic missions to return samples from the Martian surface.

A better knowledge of Mars would help us better understand Earth's history and evolution. Evidence suggests that Mars and the Earth were, long ago, even more similar than they are today. The reasons for their subsequent divergent evolution are still poorly understood. Questions remain as to whether life could have appeared on Mars or could even still exist today.

A close-up study may yield important clues about how the planet evolved from one capable of sustaining life to the barren world we see now. By studying Martian geology, weather and climate and other natural phenomena, researchers will learn not only more about Mars but also gain insight into how Earth's environment has evolved and how it may change in the future.

At present, the focus is on robotic reconnaissance and surface exploration. Drilling to collect samples will help further unravel the history of the planet and the possible evidence of life. For example, reaching under the Martian surface may reveal life forms protected from the harsh cold and radiation above.

Robotic exploration is ultimately limited, though. More effective exploration can be achieved by leveraging the insight and ingenuity of human explorers sent to Mars.

Because of its similarity to Earth, Mars is the place in the solar system where human life could most likely be sustained in the future. There are many significant technological challenges that must be overcome, but Mars also gives us something to work with. It has a thin atmosphere that partially shields the surface from radiation. Surface temperatures at low latitudes are quite harsh but not unmanageable. And it has a day-length only 37 min longer than Earth's, which makes the production of electricity from solar cells feasible. This could sustain humans and their machines until more advanced power sources are available.

The potential presence of water ice and maybe liquid water under the surface might make sustained human habitation more practical and self-supporting. It may also be possible to synthesise methane and oxygen rocket propellants from the carbon dioxide in the atmosphere and hydrogen from water ice.

When they go, humans will be able to advance the exploration of Mars in ways not possible with robots alone.

Several nations can afford to send their own robotic exploration missions to Mars but there are significant benefits in coordinating these national efforts and future human exploration missions. Groups like the International Mars Exploration Working Group are already making this happen for scientific missions. Given the enormous challenges, human exploration of Mars may only be achievable through sustained international cooperation.

The historic decision to start the human journey to Mars is still several years away. However, two important first steps are being taken: first, the engagement of more nations in space exploration; and second, the start of global coordination, as foreseen in this Framework document.

As with the Moon, the Martian environment is both fragile and special and we must protect and preserve it, even as we explore it.

Asteroids and comets left over from the formation of the solar system have high scientific interest. Robotic spacecraft have already started to explore these relics of the early solar system, which contain water and organic compounds. The first material to be returned from a comet's tail is already yielding unexpected results. Future discoveries are certain when pristine material from a comet's nucleus, and from an asteroid, can be brought to Earth.

The first sample-return mission to an asteroid is already on its way back to Earth and an attempt to land a probe on the surface of a comet is underway. Such missions could also give us a better understanding of the risk presented by a few asteroids with orbits that could cause them to hit Earth.

More distant destinations, such the moons of the giant planets Jupiter and Saturn, are extremely important scientifically. For example, Europa likely has liquid water beneath its ice crust and Titan's cold, dense atmosphere contains carbon-based molecules. These are not realistic targets for human exploration in the coming decades, but they will become more accessible as space exploration technologies improve.

In parallel with the sustained human exploration of the Moon, the robotic exploration of Mars, asteroids, and other destinations offers nations the chance to develop important skills that may later be used when humans start to explore Mars and beyond.

Chapter 6: Implementing the Global Exploration Strategy

International cooperation expands the breadth of what any one nation can do on its own, reduces risks and increases the potential for success of robotic or human space exploration initiatives. It is important to establish and sustain practical mechanisms to support exploration if humanity is to succeed in implementing long-term space exploration on a global scale.

In early 2006, 14 space agencies began discussing their common interests in space exploration. With different backgrounds, interests and capabilities, the agencies have started to develop a mutual understanding of, and language, for space exploration.

The success of the preliminary discussions suggests that the future establishment of a formal, though non-binding and voluntary, coordination mechanism among interested space agencies could assist the development and implementation of the *Global Exploration Strategy.*

Such a mechanism could help coordinate global space exploration by:

- providing a forum for participants to discuss their interests, objectives and plans in space exploration and
- promoting interest and engagement in space exploration activities throughout society.

For purposes of:

- making use of all available resources, knowledge and technological capabilities;
- leveraging each agency's individual investments;
- identifying gaps in national programmes and overlaps between them;
- sharing "lessons learnt" from national and international missions;
- improving the safety of humans in space – for example, through interoperability of life support systems and
- enhancing the overall robustness of global space exploration.

Tab. 1: *Principles of international coordination and resulting requirements*

Principles	Resulting requirements
Open and inclusive	• receives inputs from all interested agency participants that invest in and perform activities related to space exploration
	• provides for consultations among all interested agencies with a vested interest in space exploration, and also space agencies or national government agencies without specific related capabilities
Flexible and evolutionary	• takes into account and may integrate existing consultation and coordination mechanisms
	• allows consultation and coordination structures and mechanism(s) to gradually build and evolve as requirements for these activities grow
	• allows for entry of assigned representatives of governments with a vested interest and clear stake in space exploration
	• provides for different levels of consultation and coordination
Effective	• encourages participating agencies to accept the role of the coordination process and act upon the anticipated results of the coordination mechanism
Mutual interest	• contributes to common peaceful goals and benefits all participants respects the national prerogatives of participating agencies
	• allows for optional participation based on the level of each agency's interest
	• Interest

Principles of international coordination

The Table 1 below outlines key principles for international coordination for sustainable space exploration and examples of resulting requirements for the mechanism.

The way forward

Using the principles elaborated above, the 14 space agencies have agreed to pursue the establishment of a formal Coordination Mechanism for the coordination of the Global Exploration Strategy. The specific terms of reference for such a mechanism are being defined and will be described in a separate document.

Although potential areas and activities that could benefit from coordination may change over time, areas for initial consideration include:

- identification of standards to promote interoperability;
- methods for sharing scientific data and related analyses;
- identification of common services, allowing for the development of shared infrastructures;
- mechanism(s) to allow the provision of payload opportunities;
- ways and means to include broader future participation in the planning and coordination process; and
- an assessment of the requirement for any relevant international legal agreements.

The 14 participating agencies have recognized that the development of a common international exploration coordination tool will enhance the implementation of the coordination process. This coordination tool is being defined and will be described in another document.

The Coordination Mechanism will be a voluntary partnership. It will not diminish each agency's right of autonomous decision-making. However, all participants hope sharing knowledge, ideas and plans will help to optimize agency decisions.

Chapter 7: A window to tomorrow: Why we explore

Space exploration is a global partnership in service of society. It will require both human endeavour and technological innovation and it will deliver new knowledge and commercial opportunity.

Space exploration is driven by:

- *human civilization*: extending human presence to other planets to enable eventual settlement;
- *scientific knowledge*: pursuing scientific activities that address fundamental questions about the history of Earth, the solar system, and the universe – and about our place in them;
- *global partnerships*: providing a challenging, shared, and peaceful activity that unites nations in pursuit of common objectives;
- *economic expansion*: expanding Earth's economic sphere and conducting space activities that benefit life on the home planet and
- *public engagement*: using a vibrant space exploration programme to engage the public, encourage students and help develop the high-tech workforce required to address the challenges of tomorrow.

This Framework for Coordination of the Global Exploration Strategy presents a vision of tomorrow in which the human frontiers are permanently expanded into the solar system, inspiring generations of humanity to come. It foresees how the robotic and human space exploration efforts undertaken by many nations, working individually and in partnership, could be coordinated to maximise the long-term benefits for all humanity.

Each agency that has contributed to this document shares this vision and invites other agencies or institutional bodies around the world to join them in translating the vision into reality.

[486] The full text can be downloaded from NASA's website http://www.nasa.gov/pdf/178109main_ges_framework.pdf.

[487] In alphabetical order: ASI (Italy), BNSC (United Kingdom), CNES (France), CNSA (China), CSA (Canada), CSIRO (Australia), DLR (Germany), ESA (European Space Agency), ISRO (India), JAXA (Japan), KARI (Republic of Korea), NASA (United States of America), NSAU (Ukraine), Roscosmos (Russia).

[488] Some languages favour "automatic" versus "robotic" but the two should be considered interchangeable.

[489] "Space agencies" refers to government representatives that include space agencies, science organizations and groups of space agencies that have been designated by their government to represent them.

[490] "LEO" in the diagram refers to Low Earth Orbit, the location of the International Space Station.

4.5 Overview of Europe's contribution to the ISS[491]

4.5.1 The European Space Policy on the International Space Station (ISS) and Exploration of the Solar System

The international exploration endeavour has a significant political appeal in a vision of European identity, due to its potential to contribute to the creation of new knowledge, to foster innovation and to engage new companies and research organisations in space activities.[492]

The US, China and Russia have moved forward with ambitious space exploration plans. Now, Europe needs to urgently respond to these challenges.

Human spaceflight and exploration are emblematic aspects of space. The ISS offers unique opportunities for fundamental and applied research using the conditions available in space. The European participation with the Columbus laboratory module and the Automated Transfer Vehicle and the presence of European crew secures a visible European role in this venture. The knowledge and insights gained on the ISS are translated into innovative applications for the benefit of people on Earth, e.g. for the development of new materials and new therapies in medicine, and in the preparation for future planetary missions.

Europe needs to achieve optimum utilisation of the International Space Station; prepare for a visible, affordable and robust exploration programme, involving the development and demonstration of innovative technologies and capabilities and the robotic exploration of Mars, to search for evidence of life and understand the planet's habitability.

4.5.2 The Green Paper on European Space Policy and "manned space flight"[493]

Manned flights are among the most emblematic aspects of the space sector and consume an important part of the budget of ESA, which has created a European

Astronaut Corps. Forty years after the first mission in orbit, manned space flights now always take place in the co-operative framework of the international space station (ISS). The ISS combines, under US responsibility, contributions from the five main participants: the United States, Russia, Japan, Canada and Europe. It is the most ambitious and the most costly civil research infrastructure ever built (about €30 billion in development costs).

Europe's contribution via ESA to the ISS remains modest. Compared with earlier forms of co-operation with Americans and Russians in the area of manned flight (Spacelab, access to the Mir space station), it has nevertheless increased considerably in terms of both the technological and industrial dimension of the developments and enhanced possibilities for experiments offered to the European scientific community.

European participation in the ISS The European contribution amounts to about 8% of the total effort (€3 billion in investment and €300 million/year in operating costs).

It comprises:

- a component of the station, the pressurised Columbus laboratory;
- the associated scientific instrumentation;
- an automatic freight transport facility, the Automated Transfer Vehicle (ATV), which will be regularly launched by Ariane-5 towards the space station in order to meet its logistical requirements.

The European astronauts have access to the station to carry out experiments via the American space shuttle or the Russian Soyuz vehicle. The level of European effort in the field of manned space flight – principally Spacelab and ISS – has sometimes been questioned, notably concerning the scientific interest, the actual possibilities for onboard experimentation and access for European astronauts. Moreover, American decisions concerning the financing of the ISS, the onboard experimental programme, astronaut visits and access to the station must be considered in the light of Europe's objectives.

4.5.3 ISS Intergovernmental Agreement

The International Space Station is a co-operative programme between United States, Russia, Canada, Japan and eleven Member States of the European Space Agency (Belgium, Denmark, France, Germany, Italy, The Netherlands, Norway, Spain, Sweden, Switzerland, and the United Kingdom).

Fig. 4. *Artist's impression, current configuration of the ISS (source: ESA/D.Ducros).*

It is governed by an international treaty, signed by these Member States on 29 January 1998, called the ISS Intergovernmental Agreement, which provides the framework for design, development, operation, and utilisation of a permanently inhabited civil Space Station for peaceful purposes.

Furthermore, bilateral Memoranda of Understanding exist between NASA and each of the four associated space agencies: The European Space Agency (ESA), Russian Federal Space Agency (FKA, formerly Rosaviakosmos), the Canadian Space Agency (CSA) and the Japanese Space Agency (JAXA, formerly NASDA), outlining relevant ISS responsibilities, obligations and rights between the agencies.

National jurisdiction extends to the ISS elements in orbit. This applies to areas such as criminal matters, liability issues, and protection of intellectual property rights.

Utilisation rights are outlined in the Memoranda of Understanding. The European Space Agency allocation rights comprise 8.3% of the Space Station utilisation resources (e.g. communications) and 8.3% of crew time, which represent approximately 13 h per week. Concerning the user accommodations (e.g. laboratories), ESA will have 49% use of the European Columbus Laboratory, NASA 49% and the Canadian Space Agency 2%.

One important point is that ESA and the other Space Station International Partners can barter or sell their unused utilisation rights among themselves and to other non-participants to the Station's programme.

4.5.4 ISS current configuration

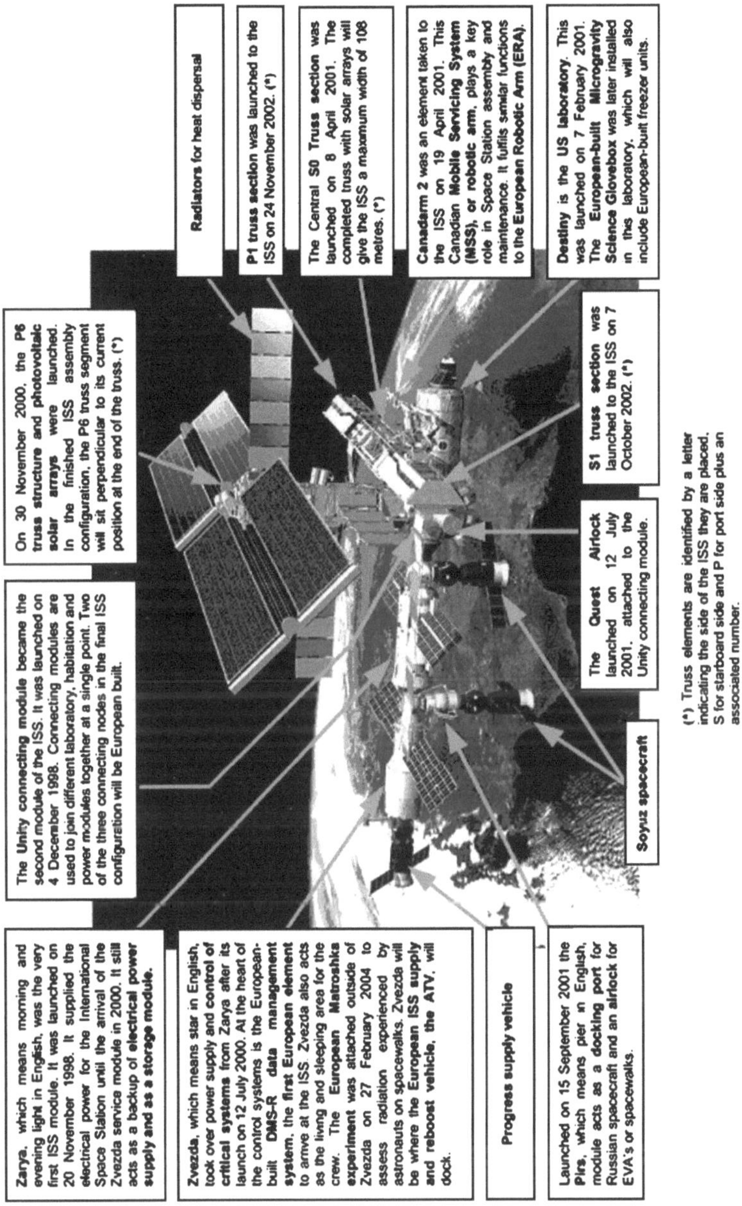

Fig. 5. *ISS current configuration (source: ESA/D.Ducros).*

4.5.5 ISS and Europe's major contributions

4.5.5.1 Columbus laboratory

Columbus is ESA's Research laboratory. It provides space for research facilities in the fields of material science, fluid physics and life science. In addition, an external payload area can accomodate experiments and applications in the fields of space science, Earth observation, technology and innovative sciences from space. Columbus will be permanently stationed at the International Space Station attached to another European-built module, Node 2.

4.5.5.2 Automated Transfer Vehicle (ATV)

The Automated Transfer Vehicle is Europe's unmanned supply vehicle for the ISS. It will take up to 9 t of cargo to the ISS, boost the station to a higher orbiting altitude and remove up to 6.5 t of waste from the station. It measures approximately 10 m long by 4.5 m in diameter, with solar arrays spanning more than 22 m

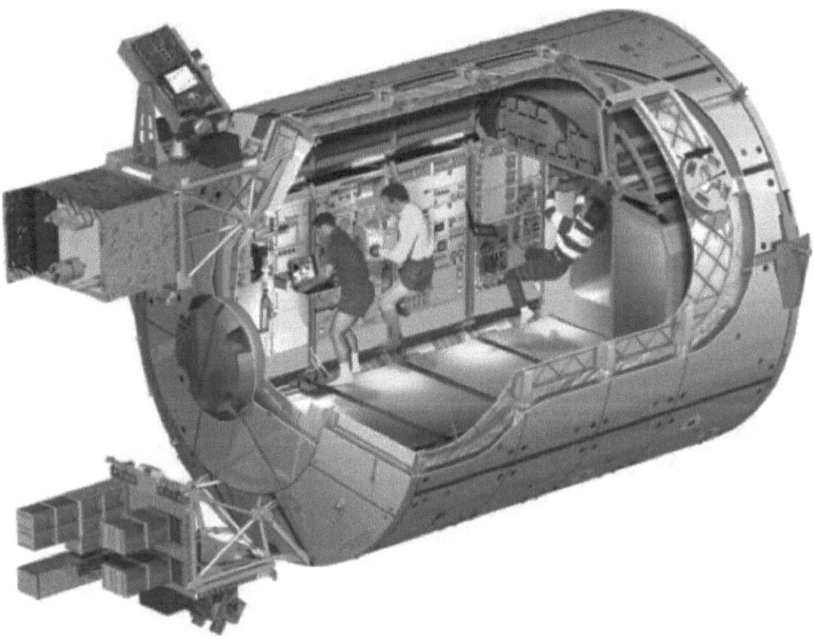

Fig. 6. *The European Columbus laboratory (source: ESA/D.Ducros).*

Fig. 7. *The Automated Transfer Vehicle (source: ESA/D.Ducros).*

for generating its electrical power. Cargo transported will include pressurized cargo, water, air, nitrogen, oxygen and attitude control propellant.

4.5.5.3 Node 2 and Node 3

Nodes are pressurised modules that interconnect the research, habitation, control and docking modules of the ISS. The Nodes are used to control and distribute resources between the connected elements. The ISS will have three Nodes. Node 1, called Unity, was developed by NASA. Node 1 became the second module of the ISS in orbit after its launch in December 1998. Nodes 2 and 3 are developed under an ESA contract with European Industry with Alenia Spazio as the prime contractor.

Node 2 has been, and Node 3 will be, transferred to NASA within the framework of a barter agreement between ESA and NASA whereby ESA will develop the Nodes for NASA as well as supplying additional hardware and

Fig. 8. *ESA-developed Node 2 (top) and Node 3 (bottom). Node 3 will be the attachment point of the Cupola (source: ESA/D.Ducros).*

services. In return NASA will launch the European Columbus Laboratory and its initial payload on the Shuttle.

Node 2 will be the first European Node launched. It will act as a connection point for the European Columbus laboratory, the US Laboratory Destiny and the Japanese Laboratory Kibo. It also will be the attachment point for the Japanese HII Transfer Vehicle, carry a docking adapter for the US Space Shuttle, and act as an attachment point for the Multi-Purpose Logistics Module (MPLM). The MPLM is a pressurised cargo container, which travels in a space shuttle cargo bay. Node 2 also provides a working base point for the Space Station Remote Manipulator System, a Canadian robotic arm called Canadarm 2.

Node 3 will be the second European node to arrive at the ISS and will be attached to the American-built Node 1, which was launched to the ISS in December 1998. The forward port of Node 3 will act as the connecting point for the European-built Cupola.

4.5.5.4 European Robotic Arm (ERA)

The European Robotic arm or ERA is a robotic arm, which serves to install solar arrays on the Russian Science and Power Platform. It further acts as an inspection tool on the Russian segment of the ISS and can carry out additional assembly and replacement tasks on the external surface of the station. The 11-m long ERA also serves to support or transfer cosmonauts carrying out tasks on spacewalks. It has an extensive range, as it is able to walk around the Russian segment of the station and while in orbit is able to manipulate up to 8000 kg of mass.

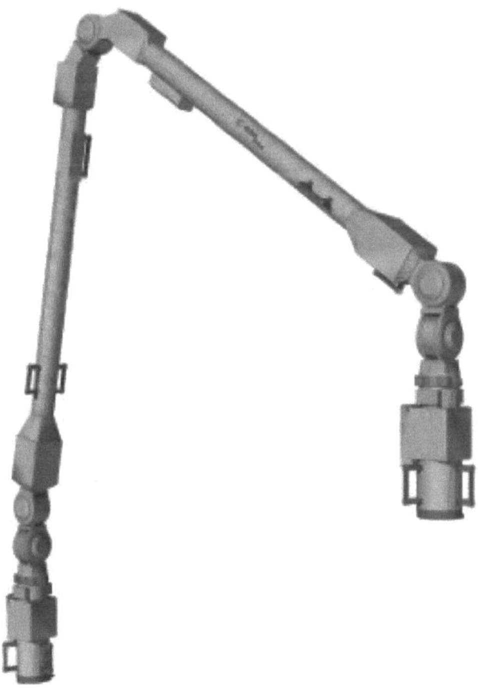

Fig. 9. *The European Robotic Arm (ERA) (source: ESA/D.Ducros).*

4.5.5.5 Data Management System (DMS-R)

Europe's DMS-R Data Management System was the first piece of European hardware on the ISS in July 2000. It includes two fault-tolerant computers and two control posts. It is the 'brain' or control centre of the Russian Segment of the ISS and carries out a great degree of the vital and fundamental functions on the station including: guidance, navigation and control of the entire ISS; failure management and recovery; and control of additional ISS systems and subsystems.

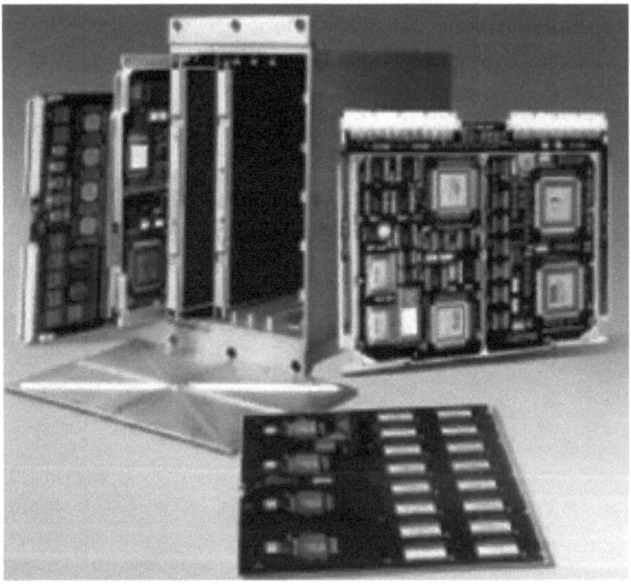

Fig. 10. *The European-built Data Management System (source: ESA).*

[491] "ISS General Information." European Space Agency. 26 May 2010. http://www.spaceflight.esa.int/documents/cupola/iss-general-information.pdf.

[492] The subsequent paragraphs are extracts from: Commission of the European Communities. European Space Policy. COM(2007) 212 final of 26 Apr. 2007. Brussels: European Communities.

[493] Extract on human space flight from: Commission of the European Communities. Green Paper – European Space Policy. COM(2003) 17 final of 21 Jan. 2003. Brussels: European Communities.

4.6 SETI's Declaration of Principles Concerning Activities Following the Detection of Extraterrestrial Intelligence[494]

Note: By resolution of the Board of Trustees on 17 August 1997, The SETI League, Inc. officially endorses the following Protocols, and respectfully requests that our members embrace them.

Adopted by the International Academy of Astronautics, 1989

We, the institutions and individuals participating in the search for extraterrestrial intelligence.

Recognizing that the search for extraterrestrial intelligence is an integral part of space exploration and is being undertaken for peaceful purposes and for the common interest of all mankind.

Inspired by the profound significance for mankind of detecting evidence of extraterrestrial intelligence, even though the probability of detection may be low.

Recalling the Treaty on Principles Governing the Activities of States in the Exploration and Use of Outer Space, Including the Moon and Other Celestial Bodies, which commits States Parties to that Treaty "to inform the Secretary General of the United Nations as well as the public and the international scientific community, to the greatest extent feasible and practicable, of the nature, conduct, locations and results" of their space exploration activities (Article XI).

Recognizing that any initial detection may be incomplete or ambiguous and thus require careful examination as well as confirmation, and that it is essential to maintain the highest standards of scientific responsibility and credibility.

Agree to observe the following principles for disseminating information about the detection of extraterrestrial intelligence:

1. Any individual, public or private research institution, or governmental agency that believes it has detected a signal from or other evidence of extraterrestrial intelligence (the discoverer) should seek to verify that the most plausible explanation for the evidence is the existence of extraterrestrial intelligence rather than some other natural phenomenon or anthropogenic phenomenon before making any public announcement. If the evidence cannot be confirmed

as indicating the existence of extraterrestrial intelligence, the discoverer may disseminate the information as appropriate to the discovery of any unknown phenomenon.

2. Prior to making a public announcement that evidence of extraterrestrial intelligence has been detected, the discoverer should promptly inform all other observers or research organizations that are parties to this declaration, so that those other parties may seek to confirm the discovery by independent observations at other sites and so that a network can be established to enable continuous monitoring of the signal or phenomenon. Parties to this declaration should not make any public announcement of this information until it is determined whether this information is or is not credible evidence of the existence of extraterrestrial intelligence. The discoverer should inform his/her or its relevant national authorities.
3. After concluding that the discovery appears to be credible evidence of extraterrestrial intelligence, and after informing other parties to this declaration, the discoverer should inform observers throughout the world through the Central Bureau for Astronomical Telegrams of the International Astronomical Union, and should inform the Secretary General of the United Nations in accordance with Article XI of the Treaty on Principles Governing the Activities of States in the Exploration and Use of Outer Space, Including the Moon and Other Bodies. Because of their demonstrated interest in and expertise concerning the question of the existence of extraterrestrial intelligence, the discoverer should simultaneously inform the following international institutions of the discovery and should provide them with all pertinent data and recorded information concerning the evidence: the International Telecommunication Union, the Committee on Space Research, of the International Council of Scientific Unions, the International Astronautical Federation, the International Academy of Astronautics, the International Institute of Space Law, Commission 51 of the International Astronomical Union and Commission J of the International Radio Science Union.
4. A confirmed detection of extraterrestrial intelligence should be disseminated promptly, openly, and widely through scientific channels and public media, observing the procedures in this declaration. The discoverer should have the privilege of making the first public announcement.
5. All data necessary for confirmation of detection should be made available to the international scientific community through publications, meetings, conferences, and other appropriate means.
6. The discovery should be confirmed and monitored and any data bearing on the evidence of extraterrestrial intelligence should be recorded and stored permanently to the greatest extent feasible and practicable, in a form that will make it

available for further analysis and interpretation. These recordings should be made available to the international institutions listed above and to members of the scientific community for further objective analysis and interpretation.

7. If the evidence of detection is in the form of electromagnetic signals, the parties to this declaration should seek international agreement to protect the appropriate frequencies by exercising procedures available through the International Telecommunication Union. Immediate notice should be sent to the Secretary General of the ITU in Geneva, who may include a request to minimize transmissions on the relevant frequencies in the Weekly Circular. The Secretariat, in conjunction with advice of the Union's Administrative Council, should explore the feasibility and utility of convening an Extraordinary Administrative Radio Conference to deal with the matter, subject to the opinions of the member Administrations of the ITU.
8. No response to a signal or other evidence of extraterrestrial intelligence should be sent until appropriate international consultations have taken place. The procedures for such consultations will be the subject of a separate agreement, declaration or arrangement.
9. The SETI Committee of the International Academy of Astronautics, in coordination with Commission 51 of the International Astronomical Union, will conduct a continuing review of procedures for the detection of extraterrestrial intelligence and the subsequent handling of the data. Should credible evidence of extraterrestrial intelligence be discovered, an international committee of scientists and other experts should be established to serve as a focal point for continuing analysis of all observational evidence collected in the aftermath of the discovery, and also to provide advice on the release of information to the public. This committee should be constituted from representatives of each of the international institutions listed above and such other members as the committee may deem necessary. To facilitate the convocation of such a committee at some unknown time in the future, the SETI Committee of the International Academy of Astronautics should initiate and maintain a current list of willing representatives from each of the international institutions listed above, as well as other individuals with relevant skills, and should make that list continuously available through the Secretariat of the International Academy of Astronautics. The International Academy of Astronautics will act as the Depository for this declaration and will annually provide a current list of parties to all the parties to this declaration.

[494] "Declaration of Principles Concerning Activities Following the Detection of Extraterrestrial Intelligence." SETI Website. 17 Aug. 1997. 26 May 2010. http://www.setileague.org/general/protocol.htm.

4.7 Extract from "Mars Life" by Ben Bova[495]

4.7.1 The atmosphere of Mars

(...)

"The atmosphere of Mars is a mere wisp, thinner than Earth's high stratosphere. It is composed mostly of carbon dioxide, with traces of nitrogen, oxygen, and inert gases such as argon and neon. The air pressure at the surface of Mars is about the same as the pressure thirty-some kilometres up in the high stratosphere of Earth's atmosphere, so thin that an uncovered glass of water will immediately boil away even when the temperature is far below zero.

Which it is most of the time. Mars is a cold world. At midsummer noon on the Martian equator, the ground temperature might get as high as 70°F. But at the height of a person's nose, the temperature would be zero, and that night it would plunge to a hundred below or even colder. The thin Martian atmosphere retains almost none of the Sun's heat: it reradiates back into space, even at noon on the equator.

There is water on Mars, however. The polar caps that can be seen from Earth even with an amateur telescope contain frozen water, usually overlain with frozen carbon dioxide: dry ice. Explorers found layers of permafrost – frozen water – beneath the surface, enough underground water to make an ocean or at least a sizable sea.

There is abundant evidence that water once flowed across the surface of Mars. The entire northern hemisphere of the planet may once have been an ocean basin. Mars was once considerably warmer and wetter than it is now.

But today the surface of Mars is a barren desert of highly oxidized iron sands that give Mars its rusty red coloration. Those sands are loaded with superoxides; the planetwide desert of Mars is more like powdered bleach than soil in which plants could grow.

Yet there is life on Mars. The First Expedition discovered lichen-like organisms living inside cracks in the rocks littering the floor of the Grand Canyon of Mars. The Second Expedition found bacteria living deep underground, extremophiles that metabolize solid rock and water leached from the permafrost.

And the human explorers discovered an ancient cliff dwelling built into a niche high up the north wall of the Valles Marineris. There were once intelligent Martians, but they were wiped out in a cataclysm that scrubbed the entire planet clean of almost all life."[496]

(...)

4.7.2 What are we getting out of exploration?

(...)

"What are we getting out of the exploration of Mars. What have you discovered that's worth the billions of dollars that have been spent on your programme?"

Jamie felt his cheeks flare with the sudden anger and hoped the cameras didn't pick it up. Forcing himself to take a calming breath before speaking, he answered, "That's sort of like asking how high is up".

"What have you found?" Samuels insisted. "After all, you've spent billions – "

"Life", Jamie said sharply. "We've found the most important thing that's ever been discovered, Rhonda. We've found that ours is not the only world on which life exists. More than that, we've found *intelligent* life. Intelligence arose on Mars, just like it has on Earth."

"But it's gone extinct."

"That's not the important point", Jamie said. "The important point is that intelligence is not rare in the universe. We've explored two planets – Earth and Mars – and found intelligence on both of them. Two for two. There's probably all sorts of intelligent species on other worlds."

"Really?" Rhonda Samuels's carefully painted face looked almost fearful.

"Really", said Jamie."[497]

(...)

[495] Bova, Ben. Mars Life. Mars Life. New York: A Tor Book, 2008.
[496] Ibid. 7–8.
[497] Ibid. 47.

4.8 Extract from "The Dream – or posthumous work on lunar astronomy" by Ludwig Kepler[498]

Fifty thousand German miles away in the ether there lies the island of Island of Levania [the Moon]. The way to it from Earth and back to it is very seldom open for passage. We demons from Levania ourselves then have quite easily access to it; for the Earthborn human willing to travel there, however, this is very difficult and fraught with the greatest danger. To accompany us, therefore, we never choose humans who only know how to sit and do nothing, fat persons or those who are only interested in sensual delights; instead we choose those who spend their lives diligently riding their hunting horses or who frequently visit India by ship and thus are used to live on hard bread, garlic, dried fish and other food despised by gourmets. We especially find gaunt old women suitable, because they are already well-versed in the art of riding broomsticks or shabby cloaks to cover never-ending distances on Earth. From Germany no men are suitable, but we don't refuse the Spaniards' skinny bodies.

To pass the whole way to Levania, long as it is, takes at most four hours. We much-occupied demons are not free to decide on the time to leave; we only come to know of it when the Moon in its Eastern part is beginning to grow dark. Before the Moon once again is fully alight, we must have reached our goal, if we don't want to loose our direction. As the opportunity for traveling thus always occurs very fast, we can only take a few of your kind, and only those who are especially devoted to us. Very many of us then converge on the chosen one and, helping him, lift him up very quickly. This first move for him is the worst, as he is catapulted upwards as if the power of gunpowder had blasted him into his flight over mountains and seas. Therefore he must be sedated with opium before and his arms and legs must be stowed away very carefully, so as not to be ripped from his body; the force of the recoil needs instead to be equally divided between his body parts. Then he will be hit by new difficulties: unimaginable cold and lack of breath, to be met by us the one through our inborn strength, the other by holding a wet sponge in front of nose and mouth. As soon as the first part is over, the journey becomes easier; then we leave our accompanying humans to their own devices – and like spiders they expand and contract and work their way on their own power, until their body mass turns to its goal on its own volition. Because, however, the force of the Moon's gravity grows as we come nearer, they are in danger of a hard crash on arrival, which is why we hurry ahead and save them from this danger. Usually, when the humans wake from their opium-induced unconsciousness, they complain of a great tiredness in all their limbs, from which they only slowly recover enough to walk.

[498] (1614); English translation of Ludwig Günther's German translation by Ulrike Landfester.

4.9 Religion and Human Space Flight

„Wer weiss, ist es ihr nicht zugedacht, dass sie dereinst jene entfernte Kugeln des Weltgebäudes und die Trefflichkeit ihrer Anstalten, die schon von weitem ihre Neugierde so reizen, von nahem soll kennen lernen? Vielleicht bilden sich darum noch einige Kugeln des Planetensystems aus, um nach vollendetem Ablaufe der Zeit, die unserem Aufenthalte allhier vorgeschrieben ist, uns in andern Himmeln neue Wohnplätze zu bereiten. Wer weiss, laufen nicht jene Trabanten um den Jupiter, um uns dereinst zu leuchten?

... wenn man mit solchen Betrachtungen und mit den vorhergehenden sein Gemüth erfüllt hat: so giebt der Anblick eines bestirnten Himmels bei einer heitern Nacht eine Art des Vergnügens, welches nur edle Seelen empfinden.“[499]

"Who knows whether it is not determined that in future the soul will get to know at close quarters those distant spheres of the cosmic structure and the excellence of their dwelling places, which already attract its curiosity from far away? Perhaps that is why some spheres of the planetary system are already developing, in order to prepare for us in other heavens new places to live after the completion of the time prescribed for our stay here on Earth. Who knows

whether those satellites do not circle around Jupiter so as to provide light for us in the future?

... when we have completely filled our dispositions with such observations and with what has been brought out previously, then the sight of a starry heaven on a clear night gives a kind of pleasure which only noble souls experience."[500]

[499] Kant, Immanuel. Allgemeine Naturgeschichte und Theorie Des Himmels oder Versuch von der Verfassung und dem mechanischen Ursprunge des ganzen Weltgebäudes, nach Newtonischen Grundsätzen abgehandelt. 1755.

[500] Kant, Immanuel. Universal Natural History and Theory of the Heavens. 1755. Translated by Ian Johnston, Vancouver Island University, Nanaimo, British Columbia, Canada.

4.10 An historian's viewpoint – Historical approaches to human space flight and the "Humans in Outer Space" project

Luca Codignola

Why do we need an historian in the introductory session of a conference that is devoted to looking forward towards the future of the Earth and indeed of the universe? Is there any real reason why we should look backward, to the past experience of humankind, in order to provide a context for papers that will deal with, among other things, interstellar communication, human impact on other planets, space governance, etc.? Is there anything else besides the fact that I, personally, have been involved in this project since its beginning, to assess whether past experiences of discovery, encounter, and contact – whatever we want to call the meeting of diverse entities – could help us provide a good practical and ideological framework for what may indeed lay ahead in the future – if not ours, then that of our progeny?

Actually, I am not really sure that we do need an historian, especially when, should you read the conclusion to my own article in the book that paved the way to this conference, *Humans in Outer Space – Interdisciplinary Odysseys*, you will note that I closed on a rather sombre note. Let me quote my own piece:

> "[I]f one really wants to know what the study of history has taught us about preparing for future encounters with entities from outer space, the truth of the matter is that there is no way humankind can be ready for them – if, and when, these encounters take place. No weapons can be readied, because we do not know our enemies. No God can be brought in, because we would not know whether it is ours or theirs. No production can be implemented, because we do not know what can be useful. No exchange can be envisaged, because we do not know what is needed. Whenever the time of encounters came in the past of humankind, neither side was ready."[501]

Still, this pessimistic assessment was somewhat sweetened by a final consideration that left some room for hope and trust:

> "[H]umankind, in the past, has proved to be able to survive, through sometimes at great costs, through adjustment, adaptation, and compromise. We trust that will happen again, when the time comes, although at what cost, no one is able to foresee."[502]

In the past two occasions in which this group has met to discuss the issue of humans in outer space, the Genoa working seminar in March 2007 and at the large Vienna conference in October 2007, what happened in the past – history, that is – has always loomed large in the mind of participants. Although scientists and space practitioners seemed to be more prone to point out the difficulties, perhaps insurmountable, of space travel, let alone interstellar contact, a number of us – historians, literary and media critics, sociologists, philosophers – took advantage of the fact that we do not have any real knowledge or familiarity with the world of space exploration to speculate what would happen if... U.S. historian James Muldoon, for one, even made the assumption that "the cost of entering space will decline to the point that small nations, alone or in concert, and wealthy private entities could afford to enter the space race".[503] Indeed, a session of this conference, as you know, is fully devoted to "Philosophy, Ethics, and Religious Beliefs".

4.10.1 Advantages of the historian's viewpoint

Historians, however, seem to enjoy an advantage over their fellow practitioners in the human and the social sciences. On account of their line of trade, historians know – or should know, if only they stopped bickering about interpretations, – what happened in the past. They know that diverse people met, sometimes by chance, sometimes by willingly pushing outside of the limits of their known world. Mind you, historians only know what happened in historical times. Indeed, what happened in pre-historical times is consigned to archaeological artifacts and biological remnants that, rather sparsely, might tell us something about major trends in the development of the human being, but will never be able to figure out what was in the mind of our common ancestor when he left Ethiopia to push towards Asia, or why she chose to cross the Bering Strait instead of staying put on the Siberian side of the earth bridge. Still, in historical times documents are able to tell us a story – albeit very imperfect – and to make available to us the thoughts and actions of real individuals – when they met, fought, struck a deal, or loved each other. Multiply these individual experiences by the number of people involved, and there you are, with patterns of what really happened in the past – patterns that were apparently repeated over and over throughout the history of humankind and could be replicated when, and if, new encounters will take place with entities from outer space.

The problem is that, in order to break even with a long list of wrongdoings for which their ancestors were deemed to be responsible, the past generation or so of western historians have embarked on a crusade of self-flagellation that tends to show only the worst sides of past encounters. Christopher Columbus's 1492 so-called "discovery of America" is the best-known case in point, but examples can be multiplied that show how Europeans, or Westerners in general, during their expansion wronged one people after another. In the context of this conference, the question is, of course, will it happen again? And what could we do to avoid it, provided that, in the foreseeable future, it will be Westerners again that might bear the responsibility and the brunt of extraterrestrial encounter?

4.10.2 Historians and the "Humans in Outer Space"-project

Let us return, however, to "Humans in Space – Interdisciplinary Odysseys", and consider the articles written by two most distinguished U.S. historians, James Muldoon and Alfred Crosby. The picture they paint is rather different. Muldoon takes the view that when Europeans discovered the new worlds there was ample room for regulatory compromise amongst the concerned European crowns. To be sure, none of them, including the papacy, had the real power to fully enforce the regulations that they issued. Still, Dr. Muldoon makes the point that

> "[e]nforcement always depend[ed] on the willingness of various states to see their own interests as supported by such enforcement. Like the regulation of the Ocean Sea, regulation of space will be rooted in and an extension of terrestrial politics. In the final analysis, it would seem that the regulation of space will fall to the most powerful states just as the regulation of the sea eventually fell to the great seaborne empires and earth-bound political realities will determine the regulation of space."[504]

Adjustment, adaptation, and compromise – an optimistic note of hope implying that the best practices of past patterns will be followed in the course of extraterrestrial encounters – even if we have no way of knowing who or what will be on the other side, the good and wise little guys of Steven Spielberg's "Close Encounters of the Third Kind" (1977) or the nasty and destructive killing machines of his later "War of the Worlds" (2005).

In his article, however, Crosby takes an opposing view. In describing the case of the Hawaiian archipelago, he shows that at the beginning of the nineteenth century this housed a population of no less than 242,000 individuals, which had fallen to 48,000 by 1878, a drop of at least four-fifths in less than a century. The Hawaiian case was far from exceptional. In an argument that he first pioneered in his "The Columbian Exchange" of 1972, Crosby showed that "[s]imilar death rates afflicted many newly contacted populations – Aztec, Incan, Maori, the indigenes of Siberia – and (...) Europeans as well when they landed on the malarial shores of tropical Africa." In Crosby's picture, there will be little room for adjustment, adaptation, or compromise when earthlings meet with entities from outer space: "In fact, such abrupt explosions and implosions of invading and indigenous organisms" Crosby concludes "may occur on planets and other heavenly bodies we land upon".[505] The question then is, I might add, not whether that catastrophic result will happen, but who of the two parties will suffer – them or us? Given the constraints and the unavoidability of Crosby's biological scenario, we are likely to set aside or forget one of the founding statements of Italian philosopher Paolo Musso's article in "Humans in Outer Space – Interdisciplinary Odysseys". According to Musso, whatever intelligent entity we meet out there will be provided with moral values, because "to be intelligent beings means also to be moral beings".[506] According to Crosby, there will be no time to assess whose moral values are superior and even less to try to integrate the two. There will be, in fact, only one way – to hope that humankind is not going to be the loser in this initial conflict, in which voluntary decisions and individual agency are not at play. Once we come out as the winner – as indeed we all hope – there will be ample time to indulge at leisure in a feeling of culpability, later to be overcome by extolling the virtues of the "others", once, of course, they have been exterminated or reduced to a non-challenging role by the microbes of our ancestors. It has, indeed, happened in the past, as shown by the several "Columbian encounters" that have taken place in the known history of humankind.

4.10.3 Getting ready to look ahead

But to return to the question with which I opened my remarks, do we really need an historian in order to provide a context for a conference that will mainly look forward into the future? Can a look at the past make us – or our offspring – more prepared to meet with extraterrestrial entities? In the strictest sense, my answer to both questions is, no. Unfortunately, if I have learned one thing from the practice of history, this is that the knowledge of history has never taught humankind to avoid past mistakes or to invent new solutions for unexpected problems or crises. Yet, in a more general, pedagogical sense, I do believe that, as much as ignorance should be avoided at all costs, knowledge should be fostered as much as possible, as there is never enough of it. And I do believe that knowledge of what happened in the past may help individuals who will find themselves in a position of power to make the right decision when (and if) the time comes.

There again, I am not suggesting that we prepare a learned manual listing behavioural rules for future encounters based on historical patterns that could be replicated when, and if, new encounters take place with entities from outer space. An exercise in this direction was attempted, about a generation ago, by a Swiss historical sociologist, Urs Bitterli, in his popular "Cultures in Conflict. Encounters Between European and Non-European Cultures, 1492–1800" (1989, German edition 1986). Bitterli made a distinction between three stages of cultural encounters, one normally following the other in a seemingly unavoidable fashion: superficial contact, sustained relation, and violent collision. The strength of the book resided in the comparative fashion through which he had examined his case studies – the Americas, Africa, Asia, the East Indies and Polynesia. Its weakness was embedded in the very idea of the existence of historical patterns. In fact, whenever one looks closely at a historical case study – or reads the portion of the Bitterli book about which one really knows something – what is really striking is the lack of depth of each chapter of "Cultures in Conflict". To me, as an historian, what really counts is not the existence of some superficial patterns involving "people", but the number of unexpected and uncontrollable variables that make the story of each individual a very different one.

While textbooks that reduce past events to neatly-recognisable behavioural patterns that are as artificial as they are useless should be avoided, historians can certainly come up with stories of past encounters that enlighten the minds of readers, increase their overall knowledge, and, when (and if) the time comes, may help people in power to make the right decision at the right time by using the many instruments that science – hard and real science – will place at their disposal at any crucial time.

I have purposefully avoided rephrasing for this audience the arguments contained in the article that I presented at the Vienna conference and that was recently edited for the "Humans in Outer Space – Interdisciplinary Odysseys" book, in spite of the fact that many people in this audience are new to this group. There I discussed the use of words such as discovery, encounter, meeting and contact, as well as the ideological issue and the biological issue that seems to be at the core of possible future encounters, at least as far as we can tell. Similarly, I have not summarised Crosby's wonderful depiction of the likely impact of micro-organisms on extraterrestrial travel, or Muldoon's legal treatise on the past allocation of unknown worlds amongst the European crowns. You will find in the book their main arguments and full references to the current debate among historians. My sole purpose, in these remarks, is to remind participants, perhaps preaching to the converted, that a good look backwards, in the direction of our past, might make each and every one of us readier to meet unexpected challenges and novelties, from wherever they come – even galaxies far, far away.

---

[501] Codignola, Luca. "Future Encounters: Learning from the Past?" Humans in Outer Space – Interdisciplinary Odysseys. Eds. Luca Codignola and Kai-Uwe Schrogl. Vienna: SpringerWien NewYork, 2008: 14–21. 20.
[502] Ibid. 21.
[503] Muldoon, James. "Inter caetera and Outer Space: Some Rules of Engagement." Humans in Outer Space – Interdisciplinary Odysseys. Eds. Luca Codignola and Kai-Uwe Schrogl. Vienna: Springer WienNewYork, 2008: 59–68. 60.
[504] Ibid. 67.
[505] Crosby, Alfred W. "Micro-Organisms and Extraterrestrial Travel." Humans in Outer Space – Interdisciplinary Odysseys. Eds. Luca Codignola and Kai-Uwe Schrogl. Vienna: SpringerWien NewYork, 2008: 6–13. 8.
[506] Musso, Paolo. "Philosophical and Religious Implications of Extraterrestrial Intelligent Life." Humans in Outer Space – Interdisciplinary Odysseys. Eds. Luca Codignola and Kai-Uwe Schrogl. Vienna: SpringerWienNewYork, 2008: 210–9. 212.

4.11 The Mars 500 isolation experiment[507]

Mars 500 simulates a mission to Mars and is therefore an important part of Europe's pathway to exploration.

4.11.1 Study overview

Following the successful completion of the initial 105-day isolation period of the Mars 500 study in July 2009, ESA and the Russian Institute for Biomedical Problems (IBMP) in Moscow are soon to embark on the full 520-day study to simulate a future mission to Mars. Human exploration of our Solar System is an important focus for the European Space Agency which has started on the path to making this a reality. Making sure that our astronauts are prepared mentally and physically for the demands of long-duration exploration missions is imperative to a mission's success.

ESA's Directorate of Human Spaceflight has a long-standing tradition of conducting research on the physiological and psychological aspects of spaceflight. ESA's bed-rest studies, in particular, are at the forefront of scientific research to understand how the human body reacts under weightless conditions, in order to devise effective countermeasures and enable humans to effectively undertake long-term missions in space.

Mars 500 is part of these scientific efforts to prepare for future human exploration missions. When preparing for long-duration space missions beyond the 6 months range currently undertaken by Expedition Crews on the International Space Station (ISS), medical and psychological aspects become an issue of major importance. When contemplating missions beyond Low Earth Orbit, such as to the Moon and Mars, daily crew life and operational capabilities may be affected by the hazardous space environment, the need for full autonomy and resourcefulness, the isolation, the interaction with fellow crewmembers and other aspects.

A better understanding of these aspects is essential for development of all necessary elements of an exploration mission. Whereas research onboard the ISS is essential for answering questions concerning the possible impact of weightlessness, radiation, and other space-specific factors, other aspects such as the affect of long-term isolation and confinement can be more appropriately addressed by the use of ground-based simulations.

The purpose of the Mars 500 study is to gather data, knowledge and experience to help prepare one day for a real mission to Mars. Obviously there will be no effect of weightlessness, but the study will help determine key psychological and physiological effects of being in such an enclosed environment for such an extended period of time.

The participants act as subjects in scientific investigations to assess the effect that isolation has on various psychological and physiological aspects, such as stress, hormone regulation and immunity, sleep quality, mood and the effectiveness of dietary supplements. The knowledge gained during the study will be invaluable in providing the basis for the potential development of countermeasures to deal with any unwanted side effects of such a mission, and also help in astronaut selection procedures, and at a modest expense. On the European side the Mars 500 programme is financed from the European Programme for Life and Physical Sciences in Space (ELIPS) and involves scientists from across Europe.

4.11.2 Life in the isolation chamber

In order to simulate a mission to Mars, six candidates (three Russian, two European and one Chinese) will be sealed in an isolation chamber in June 2010 for 520 days. This group were chosen to encompass working experience in the many fields including medicine, engineering, biology and computer engineering. Part of the chamber simulates the spacecraft that would transport them on their journey to and from Mars and another part simulates the landing module that would transfer them to and from the Martian surface. Following the successful completion of the initial 105-day isolation period in July 2009, the full 520-day study will continue from June 2010 until the end of 2011.

Once sealed into the chamber the candidates only have personal contact with each other plus voice contact with a simulated control centre and family and friends as would normally happen in a human spaceflight mission. A 20-min delay is built into communications with the control centre to simulate an interplanetary mission and the crew is given an identical diet to that used for the International Space Station. As with a human spaceflight mission, the chosen candidates are free to take certain personal items, as well as being supplied with books, movies, personal laptops and can occupy themselves with physical exercise or self studies.

During the isolation period the candidates simulate all elements of the Mars mission, travelling to Mars, orbiting the planet, landing and return to Earth. The crew have to be self reliant, organising a great deal of their daily tasks. They are responsible for monitoring and maintaining the health and psychological states of

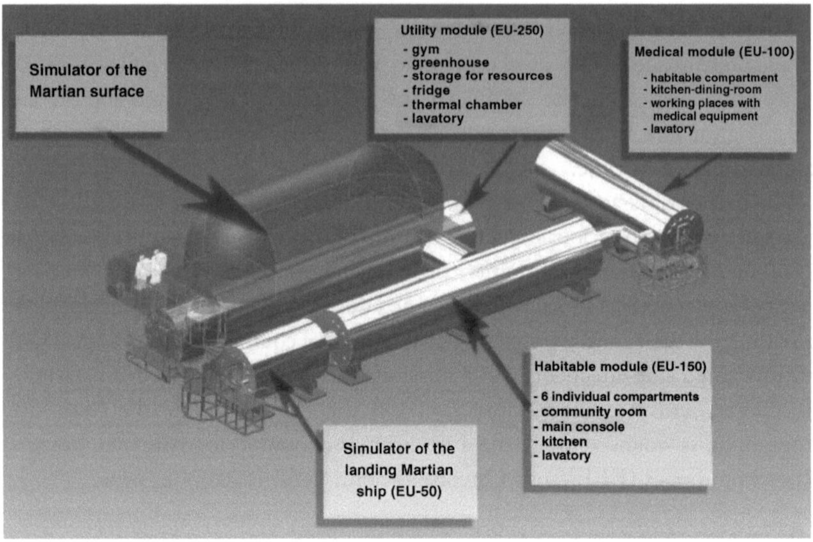

Fig. 11. *Isolation facility.*

themselves and each other, monitoring and controlling and maintaining systems including life support, control resource consumption, carrying out standard and non-standard cleaning and maintenance tasks, as well as fulfilling scientific investigations.

There is a 7-day week in place with two days off with a rotational system in place to account for night shifts. Non-standard and emergency situations are also simulated to determine the effect of a decrease in work capability, sickness, and also failures of the on-board systems and equipment. During "Mars surface operations" the crew are also divided into two groups of three people each. Once the first group exits to the Martian surface, the hatch between the Martian simulation module and the rest of the facility is closed by the second group and is only opened again when the Mars surface stay simulation has ended.

[507] European Space Agency. "Mars 500 Isolation Study – Information Kit." Noordwijk: Directorate of Human Space Flight, 2010.

About the authors

Philippe Ailleris is a Project Controller at one of the main space centers in Europe where he has worked for over 20 years. His interest in astronomy, space exploration and exobiology originated with his childhood fascination for and interest in Unidentified Aerial Phenomena (UAP), an interest which eventually led him to find work in the space industry and persists to this day. Despite the controversy surrounding UAP, he approaches the topic from a professional, rational, and scientific perspective. He has founded and currently leads the UAP Reporting Scheme Project, which he initiated under the framework of the 2009 International Year of Astronomy. The project, presented at the European Planetary Science Congress 2009 in Postdam, Germany is directed at astronomers, providing a venue for reporting unexplained sightings and resource pages documenting possible explanations for those sightings. His most recent publication is "The Lure of Local SETI: Fifty years of Field Experiments", Acta Astronautica (2010), based on his earlier presentation at UNESCO in Paris.

Berna van Baarsen is associate professor at the VU University medical centre Amsterdam where she performs research on "care at the end of life", more particularly "suffering and loneliness, dignity, euthanasia and other end-of-life decisions", and "doctor-patient relationships". She teaches clinical ethics to medical students and in post academic programmes. She gives lectures on the ethical and legal aspects of euthanasia, resuscitation, and organ donation. Since 2003 she is a member of the Medical Ethical Committee of the VU University medical centre in Amsterdam (METC-VUmc). In 2008 she has been appointed to the function of ethicist in one of the five Regional Euthanasia Review Committees in the Netherlands. She is the founding director of SELPH – Studies in Ethics, Life-issues, Psychology and Health. Starting 2009, she is visiting researcher at the University of Strasbourg, department European Centre for the Study and Teaching of Ethics (CEERE) and at the international Space University (ISU) in Strasbourg. Starting in 2010, she is visiting researcher at the VU University medical centre, department Psychiatry, in Amsterdam. She has written articles on various psychological and medical ethical topics and is a member of the editorial board of the Dutch *Journal of Health Care and Ethics*. She holds a doctorate degree in social science and has followed a post-academic education in medical and clinical ethics. Since 2007 she is active in Space research. She is the Principal Investigator of the international study on "The effects of group dynamics and loneliness on cognitive and emotional adaptation to extreme, confined environments" which is executed within the Mars520 project of the Institute for Biomedical Problems (IBMP) Moscow and the European Space Agency (ESA). Since 2008 she is member of the International Topical Team "Psychosocial and neurobehavioural aspects of human spaceflight".

Adrian Rareş Belu is a French Space Agency (CNES) Fellow at the Observatory of Bordeaux, France, since February 2010. In his previous career he worked in a human and social sciences academia-based consulting firm, for corporate and institutional top executive management. He has run for investiture in a European party for the 2005 elections, and served on the Scientific Council of the University of Nice between 2005 and 2007. He has worked within an industry-SME-academia consortia, for ESA extra-solar life-finding projects, and is currently interested in related fast-track detection and characterisation strategies. He is Member of the French Doctors in Science Association, as well as of EuroScience, He is a graduate of Ecole Centrale Paris, holds a doctorate degree in astronomy from the University of Nice, and has been invited at NASA Goddard. He is also a former American Nuclear Society Graduate Exchange Program Fellow.

Thomas Brandstetter is a postdoctoral researcher at eikones NCCR Iconic Criticism at Basel, Switzerland since October 2009. Before, he was Assistant Professor at the Institute of Philosophy at the University of Vienna. He studied Philosophy in Vienna and earned his doctoral degree in Media Studies at the Bauhaus University Weimar. His research focuses on the history of sciences. Selected publications are "*Kräfte messen. Die Maschine von Marly und die Kultur der Technik.*" Berlin: Kadmos Verlag, 2008; "Imagining Inorganic Life: Crystalline Aliens in Science and Fiction." *Imagining Outer Space: European Astroculture in the Twentieth Century.* Alexander C. T. Geppert. New York: Palgrave, 2010 (forthcoming); "Sentimental Hydraulics. Utopia and Technology in 18th Century France." *Philosophies of Technology. Francis Bacon and His Contemporaries.* Eds. Claus Zittel, Gisela Engel, Romano Nanni, and Nicole C. Karafyllis. Leiden: Brill, 2008: pp. 495–513.

Alan D. Britton is the Deputy Director of the Education for Global Citizenship Unit at the University of Glasgow, Scotland. He is involved in teaching and research on notions of national and global citizenship, educational policy-making, culture and identity. From 2007 to 2009 he co-ordinated a major EU-funded project on inter-cultural education as a vehicle for teacher's professional development. He began his pedagogical career as an outdoor education trainer before becoming a social studies and languages teacher. He was later appointed as the first Education Officer at the newly created Scottish Parliament in 1999, before gaining the prestigious post of Stevenson Lecturer in Citizenship at the University of

Glasgow in 2001–2005. His most recent major publication was a major co-edited collection: Peters, M., Britton, A. and Blee, H., eds. *Global Citizenship Education: Philosophy, Pedagogy and Practice*. Rotterdam: Sense Publishers, 2008.

Charles Cockell is a geomicrobiologist/astrobiologist at the Open University, in the United Kingdom. His academic interests encompass microbe-mineral interactions and their implications for Earth system processes and the habitability of extraterrestrial environments. He has published widely on the ethics of microbiology and the space environment in *Space Policy, Environmental Ethics, Ethics and the Environment* and other journals. He received his first degree in biochemistry and molecular biology at the University of Bristol and his PhD (DPhil) from the University of Oxford in molecular biology. He then undertook a National Research Council Associateship at the NASA Ames Research Centre in California before working at the British Antarctic Survey. He is a member of ESA's Planetary Protection and Life Sciences Working Groups. He is a Senior Editor of the journal *Astrobiology*. Popular science books include *Impossible Extinction* (CUP, 2003), which explores the tenacity of microbes on the Earth, and *Space on Earth* (MacMillan, 2006), which looks at the synergistic links between environmentalism and space exploration.

Luca Codignola-Bo (DL Rome 1970, MA Toronto 1974, DLitt hon. Saint Mary's 2003), is Head of the Institute of History of Mediterranean Europe of the Italian National Research Council. He is also Professor of North American History at the University of Genoa, and Adjunct Professor at Saint Mary's University of Halifax, NS, Canada. His main field of research is the Roman Catholic church in the North Atlantic area in the early modern era. He has also written on the early European expansion. His latest publications are *Columbus and Other Navigators* (2007); "Roman Catholic Conservatism in a New North Atlantic World, 1760–1829" (2007); "The Holy See and the Conversion of the Aboriginal Peoples in North America, 1760–1830" (2008); "The Swiss Community in Genoa from the Old Regime to the late 19th Century" (2008, with M.E. Tonizzi); *Humans in Outer Space: Interdisciplinary Odysseys* (2009, ed., with K.-U. Schrogl), "De 'Cromwell de France' à 'brigand consommé': les catholiques de la région de l'Atlantique du Nord et Napoléon (1799–1815)" (2009); "Il ruolo delle missioni religiose nella formazione dell'identità americana" (2009); "Les missionnaires spiritains à Saint-Pierre et Miquelon (1763–1816)" (2009); and "Le prime relazioni tra il Nord America e la penisola italiana, 1750–1830. Ciò che ancora non sappiamo" (2009).

David Dunér is Associate Professor in History of Science and Ideas at Lund University, Sweden, since 2008. Dunér's research concerns seventeenth and eighteenth century science, philosophy, medicine, mathematics and technology, and has resulted in numerous essays on natural history expeditions, universal languages, iatromechanics, the camera obscura, spirals, systematics, theory of matter, logical demonstrations, etc. Currently he is studying the history of mechanics and technology in the 17th and 18th century. He received his PhD in 2004 on a dissertation concerning the scientist and visionary Emanuel Swedenborg, where he proposed a cognitive history of ideas. This award-winning book is now currently being translated into English, and will be entitled *The World Machine. Emanuel Swedenborg's Natural Philosophy*. Dunér has published four monographs, edited ten books, and more then 110 articles, reports and reviews in the fields of history of science and ideas. He is editor-in-chief of *Sjuttonhundratal. Nordic Yearbook for Eighteenth Century Studies*, co-editor of the yearbook of *Swedish Linnaeus Society*, and member of the editorial board of *Lychnos*, the annual of the Swedish History of Science Society. He has been visiting scholar at Johns Hopkins and Princeton University, has been delegate to the International Society for Eighteenth-Century Studies (ISECS) Executive Committee meeting and has held numerous visiting lectures in various European countries and in North and South America. Dunér's research has been awarded by The Royal Swedish Academy of Sciences, The Royal Swedish Academy of Letters, History and Antiquities, The Bank of Sweden Tercentenary Foundation and The Swedish Foundation for International Cooperation in Research and Higher Education.

Ulrike Landfester studied German, English and Medieval Literature at the Universities of Freiburg and Munich (Germany) where she finished her dissertation thesis on the poetical function of clothing in Goethe's early works in 1993 (publ. 1995) and her habilitation thesis on Bettine von Arnim's political writings in 1998 (publ. 2000). After several visiting professorships at the universities of Frankfurt/Main, Vienna and Konstanz she is now a full professor for German Language and Literature at the University of St. Gallen (Switzerland). Since 2006, she is a member of the Standing Committee for the Humanities, and since 2008 she is member of the European Space Policy Institute's (ESPI) Advisory Council. Having published papers on diverse subjects including German Romanticism, Goethe, crime fiction, cultural body concepts and the epistemology of outer space as well as several critical-historical editions, she currently works on a book on the relationship between tattooing and modern writing culture.

Agnieszka Lukaszczyk is a space policy consultant for Secure World Foundation (SWF). A Polish/American national, Agnieszka is based in Vienna working on the European space policy and United Nations (UN) civil space activities covering the UN Committee on the Peaceful Uses of Outer Space (COPUOS) and working closely with the UN Office of Outer Space Affairs (OOSA). She has worked as the Executive Officer for the Space Generation Advisory Council (SGAC) in support of the UN Programme on Space Applications since 2006, and is now SGAC Co Chairperson. In addition, during the period of September 2006 to June 2008, Agnieszka worked at the European Space Policy Institute (ESPI). Agnieszka also serves as the Vice President – Operations for the World Space Week. She holds a Masters degree from the American University School of International Service in International Politics and a Bachelor degree in Political Science form the University of Tennessee. She also studied at the Université Catholique de Louvain in Brussels, Belgium; the Jagiellonian University in Krakow, Poland and the World Trade Institute in Berne, Switzerland. She gained professional experience at the Political Section of the Polish Embassy in Washington D.C., American Electronics Association in Brussels, European Department of the Polish Senate in Warsaw and the *Warsaw Business Journal.*

Kurt Mills is Senior Lecturer in International Human Rights at the University of Glasgow. He previously taught at Gettysburg College, James Madison University, Mount Holyoke College, and the American University in Cairo, and served as the Assistant Director of the Five College Program in Peace and World Security Studies at Hampshire College. He has published one book – *Human Rights in the Emerging Global Order: A New Sovereignty?* – and numerous articles and book chapters on human rights and humanitarian affairs. He is the founding chair of the Human Rights section of the International Studies Association. He received his PhD from the University of Notre Dame.

Martin Parker is Professor of Organisation and Culture at the School of Management, University of Leicester and Editor in Chief of the journal *Organization* (Sage). His interests are fairly wide ranging, and he has written about ethics, politics and utopias; social and organisational theory; higher education; the culture of organisations; and various aspects of the representation of organisations in contemporary culture. At the moment, he is particularly keen on thinking about non-managerial organising, and developing what might be called a "cultural studies" of organisations. Future projects include work on pirates, cowboys and smugglers, as well as a paper on art and/as work. Some of his books include *Against Management* (Polity, 2002), *The Dictionary of Alternatives* (Zed, 2007, with Valerie Fournier and Patrick Reedy) and *Space Travel and Culture* (Wiley 2009, edited with David Bell).

Anna G. Piotrowska is currently associated with the Department of Theory and Anthropology of Music at the Institute of Musicology, Jagiellonian University in Kraków, Poland and is a Fulbright Fellow in Boston University, USA. She has recently been awarded the Moritz Csaky Preis at Austrian Academy of Sciences (in 2009). In 2008 Anna G. Piotrowska participated in an international programme organised by the Bibliotheque Nationale de France. In 2007 with 20 other scholars she co-authored "A Manifesto for the Humanities in Europe. *In varietate concordia*" (sponsored by ESF). She was awarded the Mellon fellowship in Edinburgh University, UK in 2005 and in 2006 and won the CDC competition in Central European University in Budapest, Hungary. She studied musicology at Jagiellonian University in Kraków, Poland (1992–1997) and Durham University in UK (1994–1995). In 2002 she defended her PhD on the idea of nationalism in American music. Anna G. Piotrowska actively participates in many international conferences, among others in Vienna, Luxembourg, Copenhagen, Leeds, Vilnius, Thessalonica, Sheffield, Tbilisi, Prague, Brussels, Kraków. Anna G. Piotrowska is mainly interested in sociological and cultural aspects of musical life. She is an author of the book "The Idea of National Music in the Works of American Composers of the Early 20th century" (ISBN 83-7322-569-2) and about forty articles (both in Polish and English) on musical culture.

Nina-Louisa Remuss (BA, MLitt) is Associate Fellow of the European Space Policy Institute (ESPI), Vienna, Austria. Since July 2008 she has been contributing to ESPI's Research Programme Space and Security. Particularly, she co-authored a study on Europe's role in the peaceful-uses of outer space debate, led a study and a related workshop on the contribution of space applications to internal (i.e. homeland) security which was conducted under the auspices of the Czech EU Council Presidency. She also published a short policy paper on the vulnerability of space assets in the context of terrorist intended harmful interferences (all documents can be downloaded at www.espi.or.at). She has contributed numerous articles and papers to leading journals in the field and is regularly invited to speak at conferences in Europe and the U.S. In the context of her research she has been organising workshops and conference with the participation of leading personalities in the respective fields (from European institutions, agencies and industry) where she also has been acting as moderator or session chair. In 2009 she was invited to become member of the Panel of Experts of the EU Framework Programme 7 project STRAW (Security Technology Active Watch), which provides a building block towards the development of a European Security Network (ESN). Also in 2009, she was tasked with the conduct of ESPI's support of the Presidency of the European Interparliamentary Space Conference (EISC). She holds a Bachelor Degree in European Studies from the University of Maastricht (The Netherlands), a Master's Degree in International Security Studies from the University of St. Andrews (United Kingdom) and spent an exchange semester at the University of Bologna (Italy) and the Nilsson Center for European Studies of the Dickinson College (Italy). Nina-Louisa Remuss has been an intern at the Permanent Mission of Germany to the United Nations in New York as well as at the German Federal Ministry of Defence.

Michael T. Schetsche is head of the department for cultural studies and social research at the IGPP Freiburg since 1st of May 2002. Before that he was lecturer at the Institute for Sociology at the University of Bremen and thereafter at the Albert-Ludwigs-Universität in Freiburg. He holds a degree in political science (FU Berlin), finished his doctorate with the title of a Dr. rer. pol. (University of Bremen), and holds a habilitation in sociology. Hies fields of work are sociology of knowledge, sociology of media, sociology of social problems and anomalies, qualitative prognostics, exosociology. Relevant books are Schetsche, Michael. Ed. Der maximal Fremde. Begegnungen mit dem Nichtmenschlichen und die Grenzen des Verstehens [*The maximum stranger. Confrontations with non-human-actors and the limits of understanding*], Würzburg, Ergon 2004. Schetsche, Michael, and Martin Engelbrecht Ed. Von Menschen und Außerirdischen. Transterrestrische Begegnungen im Spiegel der Kulturwissenschaft [*Of men and aliens. Transterrestrian encounters as reflected by the cultural sciences*] Bielefeld: transcript, 2008.

Gustav Schörghofer is the rector of the Jesuit Church in Vienna, Austria (www.jesuitenwien1.at) since 1998. Additionally he is pastor of the artists of the archdiocese Vienna and chair of the selection committee of the "Msgr. Otto-Mauer-Preis" (http://www.otto-mauer-fonds.at/OM_Preis_Beschreibung.htm) for artists. He works on several projects with different artists such as "Jesuitenkirche", Zacherlfabrik (www.zacherlfabrik.at) and JesuitenFoyer. He studied the history of arts and classical archaeology in Salzburg before entering the Society of Jesus in 1981 and Philosophy and Theology in Munich and Rome thereafter. His Ordination to the priesthood was in 1988.

Kai-Uwe Schrogl is Director of the European Space Policy Institute (ESPI) in Vienna, Austria since 1 September 2007. Prior to this, he was the Head of the Corporate Development and External Relations Department in the German Aerospace Center (DLR). Previously he also worked with the German Ministry for Post and Telecommunications and the German Space Agency (DARA). He has been a delegate to numerous international fora and recently served as the chairman of various European and global committees (ESA International Relations Committee and two UNCOPUOS plenary working groups). He presented, respectively testified, at hearings of the European Parliament and the U.S. House of Representatives. Kai-Uwe Schrogl has written or co-edited 11 books and more than 100 articles, reports and papers in the fields of space policy and law as well as telecommunications policy. He is editor of the "Yearbook on Space Policy" and the book series "Studies in Space Policy" both published at SpringerWienNewYork. In addition he sits on editorial boards of various international journals in the field of space policy and law (*Acta Astronautica, Space Policy, Zeitschrift für Luft- und Weltraumrecht, Studies in Space Law/Nijhoff*). Kai-Uwe Schrogl is a Member of the Board of Directors of the International Institute of Space Law, Member of the International Academy of Astronautics (recently chairing its Commission on policy, economics and law) and the Russian Academy for Cosmonautics. He holds a doctorate degree in political science and lectures international relations at Tübingen University, Germany (as a Honorary Professor) and has been a regular guest lecturer including the International Space University and the European Centre for Space Law's Summer Courses.

Christopher Mark Timmins is the Director of the Fashion cluster, within the Heriot-Watt University, School of Textiles and Design. His career started as one of the top twelve textile design graduates, selected by the Royal Society of Arts, London, as Royal Society of Arts Young Designers into Industry. He spent the first two years at William Hutchinson Yarns designing yarns for a worldwide market. Following this position, he attended the St Martins School of Art in London, where he gained his postgraduate degree, a Master of Arts in fashion knitwear. The next three years were spent designing men's and women's knitwear for Marks and Spencer PLC. This invaluable commercial experience was then channelled into academia with a short lecturing spell teaching knitwear design at Plymouth College of Art and Design and then moving to his current institution, Heriot-Watt University, School of Textiles and Design, as firstly a knitwear lecturer and recently as the schools Director of the fashion cluster with responsibilities for fashion, menswear, womenswear, fashion communication and fashion marketing and retail. He has been invited to discuss couture in space at the 50th world Aeronautical Conference "Less remote" strand of paper presentations, which took place in 2008 in Glasgow and has been working with university colleagues and students since then, to define the opportunities and challenges for both clothing and couture in micro and zero gravity space environments. Alongside his lecturing and research into space couture and clothing, Mark is also a photographer and sculptor with recent exhibitions in the Scottish national fisheries museum in Anstruther and galleries in St Ives, Pitenweem, Eyemouth, Cambridge and London.

Jean-Claude Worms is Head of the Space Sciences Unit of the European Science Foundation (ESF), managing the European Space Sciences Committee (ESSC) and all space-related programmes of the ESF. He holds a PhD in physics from the University Paris 6 and was Assistant Professor in physics and astronomy in

Paris and Versailles from 1989 to 1992. He was associate researcher to several laboratories in France (LPCE, LSIIT, LSP and IPSL-SA). He worked in radiative transfer in granular media, pre-planetary aggregation and space debris, and was Principal Investigator on the PROGRA2 facility (polarimetry of dust clouds in microgravity), LIBRIS project (in-orbit optical detection of space debris), and Co-Investigator of the ESA ICAPS facility (study of particle systems on the ISS). He has been Main Scientific Organiser and Editor of solar system sessions in COSPAR Scientific Assemblies since 1998 and he is a member of the Editorial Board of the International Journal on Nanotechnologies. In 1994, he was also consultant for DASSAULT on the state-of-the-art French civilian research in Infrared and SAR imaging. Jean-Claude Worms is involved in ESA and EC high-level science advisory structures, and has participated with an observer status to ESA's Ministerial Conferences since 1999. As a result of the unique structure of the ESSC which reflects the broad spectrum of space-related disciplines, he is dealing with strategic planning, programme evaluation and reviewing, and intelligence monitoring in every sector of space sciences, including space policy and GMES.

Participants in the ESF "Humans in Outer Space – Forward Look Scoping Conference" and contributors to the book "Humans in Outer Space – Interdisciplinary Perspectives".

List of acronyms

A

A-Life: Artificial Life
ASAT: Anti-Satellite (weapon or test)
ASI: Agenzia Spaziale Italiana
ATS: Antarctic Treaty System
ATV: Automated Transfer Vehicle

B

BMVIT: Bundesministerium für Verkehr, Innovation und Technologie (Austrian Ministry for Transport, Innovation and Technology)
BNSC: British National Space Council

C

CD: Committee on Disarmament
CDC: Center for Disease Control
CEERE: European Centre for the Study and Teaching of Ethics
CNES: Centre National d'Études Spatiales (French Space Agency)
CNSA: China National Space Administration
CoC: Code of Conduct
COSPAR: Committee on Space Research
CSA: Canadian Space Agency
CSIRO: Commonwealth Scientific and Industrial Research Organisation

D

DARA: Deutsche Agentur für Raumfahrtangelegenheiten (German Space Agency)
DLR: Deutsches Zentrum für Luft- und Raumfahrt (German Aerospace Center)
DMS-R: Data Management System for the Russian segment of the ISS

E

EA: Evolutionary Algorithms
EC: European Commission
ECOSOC: UN Economic and Social Council
EELV: Evolved Expendable Launch Vehicle
EISC: European Interparliamentary Space Conference

ELIPS: European Programme for Life and Physical Sciences
EO: Earth Observation
ERA: European Robotic Arm
ESA: European Space Agency
ESF: European Science Foundation
ESPI: European Space Policy Institute
ESSC: European Space Sciences Committee

F

FKA: Russian Federal Space Agency (formerly Rosaviakosmos)
FY: Financial Year

G

GEMA: Greater Earth Metropolitan Area
GEOINT: Geospatial Intelligence
GES: Global Exploration Strategy
GMES: Global Monitoring for Environment and Security
GPS: Global Positioning System

H

HiOS: Humans in Outer Space

I

IAA: International Academy of Astronautics
IAAA: International Association of Astronomical Artists
IAASS: International Association for the Advancement of Space Safety
IBMP: Russian Institute for Biomedical Problems
ICAPS: International Conference on Automated Planning and Scheduling
ICJ: International Court of Justice
IGA: (ISS) Inter Governmental Agreement
IMINT: IMage INTelligence
IPSL: Institut Pierre Simon Laplace
ISBN: International Standard Book Number
ISECS: International Society for Eighteenth-Century Studies
ISRO: Indian Space Research Organisation
ISS: International Space Station
ISU: International Space University
IT: Information Technology
ITU: International Telecommunications Union

J

JAXA: Japan Aerospace Exploration Agency (formerly NASDA, National Space Development Agency)

K

KARI: Korea Aerospace Research Institute

L

LIBRIS: Library Information System
LPCE: Laboratoire de Physique et Chimie de l'Environnement et de l'Espace
LPS: Life and Physical Sciences in Space
LSIIT: Laboratoire des Sciences de l'Images, de l'Informatique et de la Télédétection

M

MDG: Millennium Development Goals
METC: Medical Ethical Committee (VU University medical centre in Amsterdam)
MIT: Massachusetts Institute of Technology
MOL: Manned Orbital Laboratory
MPLM: Multi-Purpose Logistics Module

N

NASA: National Aeronautics and Space Administration
NCCR: National Centres of Competence in Research
NEO: Near Earth Objects
NGO: Non-Governmental Organisation
NPoC: National Point of Contact
NRO: National Reconnaissance Office
NSAU: National Space Agency of Ukraine

O

OECD: Organisation for Economic Cooperation and Development
OST: Outer Space Treaty

P

PAROS: Prevention of an Arms Race in Outer Space

R

RCs: Regional Coordinators
Roskosmos: Russian Federal Space Agency

S

SAR: Synthetic Aperture Radar
ESF-SCH: European Science Foundation, Standing Committee for the Humanities
ESF-SCSS: European Science Foundation, Standing Committee for Social Sciences
SELPH: Studies in Ethics, Life-issues, Psychology and Health
SETI: Search for Extraterrestrial Intelligence
SFINCSS: Simulation of Flight on International Crew on Space Station
SGAC: Space Generation Advisory Council
SIGINT: Signal Intelligence
STM: Space Traffic Management
STRAW: Security Technology Active Watch
SWF: Secure World Foundation

T

TCBMs: Transparency and Confidence Building Measures

U

UAP: Unidentified Aerial Phenomena
UFO: Unidentified Flying Object
UK: United Kingdom
UN: United Nations
UNCLOS: UN Convention on the Law of the Sea
UNCOPUOS: United Nations Committee on the Peaceful Uses of Outer Space
UNESCO: United Nations Educational, Scientific and Cultural Organization
UNGA: United Nation General Assembly
UNISPACE III: Third United Nations Conference on the Exploration and Peaceful Uses of Outer Space
UNOOSA: United Nations Office for Outer Space Affairs
USA: United States of America
USSR: Union of Soviet Socialist Republics

List of figures and tables

Figures

Introduction: from "odysseys" to "perspectives" – towards new interdisciplinary approaches to humans in outer space

Chapter 1 Politics and society

Chapter 3 Culture and psychology 159

Chapter 4 Annex

Tables

Chapter 4 Annex

Acknowledgements

The book is the result of a conference which took place in La Palma in 2009. It was organised in a fabulous manner by Ms. Madelise Blumenroeder of the European Science Foundation (ESF). The editors would like to further thank Ms. Roberta Daveri, Reserach Intern at ESPI, for her support and Mr. Karim Ernst Karman from SpringerWienNewYork, who together with his colleagues from all departments of the publishing house so professionally and kindly accompanied this project. Invaluable support was received by Ms. Blandina Baranes, ESPI's Communications Manager, who efficiently handled the liaison with the publisher and provided additional assistance in numerous ways.

MIX
Papier aus verantwortungsvollen Quellen
Paper from responsible sources
FSC® C105338

If you have any concerns about our products,
you can contact us on
ProductSafety@springernature.com

In case Publisher is established outside the EU,
the EU authorized representative is:
Springer Nature Customer Service Center GmbH
Europaplatz 3, 69115 Heidelberg, Germany

Printed by Libri Plureos GmbH
in Hamburg, Germany